T0094133

SAS Programming for Elementary Statistics

SAS Programming for Elementary Statistics

Getting Started

CARLA L. GOAD

CRC Press is an imprint of the
Taylor & Francis Group, an **informa** business
A CHAPMAN & HALL BOOK

First edition published 2021
by CRC Press
6000 Broken Sound Parkway NW, Suite 300, Boca Raton, FL 33487-2742

and by CRC Press
2 Park Square, Milton Park, Abingdon, Oxon, OX14 4RN

CRC Press is an imprint of Taylor & Francis Group, LLC

ISBN: 9781138589025 (pbk)
ISBN: 9781138589094 (hbk)
ISBN: 9780429491900 (ebk)

Typeset in Palatino
by KnowledgeWorks Global Ltd.

To Mom and Dad

Contents

Preface

The primary audience for this book consists of third- or fourth-year undergraduate students and graduate students. A first semester course in statistical methods is necessary for readers. Additionally, researchers needing a refresher in SAS programming will find this book helpful. This book would be an appropriate text for a university-level SAS Programming course. For other statistical methods courses or data mining courses, it may be regarded as a supplementary book assisting with data set creation, data management skills, and elementary data analysis.

This book is an introduction to SAS Programming for elementary statistical methods written at a college level. No prior SAS experience is necessary to fully benefit from it, but the reader should have knowledge of one- and two-sample tests for population means and variances using normally distributed data. For new programmers, initial critical concepts are the creation of data sets in SAS, debugging a program, and the overall construction of a SAS program. Topics include DATA Step operations, t-tests, and confidence interval methods for population means, frequency analysis, simple linear regression, an introduction to ANOVA, and graphing. Readers are introduced to the DATA Step using simple techniques and then are led through procedures learning how to select options and ODS Graphics. Interpretation of the output is also included. After a procedure chapter or two, this process repeats itself a few more times by introducing new DATA Step operations and/or data set creation methods followed by new statistical or graphics procedures.

When reading syntax documentation produced by SAS Institute, SAS statements are in uppercase, and programmer selected variable names and SAS data set names are in italics. That practice is also in use in this book. In this way, as the reader learns introductory skills in this book, they can read SAS Institute's documentation without much adjustment to presentation style. In each of the chapters, an introduction to the SAS programming topic is motivated, programming syntax is overviewed, and for statistical methods brief coverage of statistical methods is presented. Each chapter contains objectives that are clearly stated. The SAS programming code is given for each objective. Programming notes, as needed, and the printed results or graphics are given for the objectives. All of the printed results are presented unless it was a duplication of information given earlier in the chapter. It has been noted when printed results have been suppressed. Interpretations of the results are also given. Part of becoming a SAS programmer is learning how to read error messages and other messages in the SAS Log. Attention is given to these messages in early chapters and particularly in the data set management chapters.

For the reader who is new to SAS programming, the chapters are ordered from simpler topics to those that are more advanced. The later chapters are written presuming a growing SAS programming maturity of the reader. Readers with some SAS programming experience may select chapters 2, 4, 8, 12, 13, 14, 15, 19, and 20 to enhance their SAS data set creation and data management skills. Chapter 14 covers the IMPORT and EXPORT procedures. These topics are quite useful as these are two procedures through which SAS reads and writes files associated with other software, such as Microsoft Excel and SPSS. Whenever possible, I believe well-constructed graphics support an analysis, and instructing readers how to utilize graphics tools in SAS had to be an essential part of this book. Graphics chapters are Chapters 5, 16, and 17 though there are default ODS Graphics in some of the other procedures chapters also.

In the mid-1990s, I began teaching a SAS Programming course for Statistics undergraduate and graduate majors and non-Statistics graduate students taking statistical methods courses. In addition to elementary statistical procedures in the SAS Editor, I wanted to include some SAS graphics capability and some of the new point-and-click products (SAS/ASSIST, SAS/INSIGHT, and The Analyst Application), some of which no longer exist in SAS software. At that time it was challenging to find a book that covered the topics I wanted to include in the course. That's when I began writing my own material for the course. Over the years the material has been revised to stay current with SAS versions and restructured to better assist the students in the course. Students have told me that my course notes serve as their guide when they need to create data files or begin the analysis of their research data. With the encouragement from several faculty and former students, I have written this book containing the materials I have found to be successful in the classroom for over 20 years.

I am grateful to many people for their support of this book project. To the many classes of students over the years who gave me feedback on the notes and lessons, the content and clarity of the chapters have been improved. To the students who said, "You really should publish your notes", I finally listened. To my friends and colleagues at Oklahoma State University, thank you for your support and encouragement. I am grateful to the reviewers for their thoroughness, suggestions and critique. To my family and friends, your support has been invaluable and is greatly treasured. To my husband, Dave, thank you for proofreading passages, discussing the best ways to present material, and so much more. To all of you, a heartfelt, "Thank you".

Carla L. Goad

About the Author

Carla L. Goad is a Professor of Statistics at Oklahoma State University. She holds a Ph.D. in Statistics from Kansas State University. SAS Programming has been an integral part of her work since the late 1980s. She has been teaching an introductory SAS programming course and SAS applications in graduate statistical methods courses since the mid-1990s. Her areas of expertise are experimental design and analysis and linear models. She is a statistical consultant for the Oklahoma Agriculture Experiment Station and has enjoyed working with agricultural researchers in animal and plant sciences. Through these collaborations she has coauthored many articles in agriculture research journals.

1

Introduction to SAS Programming Environment

SAS is a statistical analysis software that has international recognition. SAS Institute has products for which users must write programming syntax, and other products which are point-and-click driven. In this book, SAS 9.4 is used to demonstrate data set management skills and introductory statistical methods primarily using programming code. Many of the topics included in this book are included in the Base SAS Certification exam.

Since SAS is a Windows compatible product, it is assumed that the reader knows how to use a mouse at the computer. There is a difference between using the left and right mouse buttons. Most of the "clicking" will be with the left mouse button. Additionally, items on the SAS screen that programmers should select appear in boldface in this book, or an image may be given to assist in identifying what item(s) should be selected for the task at hand. For example, **File – Save As** identifies that "File" should be selected from the pull-down menu at the top of the SAS screen and "Save As" as the selection from the resulting pull-down menu, and identifies a Help button near the top center of the SAS screen.

The chapters of this book may briefly review statistical methods and then state particular objectives followed by the SAS programming code to achieve the stated objectives. Programming options and output results will be overviewed in these objectives. These exercises are intended to demonstrate some of the capabilities of the current SAS topic. The only objectives of this chapter are to show some very basic steps to assist in initiating a SAS session, editing and executing a SAS program, printing the results, and storing program and results files. Remaining chapters in this book address particular objectives in statistical analysis or data management and instruct the reader how to program SAS to achieve a particular goal.

1.1 Initiating a SAS Session

From the list of programs or applications on a computer, the **SAS** folder is located and from that folder **SAS 9.4 (English)** is selected.

See Figure 1.1. Each SAS installation may produce a slightly different list of SAS software products than those listed in Figure 1.1, but the current version of SAS should be selected. Selecting other options, such as, Enterprise Guide or Enterprise Miner should not be selected. These are other SAS software products not addressed in this book.

When SAS 9.4 (English) opens, a dialog box opens "in front of" the SAS programming environment. See Figure 1.2. Changes in SAS software are communicated in this dialog. For a first time SAS user, the Output Changes likely would not yet be meaningful. There is also a tutorial "Getting Started with SAS" one can select. For now, **Close** this dialog box.

When SAS first starts, there are five windows open. They are: Results, Explorer, Output – (Untitled), Log – (Untitled), and Editor – Untitled1 windows. In Figure 1.3 note the window tabs or bars for each of these appear at the bottom of the SAS screen. At the top of the SAS

SAS
SAS Operational Qualification T...
Renew SAS Software (64-bit)
SAS 9.4 (English with DBCS)
SAS 9.4 (English)
SAS 9.4 (Unicode Support)
SAS Deployment Manager 9.4
SAS Documentation Viewer 9.4 (...
SAS Enterprise Guide 6.1 (64-bit)
SAS Enterprise Miner Workstatio...
SAS IML Studio 12.3
SAS Install Qualification Tool 9.405
SAS Integration Technologies Co...
SAS Service Configuration Utility

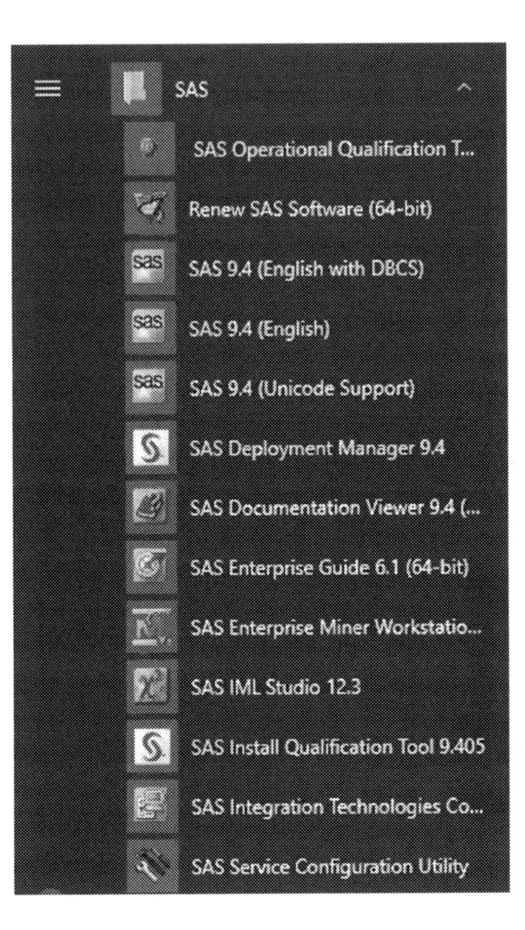

FIGURE 1.1
An example of SAS software products available in a listing of SAS folder contents (Microsoft Windows 10 operating system).

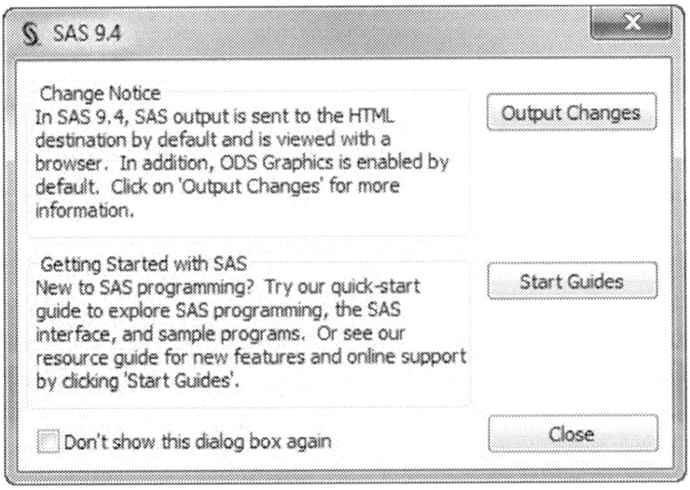

SAS 9.4
Change Notice
In SAS 9.4, SAS output is sent to the HTML destination by default and is viewed with a browser. In addition, ODS Graphics is enabled by default. Click on 'Output Changes' for more information.
Output Changes
Getting Started with SAS
New to SAS programming? Try our quick-start guide to explore SAS programming, the SAS interface, and sample programs. Or see our resource guide for new features and online support by clicking 'Start Guides'.
Start Guides
Don't show this dialog box again
Close

FIGURE 1.2
Initial dialog window that appears when SAS 9.4 opens.

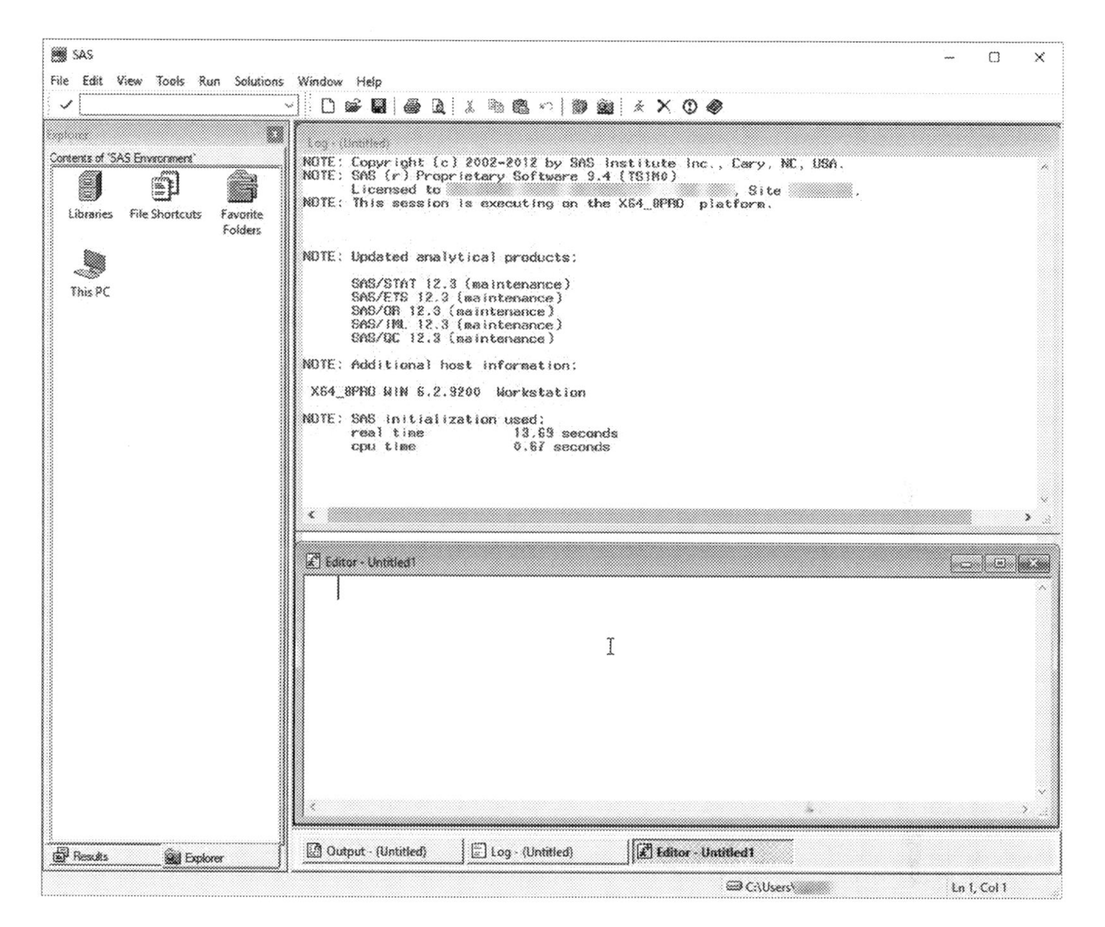

FIGURE 1.3
Appearance of the SAS Editor environment upon starting the SAS session.

screen there is a toolbar with buttons for quick actions such as Cut, Copy, Paste, etc. The blank in the upper left-hand corner of the screen is called the Command Line where SAS commands can be entered. The Command Line was utilized exclusively in older versions of SAS. The Command Line is still available in the current version though most current programmers prefer not to use it. SAS commands are generally accomplished using point-and-click actions or function keys, but a few SAS commands for the Command Line will be given in this book.

In the initial Log window the version of SAS software and maintenance release (TS#M#) information are given. The holder of the SAS site license and site license number have been obscured or *blurred* in this screen capture (Figure 1.3). Another way this pertinent license information can be accessed is by selecting **Help – About SAS 9** at the top of the SAS screen. Should SAS Technical Support need to be contacted with questions, the SAS user will be asked for site license information and other information found in this initial Log window or in the information window resulting from the **Help – About SAS 9** selection.

Below the initial Log window is the Editor window. Data and program statements for analyzing data are written in the Editor. Hidden "behind" the Log and Editor windows is the Output window. This window is one of the locations where the results from a SAS program can be displayed. To the left of these three windows are the Results and Explorer windows. More information about these two windows will be given later.

Moving from window to window in SAS can be in any of the following ways:

1. The **Window** pull-down menu at the top of the screen lists the available window environments in SAS. The computer mouse is used to select the name of the window to which to move. Window arrangement options for the Editor, Log, and Output windows are also given in this menu. Some programmers have an organizational preference for the windows in SAS.
2. Depending on the manner in which the windows are arranged, moving from window to window can be done by clicking on any exposed part of the window to be viewed.
3. At the bottom of the SAS screen there are window tabs or bars that can be selected by clicking on them to move to the desired window. The **View** pull-down menu also lists the available windows that can be viewed. On the **View** pull-down menu, the icon for the Enhanced Editor has a blue plus sign superimposed over a notepad . Though the **View** menu identifies the "Enhanced Editor" it is the same as the initial "Editor" window that opened. A notepad without the blue plus sign is the icon for the Program Editor used in other platforms and in older versions of SAS. Some current platforms may not open the Enhanced Editor but will instead open the Program Editor. Using the **View – Enhanced Editor** sequence, a new Editor window opens. If it does not, examine the log window. If, for some reason, the Enhanced Editor is not available the following error message will be displayed in the log window: ERROR: The Enhanced Editor Control is not installed. If this is the case, the programming steps can still be taken with the Program Editor. There is more on the Program Editor in Section 1.4.
4. Some of the function keys on the keyboard also move from window to window. To find out what the function keys do, from the **Tools** pull-down menu at the top of the SAS screen, **Options** and then **Keys** are selected to obtain the definitions of the function keys, that is, **Tools – Options – Keys**. For example, F6 with **Definition** "log" will move to the Log window from any of the other windows. Close the Keys window by clicking on the X in upper right-hand corner of the Keys window.

The window size can be changed using familiar point-and-click actions used in other programs.

1. **Maximize** any window by clicking on the large screen button in the upper right-hand corner of the window. When clicking on the icon in the upper left-hand corner of a window, a menu is given. One of the options available is **Maximize**.
2. A window can be returned to its original size by clicking on the button in the upper right-hand corner of the window that shows two window sizes. Or, from the menu obtained by clicking on the icon in the upper left-hand corner of the menu, **Restore** can be selected.
3. **Minimize** a window either on the menu (upper left corner) or by clicking _ (underscore) button in the upper right-hand corner of the window. When minimized, the window no longer appears on the screen. The window bar at the bottom of the SAS screen can be selected, or the correct function key can be used to restore that window.
4. Should any of the windows be closed, they can be reopened using the **View** pull-down menu and selecting the window to reopen.

1.2 Entering and Executing a SAS Program

Presently, these instructions will introduce SAS Editor. The Editor is also referred to as the Enhanced Editor in the View menu and in online SAS documentation. The (Enhanced) Editor and the Program Editor both allow data or programming statements to be entered. Moving the cursor in either editor can be done by clicking on the new location or using the directional arrow keys on the keyboard. Page Up and Page Down keys function in each of the windows in SAS as in other software.

The Program Editor is the editor utilized by all current SAS platforms and in older versions of SAS. For the Windows operating system, the Enhanced Editor is also available and is the editor that typically opens by default when a SAS session is initiated. The Enhanced Editor has a color-coded display providing assistance in debugging programs. Additionally, horizontal reference lines separate procedures, some terms are in bold face print, and there is +/– control to minimize or maximize blocks of programming statements. These additional features also assist the programmer in debugging the program.

In each chapter of this book there are clearly stated objectives. These exercises are intended to demonstrate some of the capabilities of the current SAS topic.

A simple example program appears in Objective 1.1 beginning with the DATA one; statement. Note the change in font. The font used below will be used throughout the book for programming commands. The information does not have to be entered in a case-sensitive manner as shown in the example. Case-sensitivity will be addressed in later chapters. The meaning of the lines of the example SAS program will not be explained here. The rest of this book explains the specifics of the syntax. The objective at this time is to successfully submit a SAS program, view the Log and Results Viewer windows, and to save and/or print the program and results.

OBJECTIVE 1.1: Enter the program below and its punctuation exactly as it appears. As a program is composed, one may wish to maximize the Enhanced Editor window. At least one space separates pieces of information. If two or more spaces are used instead of one, it will not affect this sample program.

```
DATA one;
INPUT Name $ Fine;
DATALINES;
Lynn 50
Evan 75
Thomas 24
Wesley 44
Marie 30
;
PROC PRINT DATA=one;
TITLE 'Objective 1.1: Outstanding Parking Fines';

PROC MEANS DATA=one;
VAR fine;
RUN;

QUIT;
```

As the program is entered, there are color changes in the statements, the appearance of boldface words, and yellow highlighting. These are features of the Enhanced Editor that will assist in finding errors. To illustrate this, create a few errors on purpose once the above program is entered in the Enhanced Editor.

1. Delete one of the A's in the word DATA in line 1. The misspelled word turns red in the Enhanced Editor, drawing attention to a problem. Restore the A before the next step. Notice after the correction the word DATA is again in boldface.
2. The punctuation used in SAS programming statements is the semicolon. Missing semicolons are a frequent source of error. Delete the semicolon after the term DATALINES and note the loss of the yellow highlighting in the next five lines. Restore the semicolon before continuing.
3. Some error indicators are a little more subtle. Delete the semicolon at the end of the PROC PRINT statement. The change in font color for the word TITLE in the following line is the result. Restore the semicolon.
4. Note the color of the text in the quotation marks of the TITLE statement. Delete the single quotation mark (or apostrophe) after the word Fines in the TITLE statement. Notice the rest of the program changes to this color indicating that the remaining lines of programming are now a part of a quotation. Restore the single quotation mark and note the color change of the text again.

Once the SAS program is entered and examined for any typographical errors, it can be "run". The SAS terminology is to ***submit*** the program. A program can be submitted in any of the following ways:

1. Click on the "running man" icon on the toolbar at the top of the screen to submit the program. If the mouse hovers over this icon for a few seconds, the word "Submit" will appear, identifying the function of this button. The definition of other buttons on this toolbar can similarly be examined.
2. While in the Enhanced Editor, clicking the *right* mouse button produces pop-up menu at the position of the cursor. Select **Submit All** with the *left* mouse button. Or select **Run-Submit** from the pull-down menus at the top of the screen.
3. Use the function key F8 to submit the program.
4. Type the word SUBMIT on the command line at the upper left of the SAS screen and press Enter.

When a SAS program executes, new content appears in the Log, Output, or Results Viewer windows. In the Log window the computer repeats back the program just submitted with some additional notes. This is also the window that will identify syntax errors if there are any. Move to the Log window. **Maximize** this window. Use the page up and page down keys to move back and forth in this window. If a programming error occurred in the submitted program, an error message will appear in the Log window. Any messages that follow the words ERROR, WARNING, or INVALID DATA mean that the program must be corrected in the *Editor window*, and it must be *submitted* again. The most common errors are missing semicolons, unclosed quotation marks, and misspellings.

To get more comfortable with the process of identifying and correcting syntax errors. Put any one of the four errors mentioned previously into the program. Submit the program and observe the error message(s) in the log window. There is often some confusion when it comes to identifying and correcting errors. Making a few errors on purpose, observing the error messages, and then correcting them is a useful learning tool.

If the program ran error free, then the requested results will appear in the Results Viewer. SAS 9.4 produces HTML output in the Results Viewer by default. SAS 9.2 and earlier versions produced Listing output in the Output window by default. There are advantages to both systems. Enabling and disabling Listing output and HTML output is overviewed in Chapter 19, Output Delivery System (ODS). If there is a minor programming error, output may still be produced in the Results Viewer or Output window. The results will not be complete nor correct however. If there is a major programming error, there will be nothing in the Results Viewer or Output window. If there are *any* errors, the program must be corrected in the Editor before it is resubmitted. These errors are syntax errors. If a typographical error in the data has been made, such as misspelling the names or incorrectly entering the fine values, then, of course, the data must be corrected and the program must be resubmitted. SAS error messages typically pertain to programming syntax.

1.3 Results and Explorer

The last two windows on the SAS screen are the **Results** and **Explorer** windows that appear at the left side of the screen and are tabbed at the bottom of the SAS screen.

The **Results** window creates an outline of the results produced by the program. By clicking on the name of the procedure in the Results window, the Results Viewer or Output window will advance to the output of that procedure. This is beneficial when output become quite lengthy and paging back and forth in the Results Viewer or Output window becomes cumbersome. However, for the short program in this chapter, the contents of the Results Viewer are quite concise. See Figure 1.4.

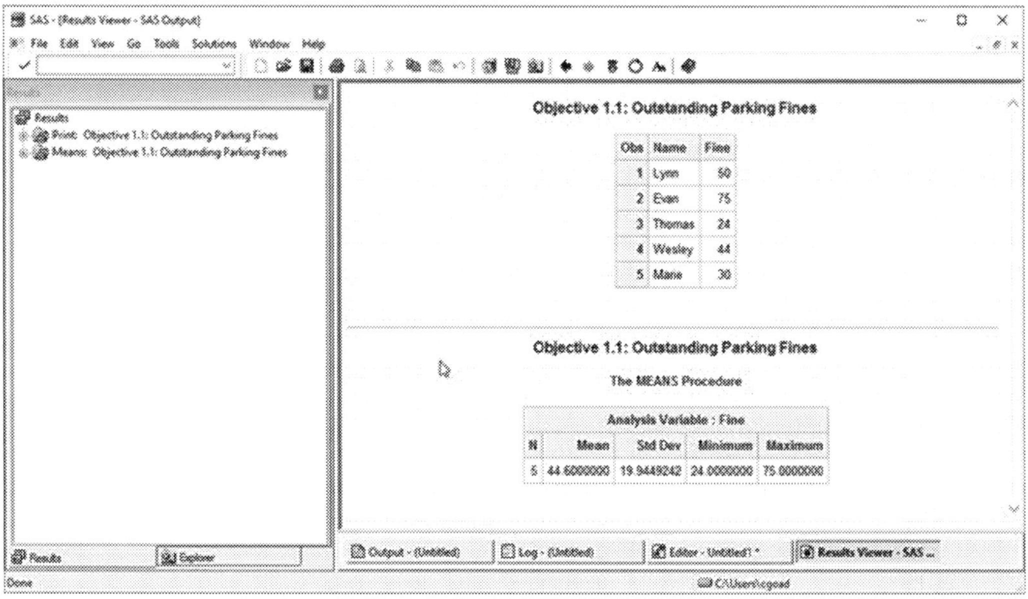

FIGURE 1.4
Results window (left) and Results Viewer (new bar at the bottom of the screen) for Objective 1.1.

Selecting the **Explorer** tab at the bottom left of the SAS screen moves to another window. The **Explorer** window lists all of the current SAS libraries, Favorite Folders, and My Computer (an index to the files on the computer). The **Explorer** window and SAS Libraries are covered in Chapter 12, SAS Libraries and Permanent SAS Data Sets.

1.4 Editing a SAS Program

The Editor window is the window in which SAS programs are written. To edit a program, simply move back to the Editor window and make the corrections or revisions to the program and **Run – Submit** it again. Be certain to check the Log window to make certain no more errors occur. Another advantage of working with the Enhanced Editor is that more than one Enhanced Editor window (more than one program) can be open at one time. After the program is submitted from the Enhanced Editor, the program will still appear in this window. Earlier versions of SAS and other current platforms use a Program Editor window. The limitation of the Program Editor is that only one Program Editor window can be open at any time. The editing process for the Program Editor is a bit different from the Enhanced Editor.

To move to the Program Editor window **View – Program Editor** is selected from the pull-down menu. Entering or copying and pasting the program from Objective 1.1 in Program Editor, it can be observed that the appearance and color features of this editor differ from those in the Enhanced Editor. The program can be submitted by the **Run – Submit** selection as before. After submitting a program from the Program Editor window, the Log and Results Viewer windows can still be viewed. However, when returning to the Program Editor, the program does not appear in this window as it did in the Enhanced Editor. SAS will store the most recently ran program in memory. The program must be recalled into the Program Editor. Once in the Program Editor, a program can be recalled in any of the following ways:

1. While in the Program Editor, **Run – Recall Last Submit** is selected from either the pull-down menu or the pop-up menu.
2. The function key F4 recalls the most recent program. Press F4 only once to recall the most recently submitted program.
3. Type RECALL (either uppercase or lowercase) on the command line and press **Enter**.

If several programs have been submitted or several revisions of the same SAS program have been submitted, pressing F4 multiple times or other multiple recall methods will "stack" up the recently submitted programs in the Program Editor. So only a single recall is necessary. SAS will only recall programs submitted during the current SAS session. If the SAS session was closed and then reopened for a new session, the submissions from the previous SAS session are no longer in memory. Once a program is recalled to the Program Editor, corrections or additions to the program can be added. After all of the changes to the program have been made, it can be submitted again. Checking the log window for any errors in the program is always recommended after running any program. If one is using the Enhanced Editor and clears the window after submitting a program, the most recently submitted program can also be recalled to the Enhanced Editor.

1.5 Clearing Log, Results Viewer, and Output Windows

When a second program or a revision of a previous program is submitted from either the Program Editor or Enhanced Editor, SAS puts the most recent log information after the previous program's log. Likewise, in the HTML Output in the Results Viewer the information is also "stacked up" or accumulates resulting in a lengthier output with redundant or incorrect results from the previous submission(s). To avoid this, the Log and Results Viewer windows can be cleared before submitting a new or corrected program. There are three ways that this can be done.

1. Move to the Log window. Select **Edit** from either the pull-down or pop-up menus. Then select **Clear All**. The Log window should then be empty. Repeat the **Edit – Clear All** process for the Results Viewer and/or Output window also.
2. Move to the Log window, type CLEAR on the command line, and then press **Enter**. Then move to other windows and repeat the CLEAR command. (CLEAR does not have to be entered in uppercase.)
3. When writing any SAS program, the following line can be used as the first line in the program.

   ```
   DM 'LOG; CLEAR; ODSRESULTS; CLEAR; ';
   ```

   When this line is at the top of the program, it will clear the Log and Results Viewer thereby clearing the history in those windows before producing a new log and new results. This avoids doing the procedures in steps 1 or 2. There are some times when those windows should not be cleared and information is allowed to accumulate. In that event, this Display Manager (DM) line should not be used in the program.

   Similarly, if the Listing output is enabled (Output window), the results will also accumulate in the Output. One can clear the Log and Output windows with the following statement:

   ```
   DM 'LOG; CLEAR; OUTPUT; CLEAR; ';
   ```

   If only the Log window is to be cleared, then the following line

   ```
   DM 'LOG; CLEAR; ';
   ```

   can be placed at the top of the program. Likewise, only the Results Viewer can be chosen to be cleared. Caution: Once these windows are cleared, the content cannot be recalled. The Recall capability is only available for submitted programs in either of the editors.

1.6 Saving the Program and Output

If the program is to be saved, the Enhanced Editor window must be the active window, and the content to be saved must appear in that window. If using Program Editor, the program may need to be recalled first. (**Run – Recall Last Submit** or F4). Either of the following two methods to save a program may be used:

1. From the pull-down menu select **File – Save As**. Browse to the folder or directory in which the program is to be saved. Enter a **File name** in the field. SAS will add the ".sas" extension to the file name as indicated in the field below **File name**. The information in the Results Viewer can similarly be saved but in webpage or text formats. Occasionally, the contents of the Log window may need to be saved. Move to the Log window and select **File – Save As** as above, then enter a **File name**. SAS will add a ".log" extension by default. This is a simple text file. Contents of the Output window (when enabled) are saved with an ".lst" extension indicating the Listing Output. The file type is also a simple text file.
2. Type FILE *a:\filename.sas* (or other appropriate extension) on the command line and press enter where the specific drive:\folder\information would be entered in place of *a:*. The extension *.sas* is used indicating that the contents of the file are a SAS program.

The item to be saved must always appear in the current or active window. A common mistake that beginners make is to save the output or results information rather than the program information. Content from any of the windows can also be copied and pasted into other applications and saved according to the procedures for that software. Thus, SAS results and/or programming can be easily included in report or presentation software.

1.7 Printing from SAS

A printout of the contents of the Editor, Log, Results Viewer, and Output windows can be obtained.

1. Select **File – Print Setup** from the pull-down menu to identify the printer, margins, font selection, and other attributes. **OK** Unless these settings are to be changed, it is not necessary to complete this process again. Then select **File** – **Print** from the pull-down menu. In the next dialog window, the printer is selected from the list of printers available, the number of copies, and the page range. Caution: If **Print to file** is enabled, a printed hardcopy of the document requested will not be produced. Disable **Print to file** by deleting the check in the box to the left of **Print to file.** From this Print dialog, a **Preview** of the printed document can be viewed or **OK** to begin printing.
2. Click on the printer button at the top of the screen to print the active window, provided that the correct printer has been previously been selected using the **File – Print Setup** sequence.

1.8 Executing a Stored SAS Program

If there is a previously stored SAS program, and it is to be edited and/or submitted, it must first be called into an editor. Ideally the file should be opened in the Enhanced Editor.

1. If the Program Editor is in use, the previous project in this window must be saved (**File – Save** as) or cleared (**Edit – Clear All**) before opening a saved file in this editor. Select **File – Open** from the menu and browse to the SAS program to be opened and select it. (**Open**) If the Program Editor is not clear, the newly opened file will be appended to the code already in that window. Only one Program Editor window can be open in SAS.
2. From any of the other windows (Enhanced Editor, Log, Output, and Results Viewer) select **File – Open Program** from the menu and browse to the SAS program to be opened and select it. (**Open**) A new Enhanced Editor window will open with this program code, and the window title will be the file name. Multiple SAS programs can be open at one time when using the Enhanced Editor.

When a new program is created or an existing program is edited in Enhanced Editor, the name of the program followed by an asterisk (*) will appear in the window bar and in the title at the top of the Enhanced Editor window. The asterisk indicates that the file has been changed since it was last saved. This is a feature of the Enhanced Editor only. The Program Editor does not do this.

If the program file to be opened was not created with a .sas extension, then changing the **Files of type** to **All Files (*.*)** in the **Open Program** dialog window must be completed before browsing to the file. Without the .sas extension, the Enhanced Editor features will not work as it does not recognize the file contents as a SAS program but a text file. To enable the features of the Enhanced Editor, the program must be resaved with a .sas extension.

1.9 Getting Help in SAS

Once familiar with the SAS programming environment, help in programming procedures can be accessed by selecting **Help – SAS Help and Documentation** from the menu at the top of the screen. Selecting the Help icon at the top of the SAS screen will also access SAS Help and Documentation. A screen capture of the **SAS Help and Documentation** window appears in Figure 1.5. Note that more information about the Enhanced Editor is available in this opening view. There are four tabs in the left pane from which to choose. In the **Contents** screen there are some general topics to choose from to help guide in an analysis. The **Index** screen is useful when seeking syntax assistance for a certain procedure as the procedure can be entered in the blank. A more general search can be done by selecting the **Search** tab.

1.10 Exiting SAS

When the SAS session is finished, one exits SAS by selecting **File – Exit** from either the pull-down or pop-up menus, or by clicking on the X in the upper right-hand corner of the SAS screen. There will be prompts to save any unsaved programs before SAS closes.

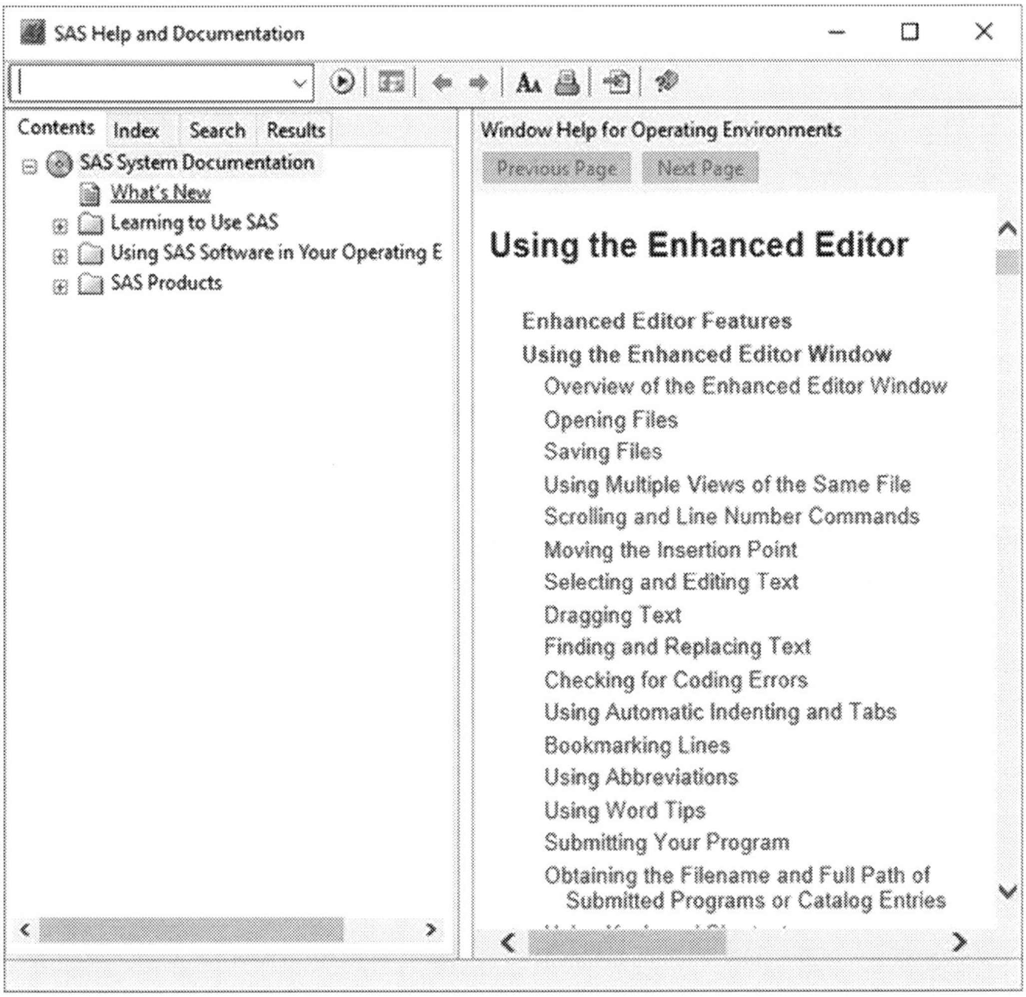

FIGURE 1.5
SAS Help and Documentation opening window.

1.11 Chapter Summary

In this chapter, a very basic introduction to the SAS Editor has been presented. It was not the purpose of this chapter to teach SAS syntax but to demonstrate how to enter a program in an editor and then run or submit the program. Editing and saving program files were additional skills overviewed. In the following chapters, programming syntax concepts are presented in more detail. As new topics are introduced, readers are encouraged to work through the objectives presented.

2

DATA Step Information 1

2.1 A Simple Form for a SAS Program

A typical SAS program may consist of a block of statements for data set creation and a block of statements for a statistical or graphical analysis. For data set creation or manipulation, the DATA step is used. Using the DATA step, SAS data sets can be created or "read" into the program, and variables in the data set can be modified. Once a data set is recognized by SAS then analysis or graphics procedures can be performed on the data set. Procedures are identified by SAS as "PROC's". A SAS program would typically include at least one DATA step and one or more procedures. This text will introduce some of the basic statistical procedures (or PROC's).

Important in SAS programming is the inclusion of RUN and QUIT statements. After DATA steps or PROC's a RUN statement can be included, or a single RUN statement can appear at the end of the SAS procedures. The last line of a SAS program is the QUIT statement.

The following rules apply when interpreting the SAS programming code printed in this book. These rules are also consistent with literature from SAS Institute. Any item that is typed in UPPER CASE is a SAS statement and must be spelled exactly as it appears. Items in lower case or *lower case italics* are items the user can define, such as variable names or optional text. When one writes programs, it is generally not necessary to use upper and lower case letters as is done in the examples. When case sensitivity is required, it will be clearly specified. The two cases are used to distinguish between SAS statements or commands and user-defined text. The punctuation in SAS is a semicolon (;). It is necessary at the end of each SAS statement in order for the program to execute. Omitting a semicolon causes run-on statements in the same way the omitting a period causes run-on sentences in written text.

As in Chapter 1, when this text refers to items on the screen that can be selected using point-and-click methods, the item will appear in boldface print. For example, to save a program, select the **File – Save As** sequence from the pull-down menus at the top of the screen.

2.2 Creating Data Sets Using INPUT and DATALINES Statements

A data set may be created within a SAS program. The simplest way to do this is:

DATA *SAS-data-set;*
INPUT *variable1 variable2 . . . variablen;*
DATALINES;

Any item typed in italics indicates variable or data set names that are up to the SAS programmer to define. When selecting names for these items, the following SAS naming conventions must be followed. The first character must be alphabetic or an underscore (_). In earlier versions of SAS these names could be no longer than eight characters, but currently up to 32 characters can be used for these names. These names **cannot** contain spaces. A single space is necessary to separate the items in each line or statement of the SAS program; therefore, spaces cannot be included in user-defined variable names.

DATA Statement

The DATA statement is where one indicates that a SAS data set is being created and assigns a name to the SAS data set adhering to the aforementioned SAS naming conventions. This statement must end in a semicolon. When coding DATA statements, one should name each SAS data set in the program code. Although naming the data set is optional since SAS will assign a default name to the SAS data set, an unnamed data set may cause confusion in debugging programs for the beginning programmer or when programs are more complicated.

INPUT Statement

The INPUT statement follows the DATA statement. In the INPUT statement one must indicate each variable that is in the data set. Each variable name corresponds to a column of information. The order in which the variables are listed in the INPUT statement must be the same order that information appears in the lines following the DATALINES statement. The values of the variables can be either numeric (prefix + and – signs allowed) or character (string). Variables are assumed to be numeric unless otherwise indicated in the INPUT statement. One identifies character variables in the INPUT statement by following the character variable name with a space and a dollar sign, $. The INPUT line is the only place the dollar sign must be specified in the program for the character variables.

DATALINES Statement

The DATALINES statement follows the INPUT statement, and indicates that the values for the variables defined in the INPUT statement will follow this statement. In later chapters statements that modify the SAS data set will be presented, and these statements are positioned after the INPUT statement and before the DATALINES statement.

OBJECTIVE 2.1: Create a SAS data set named "test" that contains student names, identification numbers, an exam score, and a letter grade. Add a RUN to create this SAS data set.

```
DATA test;
INPUT name $ id score grade $;
DATALINES;
Bill  123000000  85 B
Helen  234000000  96 A
Steven  345000000  80 B
Carla  456000000  65 C
Dana  567000000  97 A
Lisa  789000000  81 B
;
RUN;
```

The data values or records follow the DATALINES statement. Note: There are no semicolons at the ends of the lines of the data. The data values are not SAS statements and are therefore not punctuated. At least one space separates the various pieces of information in the lines of data values. Earlier versions of SAS used a CARDS statement rather than the DATALINES statement. The CARDS statement will still work as shown below.

```
DATA test;
INPUT name $ id score grade $;
CARDS;
Bill  123000000  85 B
Helen  234000000  96 A
Steven  345000000  80 B
Carla  456000000  65 C
Dana  567000000  97 A
Lisa  789000000  81 B
;
RUN;
```

From the Log window for the first DATA step code submitted:

```
1    DATA test;
2    INPUT name $ id score grade $;
3    DATALINES;

NOTE: The data set WORK.TEST has 6 observations and 4 variables.
NOTE: DATA statement used (Total process time):
  real time      0.01 seconds
  cpu time       0.00 seconds
```

When the DATA step successfully creates the SAS data set *test*, one will be able to confirm that using the feedback in the Log window. The feedback line of interest is, "Note: The data set WORK.TEST has 6 observations and 4 variables." The usage of the two-level SAS data set name, WORK.TEST, is covered in Chapter 12. For now, the SAS data set name created in the DATA statement is prefaced by "WORK." in the feedback in the Log. The number of observations in this note refers to the number of lines of data, and the number of variables is the number of variables in the INPUT statement or the number of columns of information one has collected. When one is entering data in a DATA step in the Editor window, the amount of data a researcher has collected is generally known. Thus, if anything other than six observations and/or four variables in this application were indicated, this quickly indicates that there is an error in the creation of the SAS data set *test*. There are no printed results for the DATA step operation. To obtain a printed list of the data, the PRINT procedure (Section 2.3) can be used.

When an INPUT statement is used as above, at most one observation per line must appear in each line after the DATALINES or CARDS statement. Using the above syntax, at least one space is needed between "pieces" of data. One does not use any punctuation to separate pieces of data with the syntax shown for the above INPUT statements.

2.2.1 Column and Line Pointer Controls

The number of spaces between items in the actual lines of data usually does not matter. However, the INPUT statement can be modified to precisely specify the number of spaces. To do this, one must specify which column(s) in the Editor the responses to each variable appear using column pointer controls in the INPUT statement. These column pointer controls typically appear after the variable. If the variable is a character variable, the column pointer controls appear after the $and before the next variable in the statement.

OBJECTIVE 2.2: Modify the INPUT statement with column pointer controls to precisely specify in which column(s) the value of each variable will appear.

```
DATA test;
INPUT name $ 1-6 id 9-17 score 21-22 grade $ 24;
DATALINES;
Bill    123000000   85 B
Helen   234000000   96 A
Steven  345000000   80 B
Carla   456000000   65 C
Dana    567000000   97 A
Lisa    789000000   81 B
;
```

Or, a second approach is given by

```
DATA test;
INPUT name $ 1-6 id 7-15 score 16-18 grade $ 19;
DATALINES;
Bill  123000000 85B
Helen 234000000 96A
Steven345000000 80B
Carla 456000000 65C
Dana  567000000 97A
Lisa  789000000 81B
;
```

One can add a RUN statement at the end of either of the previous DATA steps and submit the program. The contents of the Log window can be examined for errors. When the start column and the end column are specified for each variable in the INPUT statement, it is not necessary to leave spaces between pieces of information in the lines of data as shown in the second illustration. If one does not correctly specify the column numbers, incorrect values for the affected variables will result. Data loggers or data collection software may record values in a format where using the column pointer controls become necessary.

In the second illustration of Objective 2.2, score was allotted three columns. This would accommodate a student scoring 100 on an exam. In the first illustration, a score of 100 is not accommodated since only two columns, 21 and 22, are allotted to the variable score. If the column pointer controls are misspecified, this can cause a severe data error. For example, if in the second illustration score was specified in columns 16–17, this would create a severe error as each of the scores in this illustration would only be a single digit, 8, 9, 6, 8, 9, 8.

Another, less severe example, follows. If the INPUT statement were:

```
INPUT name $ 1-5 id 7-15 score 16-18 grade $ 19;
```

where only five columns are allotted to the variable name. Steven would be read as Steve.

If a character or text string has a space in it, column pointer controls must be used for that variable. For example, suppose student Bill is listed as Billy Joe. Column pointer controls can be modified so that the variable *name* can be allotted to more columns thereby picking up the space. The character variable *name* would have to be allotted at least nine columns for **all lines of data** after the DATALINES statement. That is,

```
DATA test;
INPUT name $ 1-9 id 12-21 score 22-23 grade $ 25;
DATALINES;
Billy Joe  123000000 85 B
Helen      234000000 96 A
Steven     345000000 80 B
Carla      456000000 65 C
Dana       567000000 97 A
Lisa       789000000 81 B
;
```

Instead of specifying a range of columns for the variable *name,* one could claim nine columns by specifying the number of columns followed by a period. An equivalent INPUT statement is:

```
INPUT name $ 9. id 12-21 score 22-23 grade $ 25;
```

"9." following the variable *name* indicates that *name* is allowed nine columns. It is not necessary to separate the column controls from the character string indicator, $, in the INPUT statement either.

Since *id* does not appear until column 12, the following INPUT statement is also equivalent to the previous two:

```
INPUT name $11. id 12-21 score 22-23 grade $ 25;
```

It is important to remember that SAS truncates character variables after eight characters unless column pointer controls are used. For example, longer names, such as Catherine or Christopher, would require that the *name* variable in the INPUT statement use some form of column pointer controls to capture all characters in those names.

One can also use @n to specify the cursor control to begin at column n and read the next variable. @n must appear **before** the variable in the INPUT statement. The following is another option for the INPUT statement:

```
INPUT name $ 9. @12 id score 22-23 @25 grade $;
```

This INPUT statement also correctly reads all of the variables. The character variable *name* is allotted nine columns. The cursor moves to column 12 and then reads the numeric variable *id, score* is numeric and occupies columns 22 and 23, and then the cursor moves to column 25 and reads the character variables *grade.* The *id* variable is a numeric variable, and all digits are recorded in this number. However, if *id* were identified as a character variable (since it is not likely to perform computations on an ID), it would be truncated to eight characters unless the INPUT statement identified that nine characters comprise this variable. That is,

```
INPUT name $9. @12 id $9. score 22-23  @25 grade $;
```

Not all variables in an INPUT statement have to have column pointer controls. Suppose only the variable *name* has column pointer controls. Without column pointer controls specified for the remaining variables, at least one space must separate the different values in the data lines, and the spacing for these variables does not have to be the same in each record although proofreading the data is easier when it is.

```
DATA test;
INPUT name $ 1-9 id score grade $;
DATALINES;
Billy Joe  123000000 85 B
Helen      234000000 96 A
Steven      345000000  80 B
Carla      456000000 65 C
Dana        567000000  97 A
Lisa        789000000 81 B
;
```

The lower right-hand corner of the SAS Editor screen can assist in identifying the values to specify for the column pointer controls in the INPUT statement. See Figure 2.1 in which column 9 is identified as the first column in which the variable *id* appears.

If there are a large number of variables, it may become necessary to use more than one line for each observation. In these cases, line pointer controls are needed. Line n of the data record is identified by "#n" in the INPUT statement before the first variable in that line of the record. Although there are not a large number of variables in the current example, this example can be used to illustrate the concept. A data logger or other automated data collection device may format information using more than one line per record. Column pointer controls do not have to be used when line pointer controls are used. Values in the lines of data can be separated by at least one space as they were in Objective 2.1.

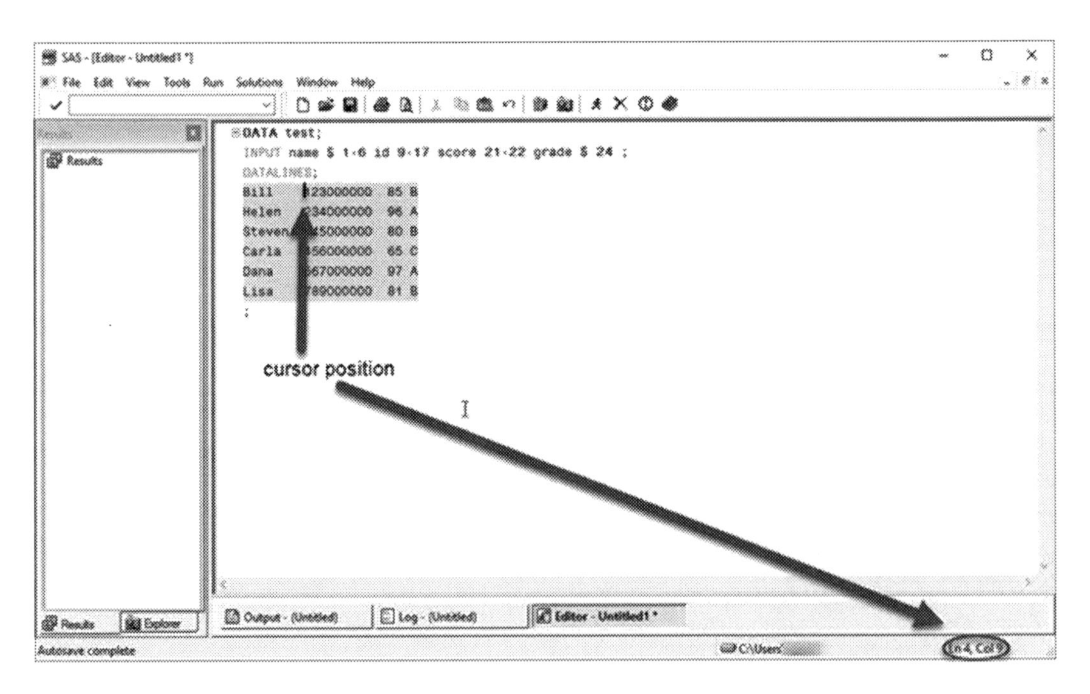

FIGURE 2.1
Identifying the column number for the cursor.

OBJECTIVE 2.3: Redo Objective 2.2 using line pointer controls and column pointer controls where line 1 of the data record contains the variables *name* and *id*, and line 2 of the data record contains *score* and *grade*.

```
DATA test;
INPUT #1 name $ 1-6 id 9-17
      #2 score 1-2 grade $ 4;
DATALINES;
Bill     123000000
85 B
Helen    234000000
96 A
Steven   345000000
80 B
Carla    456000000
65 C
Dana     567000000
97 A
Lisa     789000000
81 B
;
```

After the DATALINES statement the DATA step reads one value for each variable named in the INPUT statement by default, and line pointer controls can be used when each data record takes more than one line. When there are only a few variables, several observations can be included in the same line after the DATALINES statement with the simple addition of "@@" in the INPUT statement. SAS refers to this as "the double trailing @". The double trailing @ at the end of the INPUT statement and before the semicolon indicates that more than one observation may appear in each line after the DATALINES statement. When using the double trailing @ or "@@" convention, column pointer controls cannot be used.

OBJECTIVE 2.4: Use @@ (the double trailing @) when creating the data set *test*.

```
DATA test;
INPUT name $ id score grade $ @@;
DATALINES;
Bill  123000000  85 B  Helen   234000000  96 A  Steven   345000000  80 B
Carla  456000000  65 C Dana  567000000  97 A Lisa     789000000  81 B
;
RUN;
```

If one does not specify @@ on the INPUT line, only the first four pieces of information per row of data will be read into the data set. That is, only Bill 123000000 85 B and Carla 456000000 65 C are the only observations in Objective 2.4 that will be read in the data set *test* in this illustration.

2.2.2 Missing Data

When entering data, if the response to a variable is missing or unknown, one enters the other pieces of information in the record and puts a single period (.) in the position of the missing observation. Missing data for either character or numeric data are acknowledged the same way.

OBJECTIVE 2.5: Repeat Objective 2.1 with the modification that the name of the second student is missing, and Carla does not have values for score and grade.

```
DATA test;
INPUT name $ id score grade $;
DATALINES;
Bill    123000000  85 B
.    234000000  96 A
Steven  345000000  80 B
Carla  456000000  . .
Dana    567000000  97 A
Lisa  789000000  81 B
;
RUN;
```

If no column pointer controls are used and items that are missing from the data records are NOT indicated using a period, SAS will advance over the space(s) to read the next value. Additionally when the last item in a data record or line is missing, such as Carla's grade in Objective 2.5, omitting the period to indicate a missing value will cause an error in the creation of the SAS data set. SAS will wrap around to the next line in the record to find that response for Grade. For example,

If the DATA step did not mark missing values, SAS will create the following for *test:*

```
DATA test;
INPUT name $ id score grade $;
DATALINES;
Bill    123000000  85 B
    234000000  96 A
Steven  345000000  80 B
Carla  456000000
Dana    567000000  97 A
Lisa  789000000  81 B
;
RUN;
```

name	id	score	Grade
Bill	123000000	85	B
23400000	96	.	Steven
Carla	456000000	.	56700000
Lisa	789000000	81	B

Since no missing values were marked with a period, and there is no name given in line 2 after DATALINES, the value for ID in line 2 is read as the student's name. The value of score is read as the ID. Score is a numeric variable (no $in the INPUT statement), and the next non-missing value is the A value for Grade. Since score is numeric, this results in an "invalid data" error in the Log. By shifting all of the information in the second line after the DATALINES to accommodate the missing value, SAS wraps around to a new line to find the "Steven" character value for grade.

The error messages in the Log indicate all of the invalid data problems that exist.

```
1      DATA test;
2      INPUT name $ id score grade $;
3      DATALINES;
NOTE: Invalid data for score in line 5 19-19.
RULE:----+----1----+----2----+----3----+----4----+----5----+----6----+----7----+----8----+--
6           Steven  345000000  80 B

name=23400000 id=96 score=. grade=Steven _ERROR_=1 _N_=2
NOTE: Invalid data for score in line 8 1-4.
8           Dana    567000000   97 A
name=Carla id=456000000 score=. grade=56700000 _ERROR_=1 _N_=3
NOTE: SAS went to a new line when INPUT statement reached past the end of
a line.
NOTE: The data set WORK.TEST has 4 observations and 4 variables.
NOTE: DATA statement used (Total process time):
      real time           2.19 seconds
      cpu time            0.03 seconds
```

Near the end of the log the SAS programmer is notified of that wrap around by "Note: SAS went to a new line when INPUT statement reached past the end of a line". Additionally, this SAS data set should have six student records. The log also states that, "Note: The data set WORK.TEST has 4 observations and 4 variables". Thus, the SAS data set *test* does not contain all six of the student records, and one must work on debugging the DATA step.

When using column pointer controls, a blank space or a single period in one of the indicated columns will identify a missing value for a variable. This can be verified when the SAS data set is printed or viewed. Printing the SAS data set is covered in Section 2.3, and ViewTable is covered in Section 8.3.

It is sometimes reported that a semicolon must appear on a line by itself after the last line of data. This is not necessary for SAS to read in all of the data. SAS will correctly read in the data whether one uses the semicolon or not. This is optional, but is very commonly done and is recommended by many programmers. This use of a single semicolon was included in the previous illustrations and objectives.

Also, data **are** case sensitive. If the grade of 'a' and the grade of 'A' were entered in the lines of data, these will be regarded differently. So, while case sensitivity typically does not matter with respect to the SAS code, it **does** matter in the data.

These are just a few of the simpler ways that one can create a SAS data set. In Chapters 8 and 14 methods for reading or importing an external data file into SAS will be addressed.

When programmers and SAS literature refer to a "DATA step", the reference is to the DATA statement and the supporting lines following it. In these objectives, the three statements, DATA, INPUT, DATALINES together compose a DATA step. The INPUT statement or DATALINES statement by themselves do not have any utility. They are a part of a DATA step. Other DATA step statements and operations will be covered in later chapters.

As long as SAS statements are punctuated with a semicolon correctly, multiple statements per line or line breaks within a SAS statement are permissible. Additionally, one can choose to indent lines of SAS code or leave blank lines without creating a programming error. Indented statements and/or blank lines are style choices used by some programmers to assist in visually identifying statements that compose a block of code, such as a DATA step.

Each of the following four blocks of DATA step statements are equivalent.

```
DATA test; INPUT name $ id score grade $; DATALINES;
```

```
DATA test; INPUT name $ id
score grade $;
DATALINES;
```

```
DATA test;
    INPUT name $ id score grade $;
    DATALINES;
```

```
DATA test;


    INPUT name $ id score grade $;
    DATALINES;
```

2.3 Printing a Data Set – The PRINT Procedure

The DATA step alone does not produce any printed output. In addition to proofreading a SAS program, one of the simplest ways to examine the information in a SAS data set is to print the columns of values. The PRINT procedure can be used to do this.

The syntax of the PRINT procedure is:

PROC PRINT *<options>* ;
VAR *variable(s);* *(optional statement)*
ID *variable(s);* *(optional statement)*
BY *variable(s);* *(optional statement)*
SUM *variable(s);* *(optional statement)*

PROC PRINT statement options are:

DATA = *SAS-data-set* names the SAS data set to print. While this is optional, it is generally recommended that one specify the SAS data set name in procedures.

DOUBLE double-spaces the printed LISTING output. If using HTML output (the Results Viewer), this option has no effect.

NOOBS The PRINT procedure prints an observation number in the first column for each record in the data set by default. The column header for this is "OBS". The NOOBS option suppresses the observation number in the printed results.

UNIFORM formats each column width the same on all pages needed for the list of the data. This is useful when one has a large number of variables or columns in the data set.

LABEL uses variable labels as column headings. Without the LABEL option the column headings are the variable names defined in the SAS data set. The LABEL statement is covered in Chapter 8.

N prints the number of observations in the SAS data set. If a BY statement is used, the number of observations for each BY group is also printed. The BY statement requires that the data be sorted compatibly. The SORT procedure is covered in Section 2.6.

```
VAR variable(s);
```

The VAR statement allows one to specify which variables are to be printed and the order in which the columns (or variables) should appear. If the VAR statement is not used, all of the variables in the SAS data set will be printed in the order they were defined. This statement is useful if a data set has a large number of variables or columns, and one only wishes to view some of those columns.

```
ID variable(s);
```

When the data set is printed, the variable or variables in the ID statement will be printed in the first or left columns of the output. If one or more variables in the data set serves as an identifier, it is important to list those variables in the ID statement. The order of the variables in this statement also determines the order in which these identification columns will appear.

```
BY variable(s);
```

If one or more variables is used to define subgroups within the SAS data set, these variables should be specified in the BY statement. In order for the BY statement to function properly, the SAS data set must first be compatibly sorted by the variables in the BY statement. The SORT procedure is presented in Section 2.6.

```
SUM variable(s);
```

The variables in this statement will be summed or added, and the total will be printed at the bottom of the column. This may be useful when selected columns of the SAS data set are included in a report. See also the REPORT Procedure in Chapter 10.

The VAR, ID, BY, and SUM statements are all optional statements. The PRINT Procedure may simply consist only of the PROC PRINT statement. This single statement would print all columns of the SAS data set.

OBJECTIVE 2.6: Create a SAS data, *test*, set containing the names, identification numbers, test scores, and grades for a class of six students using Objective 2.1 for the data set creation. Print the resulting SAS data set.

```
DATA test;
INPUT name $ id score grade $;
DATALINES;
Bill   123000000  85 B
Helen  234000000  96 A
Steven   345000000  80 B
Carla  456000000  65 C
Dana   567000000  97 A
Lisa   789000000  81 B
;
PROC PRINT DATA=test;
RUN;
QUIT;
```

In this objective, the data set *test* is to be printed. When a data set is printed, SAS will assign an observation number to each line of the data. This observation number is in the

OBS column. It is not necessary to specify DATA=*SAS-data-set* in the PROC PRINT statement. If one does not, the SAS procedure will operate on the most recently created data set. However, as SAS programs become longer and include multiple SAS data sets, using the DATA=*SAS-data-set* option is strongly recommended.

The results of the PRINT procedure appear below. Note the data are case sensitive.

Obs	name	id	score	grade
1	Bill	123000000	85	B
2	Helen	234000000	96	A
3	Steven	345000000	80	B
4	Carla	456000000	65	C
5	Dana	567000000	97	A
6	Lisa	789000000	81	B

Modification: Run this objective using Objectives 2.2 – 2.5 previously given for the DATA step confirming that those variations on the INPUT statement are successful.

In the results for Objective 2.6 one may wish that the variable names had appropriate capitalization of the first letter of the variable name such as one would write in a formal report. There are a few places in SAS code where the code is case sensitive. The INPUT statement is one of those places. Typically, short but meaningful variable names in the INPUT statement are used. If more detailed text is to appear in place of the shorter variable name, a LABEL statement (Chapter 8) can be included in the DATA step. However, capitalizing a letter or two in the INPUT statement can improve the appearance of a tabular report or list.

OBJECTIVE 2.7: Create the SAS data set *test* as in Objective 2.1. Print the data set without the SAS-defined observation numbers. Additionally, capitalize the first letter of each variable name in the INPUT statement and both letters in "id" variable name.

```
DATA test;
INPUT Name $ ID Score Grade $;
DATALINES;
Bill  123000000  85 B
Helen  234000000  96 A
Steven   345000000  80 B
Carla  456000000  65 C
Dana  567000000  97 A
Lisa  789000000  81 B
;
PROC PRINT DATA=test NOOBS;
RUN;
QUIT;
```

Name	ID	Score	Grade
Bill	123000000	85	B
Helen	234000000	96	A
Steven	345000000	80	B
Carla	456000000	65	C
Dana	567000000	97	A
Lisa	789000000	81	B

When there are a large number of variables (or columns) in a SAS data set and only some of the variables are to be printed, one can specify which variables are to be printed and the order in which the variables will be printed using a VAR statement. If a VAR statement is not used, all variables in the SAS data set will be printed.

OBJECTIVE 2.8: Create the SAS data set *test* as in Objective 2.7 but only print the observation numbers, the student names, and letter grades.

```
DATA test;
INPUT Name $ ID Score Grade $;
DATALINES;
Bill   123000000   85 B
Helen   234000000   96 A
Steven   345000000   80 B
Carla   456000000   65 C
Dana   567000000   97 A
Lisa   789000000   81 B
;
PROC PRINT DATA=test;
VAR name grade;
RUN;
QUIT;
```

Obs	Name	Grade
1	Bill	B
2	Helen	A
3	Steven	B
4	Carla	C
5	Dana	A
6	Lisa	B

2.4 TITLE Statement

The TITLE statement can be used with almost all SAS procedures. The function of the TITLE statement is to label the top of each page of results with specified text for a title. The syntax of the TITLE statement is:

TITLE "text for title here";

One may use either a pair of double quotes (") or a pair of single quotes (') to designate the text of the title. Whichever symbol is used, the quotation must be closed. Unbalanced quotation marks in programs can disable all or almost all of a SAS program. For the text of a title, any letters, numbers, or symbols can be used. SAS can distinguish an apostrophe (') as in the word *can't* only if the text of the title is enclosed in double quotes ("), such as,

```
TITLE "can't";
```

If the TITLE statement reads,

```
TITLE 'can't';
```

SAS cannot distinguish between the single quotation mark and the apostrophe, and the result is an "unbalanced quotation marks" error in the log. Using the Enhanced Editor, unbalanced quotation marks can be detected. Text for titles should appear in a different color on screen. If several lines of the program appear in the same color as the title quotation, then unbalanced quotation marks are evident. The punctuation in the TITLE statement must be corrected before submitting the program.

There can be up to ten lines of title on a page. TITLE and TITLE1 both specify the first line of the title. For more than one line in the title, the TITLEn statement, where n is a number from 1 to 10, can be used. The text specified in the TITLEn statements is, by default, centered at the top of the page of output.

OBJECTIVE 2.9: Create the SAS data set *test* as in Objective 2.7. Print the contents of the SAS data set with two lines of titles. Suppress the observation numbers.

```
DATA test;
INPUT Name $ ID Score Grade $;
DATALINES;
Bill  123000000  85 B
Helen  234000000  96 A
Steven   345000000  80 B
Carla  456000000  65 C
Dana  567000000  97 A
Lisa  789000000  81 B
;
PROC PRINT DATA=test NOOBS;
TITLE 'Objective 2.9';
TITLE3 'Test Scores';
RUN;
QUIT;
```

Objective 2.9

Test Scores

Name	ID	Score	Grade
Bill	123000000	85	B
Helen	234000000	96	A
Steven	345000000	80	B
Carla	456000000	65	C
Dana	567000000	97	A
Lisa	789000000	81	B

Only the first and third lines of title are used to achieve the effect of double spacing in the titles. It's OK to skip one or more lines in the titles. Changing TITLE3 to TITLE2 in the above program will keep the lines of titles single spaced.

As programs get larger and involve more procedures (PROC's), the titles can change with each procedure. Simply type the new TITLE statements in each of the procedures.

To clear all lines of titles at any point in the program, one can include the simple statement:

```
TITLE;
```

in the program. This clears or resets all lines of titles.

```
TITLEn;
```

erases all lines of titles from the nth to the tenth. (Specify n.)

2.5 FOOTNOTE Statement

Similar to the TITLE statement is the FOOTNOTE statement. FOOTNOTE adds lines of text to the bottom of the pages of the SAS output. There are ten lines of footnotes available. The syntax of the FOOTNOTE statement is:

```
FOOTNOTEn "insert footnote text here";
```

where n can be a number from 1 to 10, and either double (") or single (') quotes can be used to define the text of the footnote. The colorization in the Enhanced Editor can again help one detect unbalanced quotations in the FOOTNOTE statements.

To clear or reset all footnotes, include

```
FOOTNOTE;
```

in the program. Or to reset the footnote lines from n through 10,

```
FOOTNOTEn;
```

where n is a number from 1 to 10.

It is important to point out that once a title or a footnote is created during a SAS session, that title or footnote stays active until one changes it, resets it, or closes the SAS session. In SAS vernacular this is stated as: TITLE and FOOTNOTE statements are global statements. Therefore, a title or footnote created in an earlier program can appear in the output of the current program if the current program does not change it. When running several programs in one SAS session, TITLE; and FOOTNOTE; statements are useful tools. By placing TITLE; FOOTNOTE; at the top of a new program, all lines of titles and footnotes generated by previous programs will be reset. Some programmers include this reset at the top of their programs after the Display Manager (DM) statement presented in Chapter 1. That is,

```
DM 'LOG; CLEAR; ODSRESULTS; CLEAR; ';
TITLE; FOOTNOTE;
```

2.6 Sorting a Data Set – The SORT Procedure

After a SAS data set is created, the data is kept in the same order as it was entered. This is evident when one creates a SAS data set and then prints the SAS data set. The lines of data can be reordered or sorted according to a specific variable in either ascending or descending order. One can also sort data using more than one variable to determine the order. The SORT procedure performs these operations. The syntax of the SORT procedure is:

```
PROC SORT DATA=SAS-data-set;
BY variable1 variable2 … variablen;
```

The lines in the SAS data set would first be placed in ascending order according to variable1. Within each level or value of variable1, the data is sorted according to variable 2 and so on. The SORT procedure **does not** produce any printed output or printed results by itself. Thus, TITLE and FOOTNOTE statements in a SORT procedure are not needed. To view the effects of sorting the data one can include a PRINT procedure after the SORT procedure.

OBJECTIVE 2.10: Sort the *test* data created in Objective 2.7 by putting the student names in alphabetical order. Print only the student names and grades.

```
DATA test;
INPUT Name $ ID Score Grade $;
DATALINES;
Bill  123000000  85 B
Helen  234000000  96 A
Steven   345000000  80 B
Carla  456000000  65 C
Dana  567000000  97 A
Lisa  789000000  81 B
;
PROC SORT DATA=test; BY name;
PROC PRINT DATA=test NOOBS;
VAR name grade;
TITLE 'Objective 2.10';
RUN;
QUIT;
```

Objective 2.10

Name	Grade
Bill	B
Carla	C
Dana	A
Helen	A
Lisa	B
Steven	B

By default, the SORT procedure sorts items in ascending order: 1 2 3 … or A B C … . This may be changed in the program. If data are to be placed in descending order according to a specified variable, type DESCENDING prior to the variable in the BY statement of the SORT procedure. Additionally, the data can be sorted using more than one variable in the BY statement of the SORT procedure.

Most SAS procedures can be modified with a BY statement. In Section 2.3, recall that the BY statement was an optional statement for the PRINT procedure. If the data are first sorted by a variable, then the PRINT procedure (and other SAS procedures) can be implemented for each level of the sorted variable. At most one BY statement can be used in any procedure.

OBJECTIVE 2.11: Sort the *test* data so that the lowest grades are listed first. For grades that are the same, list the student names in alphabetical order. Print the results two different ways. First, print only the student names and grades in the newly sorted order. Second, print the student names in alphabetical order for each grade using a BY statement on the PRINT procedure. That is, print the data by grade, lowest grade to highest grade. Suppress observation numbers in both PRINT procedures.

```
DATA test;
INPUT Name $ ID Score Grade $;
DATALINES;
Bill  123000000  85 B
Helen  234000000  96 A
Steven  345000000  80 B
Carla  456000000  65 C
Dana  567000000  97 A
Lisa  789000000  81 B
;
PROC SORT DATA=test;
BY DESCENDING grade name;

PROC PRINT DATA=test NOOBS;
VAR name grade;
TITLE 'Objective 2.11';
TITLE2 'Part 1';

PROC PRINT DATA=test NOOBS;
BY DESCENDING grade;
VAR name;
TITLE2 'Part 2';
RUN;
QUIT;
```

In the SORT procedure, DESCENDING must precede each BY variable that is to be sorted in reverse order. In the PRINT procedure, the DESCENDING modifier must be used in this BY statement also since the data have been sorted in descending order by grade. If "BY grade;" were programmed in this second PRINT procedure, then an error would occur (with an error message in the log) because the data is not sorted by ascending values of grade as the "BY grade;" statement would expect.

Objective 2.11

Part 1

Name	Grade
Carla	C
Bill	B
Lisa	B
Steven	B
Dana	A
Helen	A

Objective 2.11

Part 2

Grade=C

Name
Carla

Grade=B

Name
Bill
Lisa
Steven

Grade=A

Name
Dana
Helen

The grades are in descending order, and the names are in ascending order within each value of grade. The BY statement is used on the second PRINT procedure in the program resulting in a small table or list for each value of grade.

OBJECTIVE 2.12: Modify the VAR statement in Objective 2.11 to:

```
VAR name grade;
```

and resubmit. One can observe that grade is redundantly listed both as a subheading and in the list with the names. (This is left as an exercise for the reader.)

The SORT procedure is useful in many ways. Often calculations or an analysis (rather than just the PRINT procedure) are necessary for each level of a particular variable. If the SAS data set is first sorted by a specific variable, many SAS procedures can be invoked which can operate on the sorted data sets.

Only one SORT procedure is used in Objective 2.11. The data did not have to be sorted a second time prior to the second PRINT procedure since there was no change needed in the sorted order. One should avoid unnecessarily sorting the data. Once a SAS data set is sorted, SAS holds that sorted data in memory until the SAS session closes or until the SAS data set is sorted by some other set of variables. Throughout this book "BY processing" on other procedures requires the use of the SORT procedure.

2.7 Chapter Summary

In this chapter, using a simple example and a DATA step, it was shown how to create a SAS data set within a SAS program. Several forms of the INPUT statement were presented to accommodate the many forms in which data could be arranged. Once a SAS data set is created, one can add blocks of procedure code to a program. The PRINT procedure requests the contents of a SAS data set be printed. One likely would not want to print a SAS data set that is large, having either many columns or many lines or observations in the data set. The SORT procedure has the capability of putting the data in ascending or descending order with respect to one or more variables. The SORT procedure produces no printed output, but orders the SAS data set for further use during the current SAS session.

3

Summarizing Data Basics

In a beginning statistical methods course, first concepts include summary statistics, such as measures of central tendency and measures of dispersion. In this introduction to SAS programming, it makes sense to introduce procedures which perform these computations and many other summary statistics. Two procedures that generate statistical summary information for specified variables are the UNIVARIATE procedure and the MEANS procedure. The syntaxes for the two procedures are quite similar.

3.1 The UNIVARIATE Procedure

PROC UNIVARIATE DATA= *SAS-data-set <options>;*
BY *variable(s);* *(optional statement)*
WHERE *condition;* *(optional statement, see Chapter 4, DATA Step Information 2)*
CLASS *variable(s);* *(optional statement)*
VAR *variable1 variable2 ... variablen;*
HISTOGRAM *variable(s) </options>;* *(optional statement)*
OUTPUT OUT= *new-SAS-data-set <list options and new variable names>;*

PROC UNIVARIATE Statement

Some of the possible options for the UNIVARIATE procedure are:

ALPHA=p specifies the Type I error probability for the confidence intervals. Select p such that $0 < p < 1$. The default setting for p is 0.05 or 95% confidence.

CIBASIC requests confidence intervals for the mean, variance, and standard deviation. Confidence level is set by the ALPHA option.

FREQ requests a frequency table consisting of variable value, frequencies, percentages, and cumulative percentages.

MU0= specify the null hypothesis value for the test of the mean or location parameter; that is, H_0: $\mu = \mu_0$. MU0= *value* specifies μ_0.

NORMAL computes four test statistics and their significance levels for the test of the null hypothesis that the input data come from a normal distribution versus the alternative hypothesis that the distribution is not normal.

NOPRINT suppresses the tables of output from this procedure with the exception of those produced by the HISTOGRAM statement.

The NOPRINT option is typically used when an OUTPUT statement is included in the UNIVARIATE procedure.

This list of options is not complete. For a complete list of options consult the SAS Online Help and Documentation. It is not necessary to specify any options in the PROC UNIVARIATE statement. Several summary statistics are computed when no options are specified.

BY Statement

The UNIVARIATE procedure can operate on subgroups of the data set defined by the values of another variable. BY variables can be either numeric or character. These subgroups are identified using the BY statement. This statement is optional. Data must first be compatibly sorted BY the variables appearing in a BY statement.

CLASS Statement

The CLASS statement is also an optional statement. The CLASS statement assigns a variable or variables to form subgroups. CLASS variables can be either numeric or character. The CLASS statement has an effect similar to a BY statement. The CLASS statement does NOT require the data to be sorted first however.

VAR Statement

All numeric variables for which summary statistics are to be calculated are specified in the VAR statement. If the VAR statement is omitted, the UNIVARIATE procedure will compute summary statistics for every numeric variable in the data set.

HISTOGRAM Statement

The continuous variables for which histograms are to be produced are identified in this statement. These variables can be a smaller subset of those variables identified in the VAR statement. Histograms are available as Output Delivery System (ODS) Graphics. ODS Graphics must be enabled to view this image. See Chapter 19 for more ODS information.

HISTOGRAM statement options include:

NORMAL This option will overlay a normal curve on the histogram. There are some advanced options controlling colors and placement of the normal curve. Other distributions are also available. See SAS Help and Documentation for more information.

MIDPOINTS = *number list* SAS/UNIVARIATE will select the midpoints at which bars of the histogram are to be centered. The numbers for midpoints can be specified; this controls the number and placement of midpoints. At least one space and no commas are needed to separate the values of the midpoints. For example:

```
HISTOGRAM response/MIDPOINTS = 1.2 2.4 3.6 4.8 6.0;
```

OUTPUT Statement

Some of the information produced by the UNIVARIATE procedure can be output to a new SAS data set. This is another way to create a SAS data set outside of the

DATA step. The new SAS data set can then be used in other SAS procedures. The syntax for the OUTPUT statement identifies a new data set name and the statistics one wishes to recover for each variable in the VAR statement. The number and order of the variables in the VAR statement determine the number and order of variables to be newly created for each requested statistic. The syntax is

```
PROC UNIVARIATE DATA = SAS-data-set;
VAR variable1 variable2 … variablen;
OUTPUT OUT=new-SAS-data-set statistic1 = stat1var1 stat1var2 … stat1varn
                            statistic2 = stat2var1 stat2var2 … stat2varn
                            ⋮
                            statisticm = statmvar1 statmvar2 … statmvarn;
```

where statistic# can be any of the following:

N	the number of nonmissing values
NMISS	the number of observations having missing values
NOBS	the number of observations
MIN	the minimum or smallest value
MAX	the maximum or largest value
RANGE	the range, maximum-minimum
SUM	the sum
MEAN	the sample mean
VAR	the sample variance
STD	the sample standard deviation
STDMEAN	the standard error of the mean
CV	coefficient of variation
MEDIAN	the sample median
Q3	the upper quartile or 75th percentile
Q1	the lower quartile or 25th percentile
P1	the 1st percentile
P5	the 5th percentile
P10	the 10th percentile
P90	the 90th percentile
P95	the 95th percentile
P99	the 99th percentile
MODE	the sample mode
T	Student's t statistics for testing the hypothesis that the population mean is 0 or the value specified by the MU0 option
PROBT	observed significance level for the two-sided t-test for the population mean

3.2 Measures of Central Tendency, Dispersion, and Normality

The UNIVARIATE procedure produces several tables of summary information by default. Measures of central tendency include the mean, median, and mode. Measures of dispersion include the range, variance, standard deviation, selected percentiles, and more.

To illustrate this, two independent samples are introduced in the following example.

Two instructional programs are to be compared. Students independently participate in one of the two programs. Scores at the completion of the instructional period are recorded for each of the students. Information is given in the table below.

Program A	Program B
71 82 88 64 59 78 72 81 83 66 83 91 79 70	65 88 92 76 87 89 85 90 81 91 78 81 86 82 73 79

OBJECTIVE 3.1: Create the SAS data set *instruction* in the SAS Editor with the character variable *program* and numeric variable *score*. Examine the default information produced by the UNIVARIATE procedure for the *score* variable.

```
DATA instruction;
INPUT program $ score @@;
DATALINES;
A 71 A 82 A 88 A 64 A 59 A 78 A 72
A 81 A 83 A 66 A 83 A 91 A 79 A 70
B 65 B 88 B 92 B 76 B 87 B 89 B 85
B 90 B 81 B 91 B 78 B 81 B 86 B 82
B 73 B 79
;
PROC UNIVARIATE DATA=instruction;
VAR score;
TITLE 'Objective 3.1 - Default Information';
RUN;
QUIT;
```

Note the double trailing @ in the INPUT statement. This is one option for creating the SAS data set *instruction* (Section 2.2).

No options are selected for the UNIVARIATE procedure. The variable "program" is not needed for this analysis. It will be used in upcoming objectives.

Objective 3.1 – Default Information

The UNIVARIATE Procedure

Variable: Score

Moments

N	30	**Sum Weights**	30
Mean	79.6666667	**Sum Observations**	2390
Std Deviation	8.84476597	**Variance**	78.2298851
Skewness	–0.639265	**Kurtosis**	–0.3275697
Uncorrected SS	192672	**Corrected SS**	2268.66667
Coeff Variation	11.1022167	**Std Error Mean**	1.61482595

The name of this procedure appears at the top of the produced output, and the name of the analysis variable **Score** is also identified. If there were more than one numeric variable listed in the VAR statement, there would be another collection of tables in the output following this collection for the variable **Score.**

Note the information in the **Moments** table. Most of these measures are self-explanatory. A few of the others: "N 30" is the sample size, "Sum Observations 2390" is the total of all the scores in the set of 30. "Uncorrected SS 192672" is the sum of the squared scores, that is, $\sum_{1}^{N} y_i^2$, and "Corrected SS 2268.66667" is $\sum_{1}^{N} (y_i - \bar{y})^2$.

Basic Statistical Measures

Location		Variability	
Mean	79.66667	**Std Deviation**	8.84477
Median	81.00000	**Variance**	78.22989
Mode	81.00000	**Range**	33.00000
		Interquartile Range	14.00000

Basic Statistical Measures are measures of central tendency ("Location") and measures of dispersion ("Variability").

Tests for Location: Mu0=0

Test	Statistic		p Value	
Student's t	t	49.33452	Pr > |t|	<.0001
Sign	M	15	Pr >= |M|	<.0001
Signed Rank	S	232.5	Pr >= |S|	<.0001

The UNIVARIATE procedure will conduct tests about the population mean by default. In the table **Tests for Location** the procedure is testing H_0: $\mu = 0$ vs H_1: $\mu \neq 0$. More about this test will be covered in Objective 3.3.

Quantiles (Definition 5)

Level	Quantile
100% Max	92.0
99%	92.0
95%	91.0
90%	90.5
75% Q3	87.0
50% Median	81.0
25% Q1	73.0
10%	65.5
5%	64.0
1%	59.0
0% Min	59.0

There are default percentiles generated for the analysis variable in the **Quantiles** table.

Extreme Observations

Lowest		**Highest**	
Value	**Obs**	**Value**	**Obs**
59	5	89	20
64	4	90	22
65	15	91	12
66	10	91	24
70	14	92	17

In the table of **Extreme Observations,** the UNIVARIATE procedure will also identify the smallest five values and the largest five values for the analysis variable. Checking this list is recommended as very often typographical errors in the data may result in an extreme observation. For example, if one mistakenly entered 9 or 900 instead of 90, that error could be detected in this table.

3.3 Testing Hypotheses and Computing Confidence Intervals

Methods of inference and estimation use the sample data collected to make judgements about the population. Tests of hypotheses and confidence intervals for the population mean and variance are among the topics in an introductory statistical methods course (Freund, Wilson, and Mohr, 2010; Ott and Longnecker, 2016; Peck and Devore, 2011). For these methods the following notation is needed.

	Population Parameters	Sample Statistics
Mean	μ	$\bar{y}$
Variance	σ^2	s^2
Standard Deviation	σ	s
		sample size = n

Table 3.1 summarizes single population t-tests and confidence intervals about the mean.

TABLE 3.1

Summary of Hypothesis Testing and Confidence Intervals for a Single Population Mean

Hypotheses	**Test Statistic**	**Reject H_0 if**	**$(1-\alpha)100\%$ Confidence Interval for μ**
$H_0: \mu = \mu_0$ $H_1: \mu \neq \mu_0$	$t = \frac{\bar{y} - \mu_0}{s/\sqrt{n}}$	$\lvert t \rvert \geq t_{\alpha/2,df}$	$\bar{y} \pm t_{\alpha/2,df} \frac{s}{\sqrt{n}}$

where μ_0 is specified, $t_{\alpha/2,df}$ is the critical t-value that determines a right tail area of $\alpha/2$, and df = n – 1 for this application.

TABLE 3.2

Summary of Hypothesis Testing and Confidence Intervals for a Single Population Variance

Hypotheses	Test Statistic	Reject H_0 if	$(1-\alpha)100\%$ Confidence Interval for σ^2
$H_0: \sigma^2 = \sigma_0^2$ $H_1: \sigma^2 \neq \sigma_0^2$	$\chi^2 = \frac{(n-1)s^2}{\sigma_0^2}$	$\chi^2 \leq \chi^2_{1-\alpha/2,df}$ or $\chi^2 \geq \chi^2_{\alpha/2,df}$	$\frac{(n-1)s^2}{\chi^2_{\alpha/2,df}} \leq \sigma^2 \leq \frac{(n-1)s^2}{\chi^2_{1-\alpha/2,df}}$

where σ_0^2 is specified, $\chi^2_{p,df}$ is the critical value of χ^2 that determines a right tail area of p, and $df = n - 1$ for this application.

An observed significance level or p-value is calculated by the UNIVARIATE procedure for these tests rather than the rejection or critical region defined above. **Recall that one would reject H_0 for small p-values. That is, reject H_0 if $p \leq \alpha$ for all tests of hypotheses, not just those about μ.**

Additionally, the UNIVARIATE procedure will also compute a confidence interval for the population variance. The test procedure and confidence interval information for the population variance are in Table 3.2. The UNIVARIATE procedure does not compute the test of the variance however. A confidence interval for the standard deviation is produced in the results and is obtained by computing the square root of the endpoints of the confidence interval for the variance.

For both of the preceding test and confidence interval methods to be valid, the population the samples are drawn from must be normally distributed. The UNIVARIATE procedure has capability to assess the normality condition. For the test for normality, the null and alternative hypotheses are:

H_0: Population is normally distributed

H_1: Population is not normally distributed

There are four tests produced by the UNIVARIATE procedure when the NORMAL option is included on the PROC UNIVARIATE statement. No calculations for these test statistics are given in this text, but background information for these tests can be found in D'Agostino and Stephens (1986), Royston (1992), and Shapiro and Wilk (1965). These tests will be interpreted based upon the observed significance level.

In the UNIVARIATE procedure the CIBASIC option with the ALPHA=p option will generate confidence intervals for the mean, variance, and standard deviation using the formulas in Tables 3.1 and 3.2. This procedure will also test the mean of the population using the t-test in Table 3.1 and by two non-parametric methods: signed rank test and sign test. These tests are two-sided tests for the population mean. There is no one-sided test option in this procedure.

OBJECTIVE 3.2: Using the CLASS statement and Score as the analysis variable, compute the summary statistics, 99% confidence intervals, and tests for normality for each program and produce a HISTOGRAM with a normal curve for each program. Make certain that ODS Graphics are enabled to view the histogram.

```
PROC UNIVARIATE DATA=instruction CIBASIC ALPHA=0.01 NORMAL;
CLASS program;
VAR score;
```

```
HISTOGRAM score / NORMAL;
TITLE 'Objective 3.2 ';
RUN;
```

When using a CLASS statement within the UNIVARIATE procedure, it is not necessary to SORT the data by the class variable first.

Objective 3.2

The UNIVARIATE Procedure
Variable: score
program = A

Moments

N	14	**Sum Weights**	14
Mean	76.2142857	**Sum Observations**	1067
Std Deviation	9.40685978	**Variance**	88.489011
Skewness	−0.2825394	**Kurtosis**	−0.7691737
Uncorrected SS	82471	**Corrected SS**	1150.35714
Coeff Variation	12.3426464	**Std Error Mean**	2.51408903

< Basic Statistical Measures table has been suppressed by the author.>

Just above the **Moments** table, "program = A" is identified now. This is the result of the CLASS statement. The content of this **Moments** table summarizes those 14 scores for Program A only.

Some tables in this output for this objective are suppressed to keep output in a more manageable length.

Basic Confidence Limits Assuming Normality

Parameter	Estimate	99% Confidence Limits	
Mean	76.21429	68.64116	83.78742
Std Deviation	9.40686	6.21107	17.96323
Variance	88.48901	38.57738	322.67770

<Tests for Location: Mu0=0 table has been suppressed by the author.>

The CIBASIC and ALPHA=0.01 options produce the table **Basic Confidence Limits Assuming Normality** which contains the point estimate of the parameters and the confidence limits or intervals. If one omitted ALPHA=0.01, 95% confidence intervals are the default for CIBASIC. This table of confidence limits is not part of the default information demonstrated in Objective 3.1.

Tests for Normality

Test	Statistic		p Value	
Shapiro-Wilk	**W**	0.965895	**Pr < W**	0.8175
Kolmogorov-Smirnov	**D**	0.146708	**Pr > D**	>0.1500
Cramer-von Mises	**W-Sq**	0.04613	**Pr > W-Sq**	>0.2500
Anderson-Darling	**A-Sq**	0.251411	**Pr > A-Sq**	>0.2500

<Quantiles and Extreme Observations tables have been suppressed by the author.>

The NORMAL option on the PROC UNIVARIATE statement will compute the test statistics for four different tests for normality of the data. The consensus of the four tests is that the scores in program A do not depart from normality ($p > 0.15$).

The information tables overviewed for Program A above are also produced for Program B.

Objective 3.2

The UNIVARIATE Procedure

Variable: score

program = B

Moments

N	16	**Sum Weights**	16
Mean	82.6875	**Sum Observations**	1323
Std Deviation	7.32774408	**Variance**	53.6958333
Skewness	−0.891007	**Kurtosis**	0.69363001
Uncorrected SS	110201	**Corrected SS**	805.4375
Coeff Variation	8.86197319	**Std Error Mean**	1.83193602

<Basic Statistical Measures table has been suppressed by the author.>

Basic Confidence Limits Assuming Normality

Parameter	**Estimate**	**99% Confidence Limits**	
Mean	82.68750	77.28931	88.08569
Std Deviation	7.32774	4.95530	13.23103
Variance	53.69583	24.55503	175.06027

<Tests for Location: Mu0=0 table has been suppressed by the author.>

Tests for Normality

Test	**Statistic**		**p Value**	
Shapiro-Wilk	**W**	0.938759	**Pr < W**	0.3341
Kolmogorov-Smirnov	**D**	0.12384	**Pr > D**	>0.1500
Cramer-von Mises	**W-Sq**	0.041509	**Pr > W-Sq**	>0.2500
Anderson-Darling	**A-Sq**	0.312564	**Pr > A-Sq**	>0.2500

<Quantiles and Extreme Observations tables have been suppressed by the author.>

The panel of histograms appearing in **Distribution of score** (one histogram for each value of the CLASS variable) is the result of the HISTOGRAM statement. In Objective 3.1 where there was no CLASS variable, a HISTOGRAM request would have resulted in only one histogram for the complete data set.

Objective 3.2

The UNIVARIATE Procedure

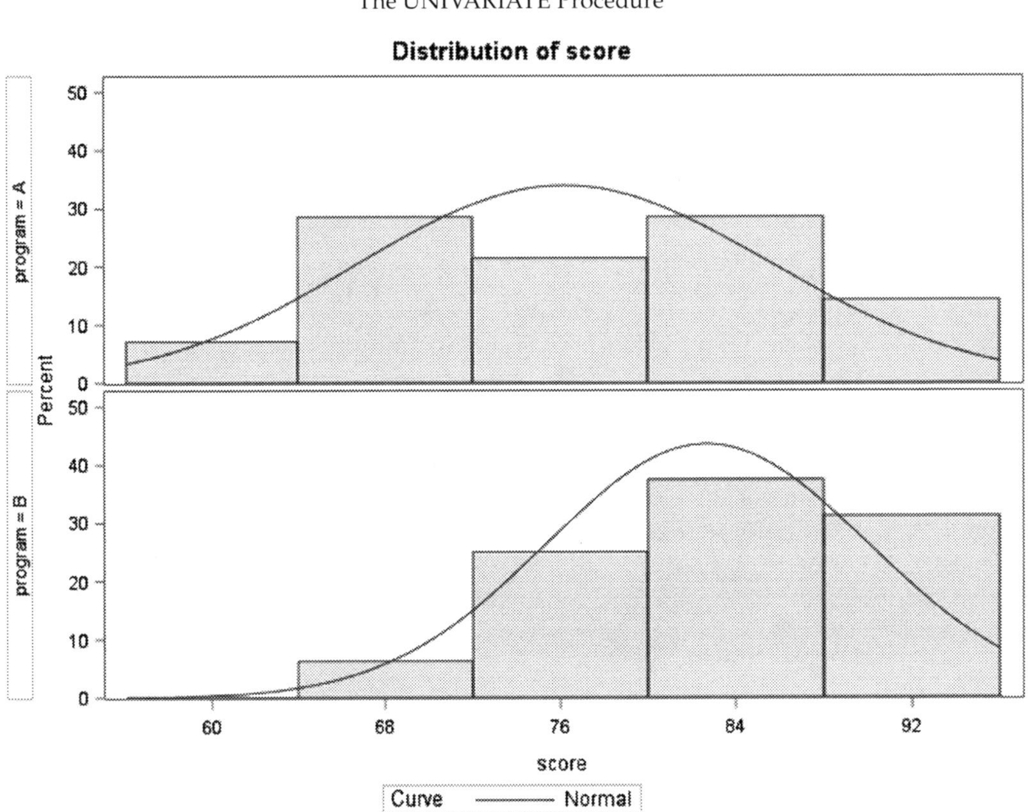

The NORMAL option on the HISTOGRAM statement resulted in the normal curves that are overlaid on the histograms. Also due to the NORMAL option on the HISTOGRAM statement are the tables **Parameters for Normal Distribution** and **Goodness-of-Fit Tests for Normal Distributions.** In **Parameters for Normal Distribution** are the sample mean and standard deviation used in the corresponding normal curves above. Notice the NORMAL option on the HISTOGRAM statement produces three of the four tests for normality (**Goodness-of-Fit Tests for Normal Distributions**) that were produced with the NORMAL option on the PROC UNIVARIATE statement.

The UNIVARIATE Procedure

program = A

Fitted Normal Distribution for score

Parameters for Normal Distribution

Parameter	Symbol	Estimate
Mean	Mu	76.21429
Std Dev	Sigma	9.40686

Goodness-of-Fit Tests for Normal Distribution

Test	Statistic		p Value	
Kolmogorov-Smirnov	**D**	0.14670782	**Pr > D**	>0.150
Cramer-von Mises	**W-Sq**	0.04612974	**Pr > W-Sq**	>0.250
Anderson-Darling	**A-Sq**	0.25141122	**Pr > A-Sq**	>0.250

Quantiles for Normal Distribution

Percent	Quantile	
	Observed	**Estimated**
1.0	59.0000	54.3307
5.0	59.0000	60.7414
10.0	64.0000	64.1589
25.0	70.0000	69.8695
50.0	78.5000	76.2143
75.0	83.0000	82.5591
90.0	88.0000	88.2697
95.0	91.0000	91.6872
99.0	91.0000	98.0979

All of the information produced above for Program A is also produced for Program B when the NORMAL option is used on the HISTOGRAM statement:

The UNIVARIATE Procedure

program = B

Fitted Normal Distribution for score

Parameters for Normal Distribution

Parameter	Symbol	Estimate
Mean	Mu	82.6875
Std Dev	Sigma	7.327744

Goodness-of-Fit Tests for Normal Distribution

Test	Statistic		p Value	
Kolmogorov-Smirnov	**D**	0.12383988	**Pr > D**	>0.150
Cramer-von Mises	**W-Sq**	0.04150937	**Pr > W-Sq**	>0.250
Anderson-Darling	**A-Sq**	0.31256404	**Pr > A-Sq**	>0.250

Quantiles for Normal Distribution

Percent	Quantile	
	Observed	Estimated
1.0	65.0000	65.6406
5.0	65.0000	70.6344
10.0	73.0000	73.2966
25.0	78.5000	77.7450
50.0	83.5000	82.6875
75.0	88.5000	87.6300
90.0	91.0000	92.0784
95.0	92.0000	94.7406
99.0	92.0000	99.7344

Objective 3.2 can be achieved in three different ways. The verification of this is left to the reader.

```
PROC UNIVARIATE DATA=instruction CIBASIC ALPHA=0.01 NORMAL;
CLASS program;
VAR score;
HISTOGRAM score;
TITLE 'Objective 3.2 - NORMAL option on PROC UNIVARIATE only';
RUN;
```

and

```
PROC UNIVARIATE DATA=instruction CIBASIC ALPHA=0.01;
CLASS program;
VAR score;
HISTOGRAM score / NORMAL;
TITLE 'Objective 3.2 - NORMAL option on HISTOGRAM only';
RUN;
```

And third
Include both NORMAL options.

3.4 Recovering Statistics in a Data Set

OBJECTIVE 3.3: Modify the program code in Objective 3.2. In the UNIVARIATE procedure test whether the mean score of each of the programs is equal to 75. That is, test H_0: $\mu = 75$ versus H_1: $\mu \neq 75$. Use $\alpha = 0.01$ in drawing a conclusion. Include a HISTOGRAM statement overlaying a normal distribution with the hypothesized mean of 75. Additionally, recover the mean, sample size, and standard error of the mean for each program in a new SAS data set, *three,* and print that new SAS data set.

```
PROC UNIVARIATE DATA=instruction MU0=75;
CLASS program;
```

```
VAR score;
HISTOGRAM score / NORMAL (MU=75);
OUTPUT OUT=three  MEAN=mnscore N=nscore STDMEAN=semscore;
TITLE 'Objective 3.3 ';
TITLE3 'Using a CLASS statement';
RUN;

PROC PRINT DATA=three;
TITLE3 'Output Data Set Identifying Each Program';
RUN;
```

To test whether or not the population mean is equal to a specified non-zero value or not, the MU0= option should be used in the PROC UNIVARIATE statement. This is "M U zero" not "M U oh". Without specifying MU0=75, one gets a test of whether or not the population mean differs from zero. This is only meaningful if zero is a plausible value for the population mean. In this context, it is not. If one specifies multiple analysis variables in the VAR statement, then the mean of each variable is tested using the value(s) in the MU0= option. In the HISTOGRAM statement a superimposed normal curve can be requested, and it is centered at the hypothesized value of the mean; 75 in this illustration.

The OUTPUT statement must occur before the next procedure (PROC) or RUN statement and must name a new SAS data set. Here that SAS data set is *three*. The statistics to be recovered must be specified in this statement, and new variable names for these values need to be specified. The OUTPUT statement creates a SAS data set with requested statistics but does not produce any printed output in the Results Viewer. To see the contents of an OUTPUT SAS data set, the PRINT procedure (Section 2.3) is used.

Objective 3.3

Using a CLASS statement

The UNIVARIATE Procedure

Variable: score

program = A

<**Moments, Basic Statistical Measures** tables are suppressed by the author.>

Tests for Location: Mu0=75

Test		Statistic	p Value	
Student's t	t	0.482992	Pr > |t|	0.6371
Sign	M	1	Pr >= |M|	0.7905
Signed Rank	S	8	Pr >= |S|	0.6323

<To avoid redundancies, several more tables of results have been suppressed by the author in this objective.>

Assuming normality, the t-test statistic in Table 3.1 is given by **Student's t 0.482992** with the observed significance level of **Pr > |t| 0.6371.** More formally, the conclusion is: The mean score for Program A does not differ from 75 ($\alpha = 0.01$, $t = 0.4830$, $p = 0.6371$). The confidence interval for Program A produced by the CIBASIC option in Objective 3.2 supports this conclusion since the 99% CI for μ: (68.64114, 83.78742) captures 75. The **Sign** and **Signed Rank** tests are non-parametric tests for the population mean and are not overviewed here.

For those cases in which there is more than one variable in the VAR statement, how should one use the MU0 option? In a case where there are three analysis variables: a, b, and c. There are multiple programming options to obtain the tests for the three population means.

Test for Location Syntax Options	Requested Tests
`PROC UNIVARIATE DATA=instruction;` `VAR a b c;`	Each of the three population means are compared to 0; that is, H_0: $\mu = 0$
`PROC UNIVARIATE DATA=instruction MU0=15;` `VAR a b c;`	The means of all three populations are compared to 15; that is, H_0: $\mu = 15$
`PROC UNIVARIATE DATA=instruction MU0=(15 27 9);` `VAR a b c;`	H_0: $\mu = 15$ for variable a H_0: $\mu = 27$ for variable b H_0: $\mu = 9$ for variable c
`PROC UNIVARIATE DATA=instruction MU0=(15 27);` `VAR a b c;`	H_0: $\mu = 15$ for variable a H_0: $\mu = 27$ for variable b H_0: $\mu = 0$ for variable c

Objective 3.3

Using a CLASS statement

The UNIVARIATE Procedure

Variable: score

program = B

<Moments, Basic Statistical Measures tables are suppressed by the author.>

Tests for Location: Mu0=75

Test	**Statistic**		**p Value**	
Student's t	**t**	4.19638	**Pr > |t|**	0.0008
Sign	**M**	6	**Pr >= |M|**	0.0042
Signed Rank	**S**	57.5	**Pr >= |S|**	0.0014

The mean score for Program B differs significantly from 75 ($\alpha = 0.01$, $t = 4.1964$, $p = 0.0008$). The confidence interval for Program B produced by the CIBASIC option in Objective 3.2 supports this conclusion. 99% CI for μ: (77.2893, 88.0857). The interval does not capture 75. One does not need the ALPHA = 0.01 option if only the test results are needed. The choice of α affects the confidence interval only. The calculated value of the test statistic and the p-value are functions of the data and are not functions of α.

Objective 3.3

Using a CLASS statement

The UNIVARIATE Procedure

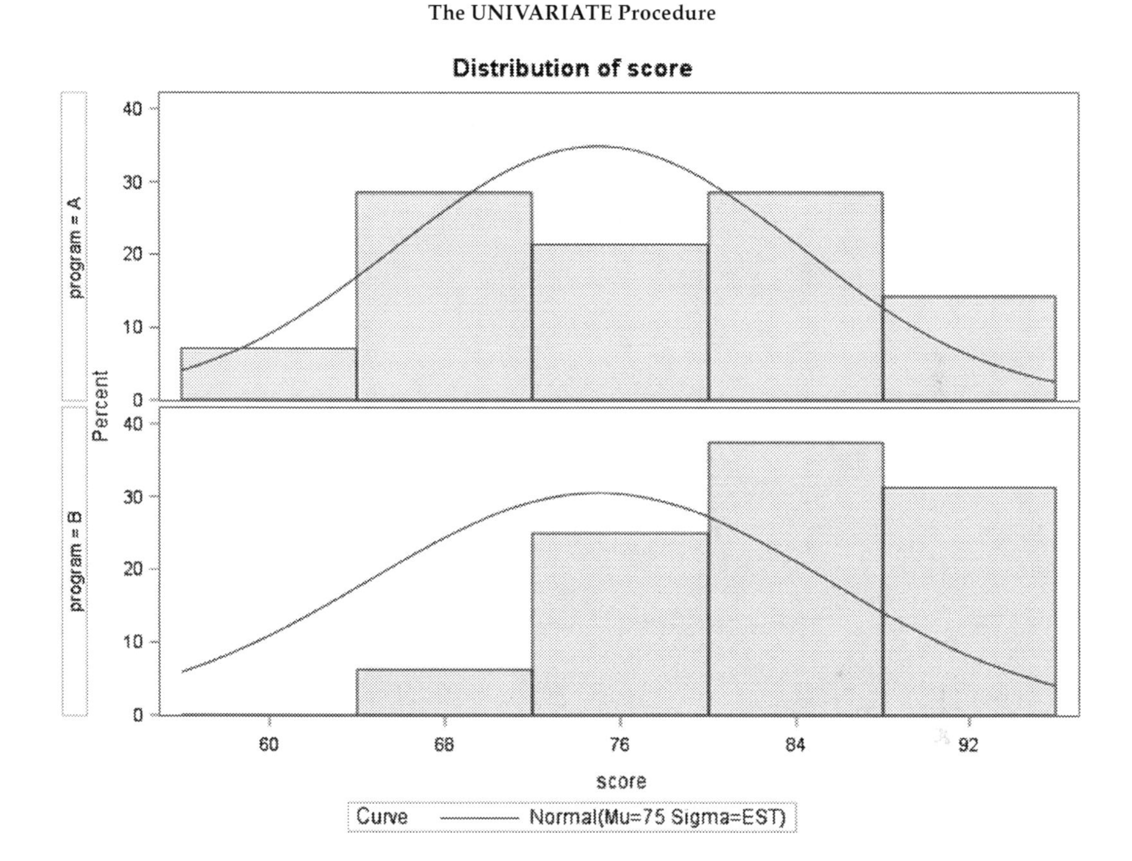

The **Distribution of score** histograms in Objective 3.3 are not the same as in Objective 3.2. The difference is that the mean of the superimposed normal distributions has been set to 75, the hypothesized mean. This is identified in the legend below the histogram. One could specify a standard deviation (SIGMA) value in addition to the value of the mean (MU) for the overlaid normal curve in the following way:

```
HISTOGRAM variable / NORMAL (MU=value SIGMA=value);
```

Objective 3.3

Using a CLASS statement

The UNIVARIATE Procedure

program = A

Fitted Normal Distribution for score

Parameters for Normal Distribution

Parameter	Symbol	Estimate
Mean	Mu	75
Std Dev	Sigma	9.145647

Note the value of 75 identified in **Parameters for Normal Distribution** for Program A and below for Program B identifying the hypothesized mean. The sample standard deviation calculated for each program (EST in the legend of the **Distribution of score** graphic) is chosen for the superimposed normal curves for both programs.

The UNIVARIATE Procedure

program = B

Fitted Normal Distribution for score

Parameters for Normal Distribution

Parameter	Symbol	Estimate
Mean	Mu	75
Std Dev	Sigma	10.46124

Objective 3.3

Output Data Set Identifying Each Program

Obs	program	nscore	mnscore	semscore
1	A	14	76.2143	2.51409
2	B	16	82.6875	1.83194

The OUTPUT statement in the UNIVARIATE procedure produced the SAS data set *three*, but the OUTPUT statement itself does not produced printed results. The PRINT procedure is used here to print and examine the contents of the newly created SAS data set. The contents of this SAS data set can be recovered for graphics or other SAS procedures.

Often a procedure such as UNIVARIATE is run only to compute the statistics that are recovered in a SAS data set, such as *three,* in this instance. If the only objective is to recover an output SAS data set, using the NOPRINT option in the PROC UNIVARIATE statement would suppress all tables except those resulting from the HISTOGRAM statement (Parameters for Normal Distribution, Goodness-of-Fit Tests for Normal Distribution, and Quantiles for Normal Distribution).

Rerun Objective 3.3 using a BY statement rather than a CLASS statement and observe the differences in the output results, specifically the output from the PRINT procedure.

3.5 The MEANS Procedure

This procedure does not generate as much output as the UNIVARIATE procedure, but is very useful for obtaining summary statistics for a data set, confidence intervals, and t-tests. As of this writing there are no ODS Graphics available in the MEANS procedure.

Basic syntax for the MEANS procedure is:

PROC MEANS DATA= *SAS-data-set* <options>;
BY *variable(s);* *(optional statement)*
WHERE *condition;* *(optional statement, see Chapter 4, DATA Step Information 2)*
CLASS *variable(s);* *(optional statement)*

VAR *variable1 variable2 ... variablen;*
OUTPUT OUT=*new-SAS-data-set* <list options and new variable names>;

PROC MEANS Statement

Some of the possible options for the PROC MEANS statement are:

N	the number of nonmissing values
NMISS	the number of observations having missing values
MIN	the minimum value
MAX	the maximum value
RANGE	the range
SUM	the sum
MEAN	the mean
VAR	the variance
STD	the standard deviation
STDERR	the standard error of the mean
CV	coefficient of variation
T	Student's t for testing the hypothesis that the population mean is 0
PRT	the observed significance level or p-value of the two-tailed test
ALPHA=p	sets the confidence limit to $(1 - p)100\%$, where $0 < p < 1$
CLM	generates a two-sided 95% confidence interval, other confidence levels can be specified by using ALPHA= p option, where n is a number between 0 and 1. $(1 - p)100\%$ specifies the confidence level.
UCLM	specifies the 95% upper confidence limit, can use ALPHA=p to set confidence level.
LCLM	specifies the 95% lower confidence limit, can use ALPHA=p to set confidence level
NOPRINT	The NOPRINT option is useful when selected items are being recovered in an output data set, and the actual output from the procedure does not need to appear in the Results Viewer. See OUTPUT statement.

It is not necessary to specify any options in the PROC MEANS statement. If no options are specified, then PROC MEANS prints the variable name and computes: N, MEAN, STD, MIN, and MAX for the variables in the VAR statement. If options are specified, only those values will be computed by the MEANS procedure, and those values will appear in the order requested in the PROC MEANS statement.

BY Statement

The MEANS procedure can operate on subgroups of the data set defined by the values of another variable. BY variables can be either numeric or character. The BY statement is optional. Data must first be sorted compatibly BY the variables appearing in the BY statement of the MEANS procedure.

CLASS Statement

The CLASS statement is also an optional statement. The CLASS statement assigns a variable or variables to form subgroups. CLASS variables can be either numeric or character. The CLASS statement has an effect similar to a BY statement. The CLASS statement does NOT require the data to be sorted first.

VAR Statement

All of the numeric variables for which summary statistics are to be calculated are specified in the VAR statement. If the VAR statement is omitted, the MEANS procedure will compute summary statistics for every numeric variable in the data set.

OUTPUT Statement

Information produced by the MEANS procedure can be output to a new SAS data set. The new SAS data set can then be used in other SAS procedures. The syntax for the OUTPUT statement identifies a new SAS data set name and the statistics one wishes to recover for each variable in the VAR statement. The number and order of the variables in the VAR statement determine the number and order of variables to be created for each requested statistic. The syntax for these two statements is:

```
VAR variable1 variable2 … variablen;
OUTPUT OUT=new-SAS-data-set  statistic1 = stat1var1 stat1var2 … stat1varn
                             statistic2 = stat2var1 stat2var2 … stat2varn
                             :
                             statisticm = statmvar1 statmvar2 … statmvarn;
```

Specify the statistic and assign it a variable name, such as N = *size*. The list of statistics which can be output is identical to the list of statistics options for the PROC MEANS statement. This is a shorter list of statistics than those one can output from the UNIVARIATE procedure.

OBJECTIVE 3.4: Run the MEANS procedure for the same SAS data set, *instruction,* the UNIVARIATE procedure was analyzing. Examine the default output, that is, select no options.

```
PROC MEANS DATA=instruction;
TITLE 'Objective 3.4';
RUN;
```

Objective 3.4

The MEANS Procedure

Analysis Variable : score

N	Mean	Std Dev	Minimum	Maximum
30	79.6666667	8.8447660	59.0000000	92.0000000

Only the SAS data set was specified in the PROC MEANS statement, and no VAR statement was used. Since there is only one numeric variable in the data set, the summary statistics for that one numeric variable are calculated in the **Analysis Variable** table. The MEANS procedure computes the five indicated statistics for the data set *instruction*. These results are not separated into the two programs because that was not requested in this objective. If there were more than one numeric variable, then there would be a row of the **Analysis Variable** table for each variable analyzed.

OBJECTIVE 3.5: Compute the minimum, maximum, mean, standard deviation, sample size, and the coefficient of variation for each of the programs in the *instruction* data set. Use a CLASS statement to identify the levels of program as the subgroups.

```
PROC MEANS DATA=instruction MIN MAX MEAN STD N CV;
CLASS program;
VAR score;
TITLE 'Objective 3.5';
RUN;
```

Objective 3.5

The MEANS Procedure

Analysis Variable : score

program	N Obs	Minimum	Maximum	Mean	Std Dev	N	Coeff of Variation
A	14	59.0000000	91.0000000	76.2142857	9.4068598	14	12.3426464
B	16	65.0000000	92.0000000	82.6875000	7.3277441	16	8.8619732

Notice that the **Analysis Variable** table contains the statistics requested in the PROC MEANS statement in the order that they were specified. If, say, only CV option was requested in the PROC MEANS statement, then only the coefficient of variation will be computed for each program.

OBJECTIVE 3.6: Rerun Objective 3.5 but this time add an OUTPUT statement to recover the minimum, maximum, and range in a SAS data set named *six*. Examine the contents of the new SAS data set by adding a PRINT procedure.

```
PROC MEANS DATA=instruction MIN MAX MEAN STD N CV;
CLASS program;
VAR score;
TITLE 'Objective 3.6';
OUTPUT OUT=six MIN=score_min MAX=score_max RANGE=score_range;
RUN;

PROC PRINT DATA=six;
RUN;
```

The statistics requested in the PROC MEANS statement do not have to be the same as those statistics recovered in a SAS data set. In the OUTPUT statement a variable naming convention that assists in identifying the values in the new data set is recommended.

Objective 3.6

The MEANS Procedure

Analysis Variable : score

program	N Obs	Minimum	Maximum	Mean	Std Dev	N	Coeff of Variation
A	14	59.0000000	91.0000000	76.2142857	9.4068598	14	12.3426464
B	16	65.0000000	92.0000000	82.6875000	7.3277441	16	8.8619732

Obs	program	_TYPE_	_FREQ_	score_min	score_max	score_range
1		0	30	59	92	33
2	A	1	14	59	91	32
3	B	1	16	65	92	27

The contents of the **Analysis Variable** table contain only those statistics requested in the PROC MEANS statement, and the table following that is the result of the PRINT procedure. The CLASS statement not only produced a row of results for each program (Obs 2 and 3) but also computed the statistics for the entire data set (Obs 1). _FREQ_ identifies the number of observations for each row in the table.

If Objective 3.6 is run a second time with the NOPRINT option and no specific statistics requested in the PROC MEANS statement, the effect is that a SAS data set *six* is created with the variables named in the OUTPUT statement but the MEANS procedure produces no printed output.

The data set *instruction* has only one response variable, score, to be analyzed. If multiple variables are analyzed in the VAR statement, one would also have to provide multiple variable names for each statistic in the OUTPUT statement. The names selected correspond to the order of the variables in the VAR statement. For example, if additionally there were numeric variables x, y, and z in the analysis, the program code would be:

```
PROC MEANS DATA=instruction MIN MAX MEAN STD N CV;
CLASS program;
VAR score x y z;
OUTPUT OUT=six  MIN=score_min xmin ymin zmin  MAX=score_max xmax ymax zmax
                RANGE=score_range xrange yrange zrange;
RUN;
```

OBJECTIVE 3.7: Repeat Objective 3.6 using a BY statement rather than a CLASS statement. If data set *instruction* has not been sorted previously, a SORT procedure will need to be included. Name the output SAS data set *seven* and print it.

It is not necessary to sort the data by the variables in the VAR statement, only those variables in the BY statement.

```
PROC SORT DATA=instruction; BY program;
PROC MEANS DATA=instruction MIN MAX MEAN STD N CV;
BY program;
VAR score;
TITLE 'Objective 3.7';
OUTPUT OUT=seven MIN=score_min MAX=score_max RANGE=score_range;
RUN;

PROC PRINT DATA=seven;
RUN;
```

Objective 3.7

The MEANS Procedure

program=A

Analysis Variable : score

Minimum	Maximum	Mean	Std Dev	N	Coeff of Variation
59.0000000	91.0000000	76.2142857	9.4068598	14	12.3426464

program=B

Analysis Variable : score

Minimum	Maximum	Mean	Std Dev	N	Coeff of Variation
65.0000000	92.0000000	82.6875000	7.3277441	16	8.8619732

The BY statement on the MEANS procedure results in an **Analysis Variable** table for each value of the "BY variable". When using a CLASS statement, the output in Objectives 3.5 and 3.6 was a single table in each case.

Objective 3.7

Obs	program	_TYPE_	_FREQ_	score_min	score_max	score_range
1	A	0	14	59	91	32
2	B	0	16	65	92	27

The SAS data set *seven* created by the MEANS procedure has only two rows, one for each program. This is different from the SAS data set *six* created when there is a CLASS statement. (Objective 3.6) The number of rows produced in these SAS data sets *six* and *seven* may influence the decision of whether to use BY statement or a CLASS statement in this procedure.

The UNIVARIATE procedure differs from the MEANS procedure in the form of the output data sets. Using either a CLASS variable or a BY variable, the SAS data set output from UNIVARIATE will only have rows of data for each level of the class variable as in Objective 3.7 results.

The MEANS procedure can also be used to test hypotheses about the population mean using a t-test, assuming normality, of course.

OBJECTIVE 3.8: In the MEANS procedure, for each level of program compute the default test for the population mean and include a 99% confidence interval for the mean. Use $\alpha = 0.01$ in the conclusions.

```
PROC MEANS DATA=instruction MEAN STDERR T PRT CLM ALPHA=0.01;
CLASS program;
VAR score;
TITLE 'Objective 3.8';
RUN;
```

The T and PRT options on the PROC MEANS statement produce the t-statistic and p-value for testing H_0: $\mu = 0$ versus H_1: $\mu \neq 0$. CLM and ALPHA = 0.01 produce the 99% confidence interval for the mean only. The MEAN and STDERR options are not required, but generally one would request more information than just T and PRT. The MEAN divided by the STDERR is the value of T in the output.

Objective 3.8

The MEANS Procedure

program	N Obs	Mean	Std Error	t Value	Pr > |t|	Lower 99% CL for Mean	Upper 99% CL for Mean
A	14	76.2142857	2.5140890	30.31	<.0001	68.6411561	83.7874154
B	16	82.6875000	1.8319360	45.14	<.0001	77.2893105	88.0856895

The t-values and Pr>|t| are useful only if zero is a plausible value for the mean, and in this example of test scores, it is not. Because zero is not a plausible value, the values of t are quite large in magnitude, and the associated p-values are less than 0.0001. The confidence intervals computed are interpretable. A test of H_0: $\mu = 75$ versus H_1: $\mu \neq 75$ is possible using the MEANS procedure, but information from Chapter 4 is needed to do that. Testing the mean for a difference from a non-zero value using the MEANS procedure will be revisited in Chapter 6.

The confidence intervals produced here are the same as those in Objective 3.2 differing only in the number of decimal places between the UNIVARIATE and MEANS procedures. The interpretation of the confidence intervals with respect to the test of H_0: $\mu = 75$ versus H_1: $\mu \neq 75$ would therefore be the same as done in Objective 3.3.

3.6 Chapter Summary

In this chapter, summary statistics from both the UNIVARIATE and MEANS procedures were presented. The UNIVARIATE procedure produces more tables of output than the MEANS procedure. Both procedures can output a variety of statistics to a new SAS data set. Both procedures have confidence interval and t-test capability for the population mean. Only the UNIVARIATE procedure produces confidence intervals for a population variance, tests for normality, and has ODS Graphics available.

4

DATA Step Information 2

4.1 New Variable Assignment or Data Transformations

Data sets may be created using a DATA step with an INPUT and DATALINES statements in the program. Oftentimes, variables of interest are functions of the variables that have been measured and recorded in the original data set.

Example:

Suppose the lengths (in millimeters) of fish in a study are recorded along with other characteristics and a SAS data set, *fish*, has been created in a DATA step; such as,

```
DATA fish;
INPUT ID Location $ Length Weight Age Gender $;
DATALINES;
23   Payne   75    24   2.5   f
41   Payne   68    16   2     m
17   Payne   57    12   1.5   F
33   payne   45    14   0.5   m
18   Payne   71    20   3     F
77   Payne   60    19   2.5   f
;
```

The SAS data set named *fish* will contain six variables: an ID number, location or county in which the fish was caught, length in millimeters, weight, age, and gender. An inconsistency or errors in the usage of capital letters is present in the data. Those corrections will be addressed in Objectives 4.4 and 4.5. Suppose that the length in millimeters is to be converted to length in inches. A new variable in data set *fish* that is a measure in inches can be created in the following way.

```
DATA fish;
INPUT ID Location $ Length Weight Age Gender $;
Length_in = length / 25.4;
DATALINES;
23   Payne   75    24   2.5   f
41   Payne   68    16   2     m
17   Payne   57    12   1.5   F
33   payne   45    14   0.5   m
18   Payne   71    20   3     F
77   Payne   60    19   2.5   f
;
```

A new variable called "Length_in" is created. The values of this new variable are obtained by dividing each of the values read in for the length variable by the conversion factor of 25.4. The values input after the DATALINES (or CARDS) statement are the same as before. When the DATA step executes, there will be seven variables in the resulting SAS data set: six of them originally input (ID, location, length, weight, age, and gender) and a new variable (Length_in) that is a function of those input values.

OBJECTIVE 4.1: Clear the log window and the Results Viewer, and reset any active lines of titles. Create the data set *fish* with seven variables and print the contents of the data set.

```
DM 'LOG; CLEAR; ODSRESULTS; CLEAR; ';
TITLE;

DATA fish;
INPUT ID Location $ Length Weight Age Gender $;
Length_in = length / 25.4;
DATALINES;
23   Payne   75    24   2.5   f
41   Payne   68    16   2     m
17   Payne   57    12   1.5   F
33   payne   45    14   0.5   m
18   Payne   71    20   3     F
77   Payne   60    19   2.5   f
;
PROC PRINT DATA=fish;
TITLE 'Objective 4.1';
RUN;
QUIT;
```

Objective 4.1

Obs	ID	Location	Length	Weight	Age	Gender	Length_in
1	23	Payne	75	24	2.5	f	2.95276
2	41	Payne	68	16	2.0	m	2.67717
3	17	Payne	57	12	1.5	F	2.24409
4	33	payne	45	14	0.5	m	1.77165
5	18	Payne	71	20	3.0	F	2.79528
6	77	Payne	60	19	2.5	f	2.36220

In the output for Objective 4.1 the new variable Length_in is calculated and is the new or seventh variable in the SAS data set. Using options in Section 2.3 the observation numbers could be suppressed (NOOBS option), and the order of the columns or variables can be controlled with a VAR statement. The numeric data values in the printed results are right-justified under their respective variable names or column headers, and character data are left justified.

Objective 4.1 illustrates how a single variable can be added to a data set using a variable assignment statement. Variable assignment or data transformation statements must be placed **after** the INPUT statement and **before** the DATALINES (or CARDS) statement.

TABLE 4.1

Code for Arithmetic Operations

Symbol	Operation
+	addition
–	subtraction
*	multiplication
/	division
**	exponentiation
( )	parenthesis as needed

Any number of assignment statements can be used in one DATA step as long as each statement has correct syntax and is correctly punctuated. The form of the variable assignment statement is as follows:

```
newvariablename = function of variables input or created in prior
statements;
```

The new variable name must follow the SAS naming convention rules for variable names in the INPUT statement (Section 2.2). Also, as in the INPUT statement, when a new variable name is initialized in a variable assignment statement, the uppercase or lowercase format used in that initialization will be used in the output.

Table 4.1 lists the arithmetic operations that can appear in these variable assignment statements of the DATA step. Parenthesis are the only grouping symbol one can use. That is, square brackets [ ] or braces { } in algebraic expressions are not recognized in these statements. Parenthesis should be nested when necessary. So, an algebraic expression creating a new variable y, such as, $3\,[(x+2)^2-7]=y$ would be coded as

```
y = 3*( (x + 2)**2 - 7 );
```

in a DATA step. One could omit the spaces, of course.

There are also many, many numeric, text, time, and date functions available. Some of the more commonly used functions are given in Table 4.2. The word *argument* in the syntax column refers to a variable name in the data set, numeric values, or a function of variables.

For more SAS functions a list can be obtained from SAS Help and Online Document. From the pull-down menu select **Help – SAS Help and Documentation** (version 9). After SAS Help and Documentation opens,

1. expand the folder **SAS Products** by clicking on the + to the left of the folder icon
2. expand the **Base SAS** folder
3. expand the **SAS 9.4 DS2 Language Reference** folder
4. expand the **DS2 Language Reference** folder
5. expand the **DS2 Functions** list

See Figure 4.1 for this arrangement as it appears in SAS 9.4.

TABLE 4.2

Commonly Used SAS DATA Step Functions

Function	Syntax	Description
Algebraic Functions		
MOD	y = MOD(dividend, divisor)	returns the remainder when dividing the dividend by the divisor
SIGN	y = SIGN(argument)	returns the sign of the value or 0
SQRT	y = SQRT(argument)	computes a square root
ABS	y = ABS(argument)	computes the absolute value
INT	y = INT(argument)	computes the greatest integer of the value
ROUND	y = ROUND(argument)	rounds the value to the nearest unit
EXP	y = EXP(argument)	computes e^x where x is specified argument
LOG	y = LOG(argument)	computes the natural log or log base e of the argument
LOG2	y = LOG2(argument)	computes the log base 2 of the argument
LOG10	y = LOG10(argument)	computes the log base 10 or common log of the argument
Statistical Functions		
MEAN	y = MEAN(argument1, . . . , argumentn)	computes the arithmetic mean of the arguments listed
STD	y = STD(argument1, . . . , argumentn)	returns the standard deviation of the arguments listed
MAX	y = MAX(argument1, . . . , argumentn)	returns the largest value of those variables or arguments specified in parenthesis
MIN	y = MIN(argument1, . . . , argumentn)	returns the smallest value of those variables or arguments specified in parenthesis
N	y = N(argument1, . . . , argumentn)	counts all nonmissing numeric values
NMISS	y = NMISS(argument1, . . . , argumentn)	counts all missing numeric values
Trigonometric Functions		*All angles are measured in radians.*
SIN	y = SIN (argument)	computes the sine of an angle (radian)
COS	y = COS(argument)	computes the cosine of an angle
TAN	y = TAN(argument)	computes the tangent of an angle
ARCOS	y = ARCOS(argument)	computes the inverse cosine
ARSIN	y = ARSIN(argument)	computes the inverse sine
ATAN	y = ATAN(argument)	computes the inverse tangent
Text Functions		
UPCASE	y = UPCASE(argument)	converts the character value to uppercase
LOWCASE	y = LOWCASE(argument)	converts the character value to lowercase
PROPCASE	y = PROPCASE(argument)	converts the character value to proper case; all lowercase except for first letter of the word
Miscellaneous Functions		
_n_	y = _n_	sets y equal to the value of the observation number. The observation number SAS displays is not a variable in the data set otherwise.
LAGn	y = LAGn(argument)	returns the value occurring n observations before this one. n must be a whole number between 1 and 100. The lines of the data must be in a meaningful order prior to using the LAG function. LAG and LAG1 are the same function.

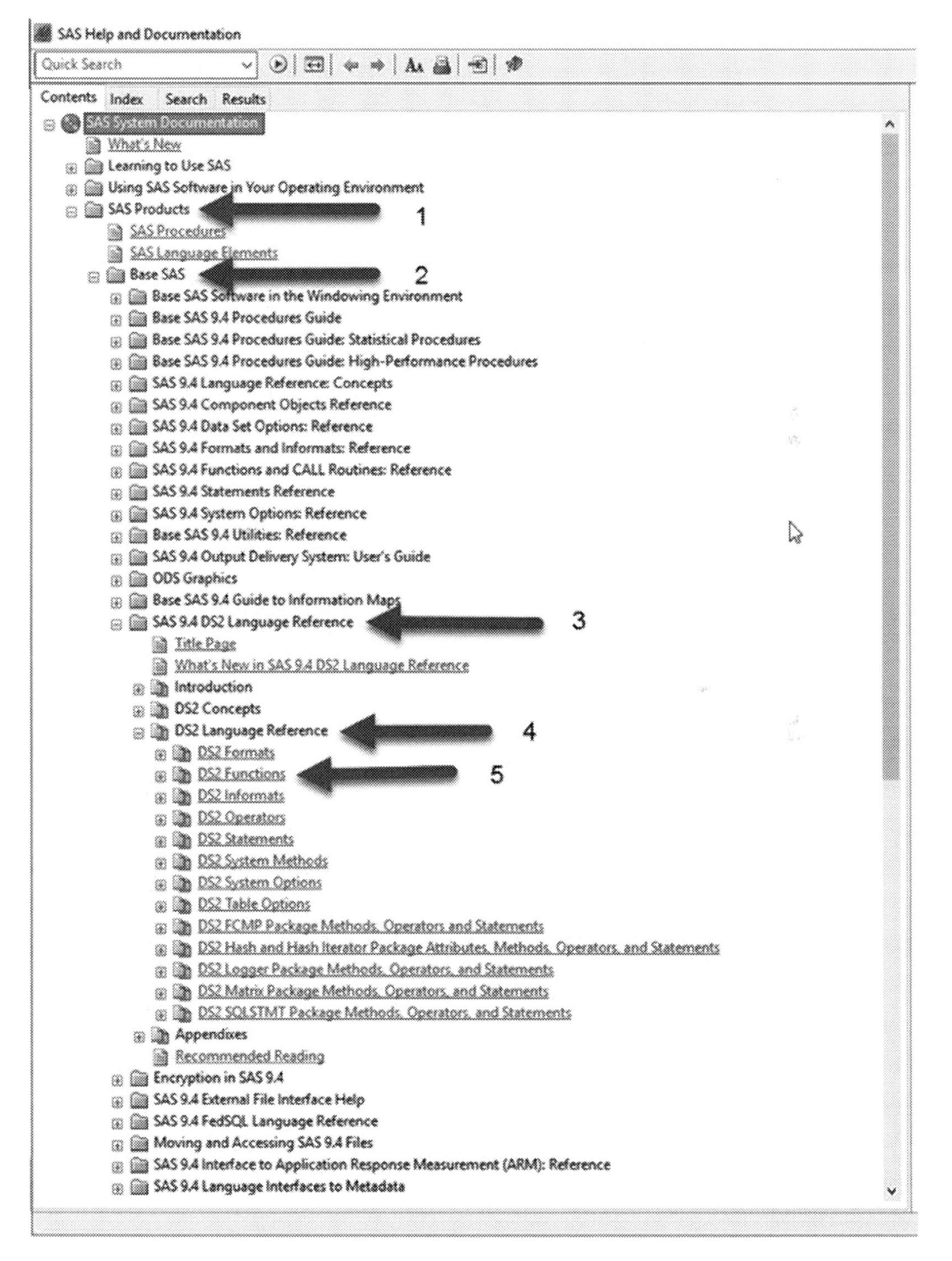

FIGURE 4.1
SAS Help and Documentation for identifying available SAS DATA step functions.

Consider an updated version of the list of students and grades used in Chapter 2. In this grade record are three exam scores where each exam is worth 100 points, and five quiz scores each worth 25 points. Bill is missing a score for Quiz 4, and Dana is missing a score for Exam 3.

Name	ID	Exam 1	Exam 2	Exam 3	Quiz 1	Quiz 2	Quiz 3	Quiz 4	Quiz 5
Bill	123000000	85	88	84	20	22	16		21
Helen	234000000	96	90	89	16	25	20	18	22
Steven	345000000	80	92	82	19	24	19	20	21
Carla	456000000	65	78	74	18	20	23	20	24
Dana	567000000	97	94		22	17	24	18	20
Lisa	789000000	81	88	92	15	20	22	18	19

OBJECTIVE 4.2: Create a SAS data set *grades* containing the three exams scores and the five quiz scores. Using SAS functions or arithmetic operations in a DATA step, for each student compute the exam score total, the number of exam scores recorded for each student, exam average, the minimum quiz score, and the total of the four best quizzes. Print the exam results separate from the quiz results. Do not print the student ID numbers nor the observation numbers in either list.

```
DATA grades;
INPUT Name $ ID Exam1 Exam2 Exam3 Q1 Q2 Q3 Q4 Q5;

ExamTotal1 = SUM(exam1, exam2, exam3);
ExamTotal2 = exam1 + exam2 + exam3;
ExamsN = N(exam1, exam2, exam3);              *number of exams completed;
ExamAvg1 = MEAN(exam1, exam2, exam3);
ExamAvg2 = (exam1 + exam2 + exam3) / ExamsN;
MinQuiz = MIN(q1, q2, q3, q4, q5);
Best4QuizTotal = SUM(q1, q2, q3, q4, q5) - MIN(q1, q2, q3, q4, q5);

DATALINES;
Bill      123000000  85 88 84 20 22 16 . 21
Helen     234000000  96 90 89 16 25 20 18 22
Steven    345000000  80 92 82 19 24 19 20 21
Carla     456000000  65 78 74 18 20 23 20 24
Dana      567000000  97 94 . 22 17 24 18 20
Lisa      789000000  81 88 92 15 20 22 18 19
;
PROC PRINT DATA=grades NOOBS;
VAR name exam1 exam2 exam3 examtotal1 examtotal2 examsn examavg1 examavg2;
TITLE 'Objective 4.2';
TITLE3 'Exam Info Only';

PROC PRINT DATA=grades NOOBS;
VAR name q1 q2 q3 q4 q5 minquiz best4quiztotal;
TITLE3 'Quiz Info Only';

RUN;
QUIT;
```

The total and the average of the three exams were each computed two ways, using the appropriate SAS function and using arithmetic expressions. In the line for "ExamsN", a single line comment (`*number of exams completed;`) appears to the right of the statement to identify what this statement is producing. It is a good practice to annotate all programs. Comments are overviewed in Section 20.2.

In this objective, when multiple words identify the computation, and the variable name cannot contain spaces, spaces are omitted in variable names, such as "ExamTotal1". Variable labels are presented in Chapter 8 and provide a strategy for avoiding long variable names.

Objective 4.2

Exam Info Only

Name	Exam1	Exam2	Exam3	ExamTotal1	ExamTotal2	ExamsN	ExamAvg1	ExamAvg2
Bill	85	88	84	257	257	3	85.6667	85.6667
Helen	96	90	89	275	275	3	91.6667	91.6667
Steven	80	92	82	254	254	3	84.6667	84.6667
Carla	65	78	74	217	217	3	72.3333	72.3333
Dana	97	94	.	191	.	2	95.5000	.
Lisa	81	88	92	261	261	3	87.0000	87.0000

Objective 4.2

Quiz Info Only

Name	Q1	Q2	Q3	Q4	Q5	MinQuiz	Best4QuizTotal
Bill	20	22	16	.	21	16	63
Helen	16	25	20	18	22	16	85
Steven	19	24	19	20	21	19	84
Carla	18	20	23	20	24	18	87
Dana	22	17	24	18	20	17	84
Lisa	15	20	22	18	19	15	79

In the first table of the output for Exam Info Only, note that ExamTotal1 and ExamTotal2 are the same values for each student except for Dana. Dana's Exam 3 score is missing. ExamTotal1 used the SUM function. When the SUM function encountered a missing value, it added the nonmissing values, whereas the arithmetic operations used in ExamTotal2 produced a missing value since one of the operands is missing. This is an important difference between the two methods. Similarly, the ExamAvg1 and ExamAvg2 are in agreement unless one or more of the operands are missing values. A common error is for researchers to record a zero when, in fact, the observation is missing. Knowing how programming statements are affected by missing values is important.

A similar result is observed in the Quiz Info Only. Notice that the missing Quiz 4 score for Bill is not the minimum value, but 16 is. Thus, Bill's Best4QuizTotal is 20 + 22 + 21 = 63. If the missing score should be regarded as a zero when adding up the four largest quiz scores, then Bill's QuizTotal should be 20 + 22 + 16 + 21 = 79. If for some reason Bill's (nonzero) score has not been recorded, a zero score is not appropriate. Notice that Steven has two quiz scores of 19, and 19 is also his minimum score. In the computation of his QuizTotal, one of the 19's was correctly subtracted.

The MEANS procedure and UNIVARIATE procedure both compute sums, means, and various other statistics for **specified variables or columns of data**. In Objective 4.2 the sum and mean **for each of the subjects or rows of data** are computed.

OBJECTIVE 4.3 For the *fish* data in Objective 4.1, sort and print the data by gender. Observe the effects of the typographical errors when entering gender using a mix of uppercase and lowercase letters.

```
DATA fish;
INPUT ID Location $ Length Weight Age Gender $;
Length_in = length / 25.4;
DATALINES;
23   Payne   75    24   2.5   f
41   Payne   68    16   2     m
17   Payne   57    12   1.5   F
33   payne   45    14   0.5   m
18   Payne   71    20   3     F
77   Payne   60    19   2.5   f
;
PROC SORT DATA=fish; BY gender;
PROC PRINT DATA=fish; BY gender;
TITLE 'Objective 4.3';
RUN;
QUIT;
```

Objective 4.3

Gender=F

Obs	ID	Location	Length	Weight	Age	Length_in
1	17	Payne	57	12	1.5	2.24409
2	18	Payne	71	20	3.0	2.79528

Gender=f

Obs	ID	Location	Length	Weight	Age	Length_in
3	23	Payne	75	24	2.5	2.95276
4	77	Payne	60	19	2.5	2.36220

Gender=m

Obs	ID	Location	Length	Weight	Age	Length_in
5	41	Payne	68	16	2.0	2.67717
6	33	payne	45	14	0.5	1.77165

In the output for Objective 4.3, the females are indicated by "f" and "F". It is important to remember that **data are case sensitive**, and this is an example of that. "f" and "F" are regarded as two different values when, in fact, they both are meant to identify the category of females. Clearly in this small data set these typographical errors can be easily corrected, but what if a large data set contains these errors? Finding each error and correcting them may be quite time consuming if it is even feasible. SAS functions can be used to set all of the values to the same case.

OBJECTIVE 4.4 For the SAS data set *fish*, correct the data so that all of the responses to the Gender variable are lowercase. Print the data by gender.

```
DATA fish;
INPUT ID Location $ Length Weight Age Gender $;
Length_in = length / 25.4;
gender = LOWCASE(gender);
DATALINES;
23    Payne    75    24    2.5    f
41    Payne    68    16    2      m
17    Payne    57    12    1.5    F
33    payne    45    14    0.5    m
18    Payne    71    20    3      F
77    Payne    60    19    2.5    f
;
PROC SORT DATA=fish; BY gender;
PROC PRINT DATA=fish; BY gender;
TITLE 'Objective 4.4';
RUN;
QUIT;
```

The variable *gender* is corrected when the existing variable is used on the left side of the equal sign in this variable assignment statement. One could have written

```
gender2 = LOWCASE(gender);
```

This would produce a new variable "gender2" that has all of the lowercase values of "f" and "m" identifying females and males. The original "gender" variable would remain in SAS data set *fish* as it was originally created. Similarly, the UPCASE or PROPCASE functions could have been used to make this correction.

Objective 4.4

Gender=f

Obs	ID	Location	Length	Weight	Age	Length_in
1	23	Payne	75	24	2.5	2.95276
2	17	Payne	57	12	1.5	2.24409
3	18	Payne	71	20	3.0	2.79528
4	77	Payne	60	19	2.5	2.36220

Gender=m

Obs	ID	Location	Length	Weight	Age	Length_in
5	41	Payne	68	16	2.0	2.67717
6	33	payne	45	14	0.5	1.77165

In the output for Objective 4.4, the two corrected values of *Gender* are the headings for each of the printed data tables.

How could the values for Location be corrected? The PROPCASE function would set all of the locations in *fish* to "Payne".

One of the simplest algebraic functions one can include in a DATA step, is setting the values of a variable equal to a single value, a constant function. That is, a new variable assignment statement could be of the forms:

```
newvariable1 = number;
```

or

```
newvariable2 = "text";
```

OBJECTIVE 4.5: Modify the DATA step in Objective 4.4. Use PROPCASE to correct the location variable. Include a new variable, State, in the SAS data set, and set it equal to "OK" identifying the state of Oklahoma for all lines of data. Print only the location, state, length in inches, weight, and gender variables, and print them in this order.

```
DATA fish;
INPUT ID Location $ Length Weight Age Gender $;
Length_in = length / 25.4;
gender = LOWCASE(gender);
State = "OK";
Location = PROPCASE(location);
DATALINES;
23   Payne   75    24   2.5   f
41   Payne   68    16   2     m
17   Payne   57    12   1.5   F
33   payne   45    14   0.5   m
18   Payne   71    20   3     F
77   Payne   60    19   2.5   f
;
PROC PRINT DATA=fish NOOBS;
VAR location state length_in weight gender;
TITLE 'Objective 4.5';
RUN;
QUIT;
```

Objective 4.5

Location	State	Length_in	Weight	Gender
Payne	OK	2.95276	24	f
Payne	OK	2.67717	16	m
Payne	OK	2.24409	12	f
Payne	OK	1.77165	14	m
Payne	OK	2.79528	20	f
Payne	OK	2.36220	19	f

Perhaps the constant function may not make much sense yet. But suppose data from other locations (counties) and/or states are collected, and each of these locations is in a separate SAS data set. Before the end of this chapter, it will be shown how to combine SAS data sets in a SAS program. Having one or more variables such as, State, becomes an important part of the information as these variables may help identify the origin of the data records. Another advantage of the constant function is that typographical errors, such as "payne", are avoided. If Location had not been a part of the INPUT statement, one could have used a constant function, `Location = "Payne";` and avoided the error.

4.2 IF - THEN Statement

Variables can also be created, kept or deleted by using an IF – THEN statement in a DATA step. An IF – THEN statement also must appear after the INPUT line and before the DATALINES (or CARDS) statement. Multiple IF – THEN statements can appear in a single DATA step. IF – THEN statements cannot be placed within any procedures (PROCs).

The form of the IF – THEN statement is:

```
IF condition THEN assignment1;
```

If the condition specified is true, then assignment1 will be performed. If the condition specified is not true, then the assignment1 is not done. This may create a missing value in the SAS data set. The assignment can follow the variable assignment guidelines presented in Section 4.1. There are some other assignments that can be programmed. Section 15.2 illustrates the OUTPUT assignment, another type of assignment resulting from a true condition.

Another form of the IF – THEN statement includes an ELSE statement.

```
If condition THEN assignment1; ELSE assignment2;
```

The ELSE statement is optional. If the condition specified is true, then assignment1 will be performed. If it is not true, then assignment2 will be performed. Table 4.3 lists the syntax for selected variable conditions.

TABLE 4.3

Syntax for Specifying Conditions

Condition	Symbol Text	Sample Syntax	Comments
equal	=	x = 5	
	EQ	x EQ 5	
not equal	^=	color ^= "Red"	Single or double quotes can be used to identify character strings.
	NE	color NE 'Red'	
less than	<	6.7 < x < 8.2	
	LT	6.7 LT x LT 8.2	
less than or equal to	<=	y <= 15.8	
	LE	y LE 15.8	
greater than	>	t > 10.2	
	GT	t GT 10.2	
greater than or equal to	>=	r >= 14.3	
	GE	r GE 14.3	
inclusion in a group	IN	time IN (3, 6, 9)	
		city IN ("KC", "Omaha")	
or	OR	x < 3 OR x GE 7	Logical OR implies that either one or both of the conditions must be true for the assignment in the THEN clause to be made.
and	AND	y > 12 AND t NE 13	Logical AND implies that **both** of the conditions must be true for the assignment in the THEN clause to be made.

Example:

Consider the *fish* data set from Objective 4.5. Some examples of IF – THEN statements appear in Table 4.4. The text in italics in the right column describes what the code is doing. It is NOT a part of the SAS code.

There are two other IF – THEN statements that can be also be useful. The first is

```
IF condition THEN DELETE;
```

Notice the assignment is to remove all the observations or lines of data that satisfy the given condition. Conversely,

```
IF condition;
```

retains the observations or lines of data satisfying the given condition. The author recommends the usage of these statements with care as one should be absolutely certain that data records are not needed before they are deleted.

TABLE 4.4

Examples of IF – THEN statements

Syntax	Remarks
`DATA fish;`	

`INPUT ID Location $ Length Weight Age Gender $;`
`Length_in = length / 25.4;`
`gender = lowcase(gender);`
`State = "OK";`
`Location = PROPCASE(location);`
	DATA and INPUT statements as they appeared in Objective 4.5

`IF length < 60 THEN Size = 'Small';`	*All lengths smaller than 60 are classified as small.*
`ELSE size = 'Large';`	*All items greater than or equal to 60 are classified as large.*
`IF age LE 2.0 THEN group = 1;`	*Three groups are created.* *Group 1 contains the youngest specimens.*
`IF 2.0 < age < 3.0 THEN group = 2;`	*All specimens with ages between 2 and 3 are in group 2.*
`IF age GE 3.0 THEN group = 3;`	*The oldest specimens are in group 3.*
`IF location = "payne" then lake='CB';`	*Because data are case-sensitive, lake is missing for all observations. Correctly using "Payne" would set lake equal to CB for all observations.*

`DATALINES;`
 	
`23 Payne 75 24 2.5 f`
`41 Payne 68 16 2 m`
`17 Payne 57 12 1.5 F`
`33 payne 45 14 0.5 m`
`18 Payne 71 20 3 F`
`77 Payne 60 19 2.5 f`
`;`
	After inserting all of the modification statements, the DATALINES statement is still needed.

4.3 WHERE Statement

The WHERE statement is not a type of variable assignment but is a convenient tool for creating a subgroup on which a procedure is to operate. The WHERE statement also requires that a condition be specified. The WHERE statement is not limited to usage in the DATA step. It may be necessary to have a procedure perform only with data that satisfy one or more conditions. The *condition* specified in the WHERE statement follows the same syntax as the IF – THEN statement conditions in Section 4.2; hence it is introduced in the same chapter as the IF – THEN statement.

```
WHERE condition;
```

For the PRINT, UNIVARIATE, and MEANS procedures it was stated that the WHERE statement was an optional statement to include in the block of code for that procedure. Using a WHERE statement, a procedure can be controlled to perform only when a specified condition is met.

OBJECTIVE 4.6 Use the DATA step from Objective 4.5. First, print the observations having a weight of 15 or lower. Second, compute the mean and the other default sample statistics for the variables length (in both millimeters and inches) and weight only for the males using the MEANS procedure.

```
DATA fish;
INPUT ID Location $ Length Weight Age Gender $;
Length_in = length / 25.4;
gender = LOWCASE(gender);
State = "OK";
Location = PROPCASE(location);
DATALINES;
23    Payne    75    24    2.5    f
41    Payne    68    16    2      m
17    Payne    57    12    1.5    F
33    payne    45    14    0.5    m
18    Payne    71    20    3      F
77    Payne    60    19    2.5    f
;
PROC PRINT DATA=fish;
WHERE weight le 15;
TITLE 'Objective 4.6';
TITLE2 'Observations with Weight <= 15';

PROC MEANS DATA=fish; WHERE gender='m';
VAR length length_in weight;
TITLE2 'Summary Statistics for the Males';
RUN;
QUIT;
```

Like all SAS statements, the WHERE statement does not have to appear on a line by itself so long as the correct semicolon punctuation is used to separate each of the commands.

Objective 4.6

Observations with Weight <= 15

Obs	ID	Location	Length	Weight	Age	Gender	Length_in	State
3	17	Payne	57	12	1.5	f	2.24409	OK
4	33	Payne	45	14	0.5	m	1.77165	OK

Objective 4.6

Summary Statistics for the Males

The MEANS Procedure

Variable	N	Mean	Std Dev	Minimum	Maximum
Length	2	56.5000000	16.2634560	45.0000000	68.0000000
Length_in	2	2.2244094	0.6402935	1.7716535	2.6771654
Weight	2	15.0000000	1.4142136	14.0000000	16.0000000

When a WHERE statement is used, a smart choice of TITLE statements is recommended to identify the subgroups on which the PRINT and MEANS procedures are operating.

4.4 SET Statement

One of the extremely useful DATA step commands is the SET command. Up to this point SAS data sets have been created using the DATA – INPUT – DATALINES syntax or an OUTPUT statement available in some procedures. The SET commands allows the programmer to operate on a SAS data set previously created in the SAS session without beginning all over again with the DATA – INPUT – DATALINES syntax. The SET statement can be used within a DATA step operation. The SET command can be used in two ways, modifying an existing SAS data set and concatenating two or more SAS data sets.

4.4.1 Modify an Existing SAS Data Set

SAS data sets created in the current session are available and active until the SAS session closes. An existing SAS data set can be called into a DATA step to be modified. This new SAS data set can be saved by a new name, or can overwrite the old SAS data set.

Option 1: Creating a new SAS data set while keeping the original data set.

```
DATA  new-SAS-data-set-name;
SET oldname;
< DATA Step operations>
```

The first line creates a new SAS data set.

`SET oldname;` calls an existing SAS data set from the current SAS session into this DATA step.

The DATA Step operations can include new variable assignment statements, IF – THEN statement(s), WHERE statement, and other DATA Step commands that are presented in Chapters 8, 12, 13, and 15. These statements must follow the SET statement and occur before any new DATA step or procedure (PROC) code.

This is useful if one wishes to modify variables in an existing SAS data set and keep the original SAS data set intact. The SAS data set specified in the SET statement must have been created earlier in the SAS session but not necessarily in the same program as this second DATA step. Chapter 12 contains information about SAS data sets that are permanent files.

OBJECTIVE 4.7: Create a new SAS data set containing only the data for the male specimens in the SAS data set *fish*.

When the SAS data set *fish* was created in Objective 4.5, the following message appeared in the log window.

```
NOTE: The data set WORK.FISH has 6 observations and 8 variables.
```

In a new Enhanced Editor window the following brief DATA step code can be entered and submitted:

```
DATA males_fish;
SET fish;
IF gender = "f" THEN DELETE;
RUN;
```

All females will be deleted from this new SAS data set. The original SAS data set *fish* is unchanged. Observe the message lines in the log window referencing the number of observations in the original SAS data set *fish* and in the new SAS data set *males_fish*.

```
NOTE: There were 6 observations read from the data set WORK.FISH.
NOTE: The data set WORK.MALES_FISH has 2 observations and 8 variables.
```

Option 2: Updating an existing SAS data set keeping the original SAS data set name.

```
DATA oldname;
SET oldname;
< DATA Step operations>
```

This DATA step calls in an existing SAS data set, and modifies it. All modifications are saved to the original SAS data set. When this DATA step is completed, only one SAS data set, *oldname*, will exist. It will contain all of the new information specified in the DATA step operations and will replace the previous version of the SAS data set *oldname*.

OBJECTIVE 4.8: Add the additional variable "Species" to the SAS data set *fish* (Objective 4.5) using a SET command for the modification. Set the value of Species equal to "darter" for all observations. Print the id, location, species, and weight. (Clearly this could also be done by modifying the DATA step code in Objective 4.5.)

```
DATA fish;
SET fish;
Species = "darter";

PROC PRINT DATA=fish;
VAR ID location species weight;
```

```
TITLE 'Objective 4.8';
RUN;
QUIT;
```

Though only four of the variables were requested to be printed, there are eight variables initially in the SAS data set *fish,* and nine variables after the addition of Species. The log message for this DATA step is:

```
254   DATA fish;
255   SET fish;
256   Species = "darter";
257

NOTE: There were 6 observations read from the data set WORK.FISH.
NOTE: The data set WORK.FISH has 6 observations and 9 variables.
```

Objective 4.8

Obs	ID	Location	Species	Weight
1	23	Payne	darter	24
2	41	Payne	darter	16
3	17	Payne	darter	12
4	33	Payne	darter	14
5	18	Payne	darter	20
6	77	Payne	darter	19

The Species variable is now included in the SAS data set *fish.*

4.4.2 Concatenating SAS Data Sets

More than one SAS data set can be a part of a SAS session or program. Sometimes it is necessary to combine the information in two or more SAS data sets. This manner of combining SAS data sets "stacks" them vertically resulting in a greater number of data records (rows of data). If two or more SAS data sets are to be vertically concatenated the SET statement in a DATA step can do this. The syntax of this command is:

```
DATA newsetname;
SET SAS-data-set1 … SAS-data-setn;
```

When using the SET command, the SAS data sets must have most, if not all, of the same variable names. The SET command vertically concatenates (or stacks) the data sets. The number of observations in the resulting SAS data set is increased when concatenating two or more SAS data sets. In this vertical concatenation, the order of the data top to bottom is the order of the SAS data sets specified in the SET statement.

OBJECTIVE 4.9: Form one SAS data set from the Payne County *fish* data set in Objective 4.5 and a new set of data from Noble County by using the SET statement. Print the resulting SAS data set. Using this combined SAS data set compute summary statistics for each location (county) using the MEANS procedure.

```
DATA fish_noble;
INPUT ID Lake $ Gender $ Weight Length Age;
```

```
Length_in = length / 25.4;
gender = LOWCASE(gender);
Location="Noble";
State = "OK";
DATALINES;
83 PRY f 20 61 2
72 MCM m 24 80 3
30 MCM m 19 69 1.5
46 PRY f 18 50 2.5
78 MCM f 19 54 2
;
DATA combine;
SET fish  fish_noble;

PROC PRINT DATA=combine;
TITLE 'Objective 4.9';

PROC MEANS DATA=combine;
CLASS location;
    *** No VAR statement used. All numeric variables summarized;
RUN;
QUIT;
```

The variables or columns in the SAS data set *fish_noble* are not in the same order as the variables in the SAS data set *fish*. The two sets have the variables ID, Location, Length, Weight, Age, Length_in, and State in common. SAS data set *fish* does not have the Lake variable. It was in the example in Table 4.4 but was not a part of Objective 4.5. In Objective 4.5, Location was in the input data values and corrected with the PROPCASE function. In *fish_noble* Location was created using the constant function, Location = "Noble". For the computation of the summary statistics, a CLASS statement within the MEANS procedure does not require that the new combined data set be sorted first. If one had used "BY location" instead of the CLASS statement, sorting the combined data by location would be necessary. The single line comment in the MEANS procedure only needs a single * to the left to identify the line as a comment. Multiple *'s are the programmers choice. See Section 20.2 for more about the syntax used in comments.

Objective 4.9

Obs	ID	Location	Length	Weight	Age	Gender	Length_in	State	Lake
1	23	Payne	75	24	2.5	f	2.95276	OK	
2	41	Payne	68	16	2.0	m	2.67717	OK	
3	17	Payne	57	12	1.5	f	2.24409	OK	
4	33	Payne	45	14	0.5	m	1.77165	OK	
5	18	Payne	71	20	3.0	f	2.79528	OK	
6	77	Payne	60	19	2.5	f	2.36220	OK	
7	83	Noble	61	20	2.0	f	2.40157	OK	PRY
8	72	Noble	80	24	3.0	m	3.14961	OK	MCM
9	30	Noble	69	19	1.5	m	2.71654	OK	MCM
10	46	Noble	50	18	2.5	f	1.96850	OK	PRY
11	78	Noble	54	19	2.0	f	2.12598	OK	MCM

Objective 4.9

The MEANS Procedure

Location	N Obs	Variable	N	Mean	Std Dev	Minimum	Maximum
Noble	5	**ID**	5	61.8000000	22.7859606	30.0000000	83.0000000
		Length	5	62.8000000	12.0291313	50.0000000	80.0000000
		Weight	5	20.0000000	2.3452079	18.0000000	24.0000000
		Age	5	2.2000000	0.5700877	1.5000000	3.0000000
		Length_in	5	2.4724409	0.4735878	1.9685039	3.1496063
Payne	6	**ID**	6	34.8333333	22.6310995	17.0000000	77.0000000
		Length	6	62.6666667	10.9666160	45.0000000	75.0000000
		Weight	6	17.5000000	4.3703547	12.0000000	24.0000000
		Age	6	2.0000000	0.8944272	0.5000000	3.0000000
		Length_in	6	2.4671916	0.4317565	1.7716535	2.9527559

In the first table of the output for Objective 4.9, the order of the variables or columns are determined by the SAS data set specified first in the SET statement. The missing values for Lake in the Payne County data are also evident. What if the Noble County column for the weight variable was entered as "Wt" rather than as "Weight"? How would that affect the concatenation? Without restating the programming, the output from the PRINT procedure would look like this:

Obs	ID	Location	Length	Weight	Age	Gender	Length_in	State	Lake	Wt
1	23	Payne	75	24	2.5	f	2.95276	OK		.
2	41	Payne	68	16	2.0	m	2.67717	OK		.
3	17	Payne	57	12	1.5	f	2.24409	OK		.
4	33	Payne	45	14	0.5	m	1.77165	OK		.
5	18	Payne	71	20	3.0	f	2.79528	OK		.
6	77	Payne	60	19	2.5	f	2.36220	OK		.
7	83	Noble	61	.	2.0	f	2.40157	OK	PRY	20
8	72	Noble	80	.	3.0	m	3.14961	OK	MCM	24
9	30	Noble	69	.	1.5	m	2.71654	OK	MCM	19
10	46	Noble	50	.	2.5	f	1.96850	OK	PRY	18
11	78	Noble	54	.	2.0	f	2.12598	OK	MCM	19

Weight and Wt would be identified as two different variables, and one location would have missing values for one of those columns. In this small data set, this is relatively easy to identify. When there are more variables and larger numbers of observations, identifying this issue can be more difficult. To correct this issue, one could recheck each of the INPUT statements and make certain the spelling of the variables in common are spelled the same, or using variable assignment statements, create Weight = wt; in the *fish_noble* SAS data set.

What would change if the two SAS data sets were reversed in the SET statement in the programming code for Objective 4.9? That is,

```
DATA combine;
SET fish_noble fish;
```

The SET command "stacks" the SAS data sets in the order specified in the SET statement. The order in the SET statement is affected whether concatenating two data sets or several. The order of the columns would also change in the default PRINT procedure output since the SAS data set *noble_fish* variables have a different order than the SAS data set *fish*, and the first SAS data set identified in the SET statement determines the initial order of the column variables.

Obs	ID	Lake	Gender	Weight	Length	Age	Length_in	Location	State
1	83	PRY	f	20	61	2.0	2.40157	Noble	OK
2	72	MCM	m	24	80	3.0	3.14961	Noble	OK
3	30	MCM	m	19	69	1.5	2.71654	Noble	OK
4	46	PRY	f	18	50	2.5	1.96850	Noble	OK
5	78	MCM	f	19	54	2.0	2.12598	Noble	OK
6	23		f	24	75	2.5	2.95276	Payne	OK
7	41		m	16	68	2.0	2.67717	Payne	OK
8	17		f	12	57	1.5	2.24409	Payne	OK
9	33		m	14	45	0.5	1.77165	Payne	OK
10	18		f	20	71	3.0	2.79528	Payne	OK
11	77		f	19	60	2.5	2.36220	Payne	OK

4.5 MERGE Statement

SAS data sets that have one or more common variables can be merged or concatenated "side by side". The MERGE statement increases the number of variables or columns in the resulting SAS data set. **Each of the SAS data sets must first be sorted by the key variable(s) that they have in common.** If the SAS data sets do not have at least one common key variable, they should not be merged. The general approach for merging two SAS data sets is given by:

```
PROC SORT DATA=a; BY variablelist;
PROC SORT DATA=b; BY variablelist;
DATA new;
MERGE a b; BY variablelist;
```

The above SAS code can be modified for more than two SAS data sets. Each SAS data set must contain the same common key variable information by which the tables are to be merged. Using the *grades* SAS data set created in Objective 4.2, consider the data set *finals* for the same students. That is, the following new information for the six students is available.

Name	ID	Final Project Score	Final Project Letter Grade	Final Exam Score (%)
Bill	123000000	83	B	88
Helen	234000000	91	A	84
Steven	345000000	94	A	85
Carla	456000000	77	C	92
Dana	567000000	86	B	79
Lisa	789000000	79	C	100

OBJECTIVE 4.10: Combine the two SAS data sets "grades" and "finals" into a single SAS data set "all_grades" so that each of the six student records are correctly aligned. Print the resulting SAS data set without observation numbers.

```
DATA grades;
INPUT Name $ ID Exam1 Exam2 Exam3 Q1 Q2 Q3 Q4 Q5;
DATALINES;
Bill      123000000  85 88 84 20 22 16 . 21
Helen     234000000  96 90 89 16 25 20 18 22
Steven    345000000  80 92 82 19 24 19 20 21
Carla     456000000  65 78 74 18 20 23 20 24
Dana      567000000  97 94 . 22 17 24 18 20
Lisa      789000000  81 88 92 15 20 22 18 19
;

DATA finals;
INPUT Name $ ID FPScore FPGrade $ FExam;
DATALINES;
Bill      123000000  83  B 88
Helen     234000000  91  A 84
Steven    345000000  94  A 85
Carla     456000000  77  C 92
Dana      567000000  86  B 79
Lisa      789000000  79  C 100
;
PROC SORT DATA=grades; BY name id;
PROC SORT DATA=finals; BY name id;
DATA all_grades;
MERGE grades finals;
BY name id;

PROC PRINT DATA=all_grades NOOBS;
TITLE 'Objective 4.10' ;
RUN;
QUIT;
```

In this instance the order of the students was the same in both files. Very often this is not the case.

Objective 4.10

Name	ID	Exam1	Exam2	Exam3	Q1	Q2	Q3	Q4	Q5	FPScore	FPGrade	FExam
Bill	123000000	85	88	84	20	22	16	.	21	83	B	88
Carla	456000000	65	78	74	18	20	23	20	24	77	C	92
Dana	567000000	97	94	.	22	17	24	18	20	86	B	79
Helen	234000000	96	90	89	16	25	20	18	22	91	A	84
Lisa	789000000	81	88	92	15	20	22	18	19	79	C	100
Steven	345000000	80	92	82	19	24	19	20	21	94	A	85

The variable names were kept short. One could have used a lengthier name, such as Final_Project_Letter_Grade instead of FPGrade. Since variable names cannot contain spaces, the usage of the underscore achieves the multiple word appearance in a single variable name.

See Section 2.2 for SAS variable naming restrictions. Longer variable names, such as this, are not typically recommended however, as longer names become cumbersome and more susceptible to typographical errors in later programming lines. Keeping shorter variable names and developing more detailed variable labels are covered in Chapter 8.

There are multiple ways one could have programmed this objective. Both SAS data sets have Name and ID in common as key variables. Since all of the names, the BY statements on the SORT procedures and in the DATA step could have simply been "BY name". That is, the SAS data sets are compatibly sorted for any action or procedure to be completed "BY name". If one used "BY id" in the DATA step, the merging process would have failed since the data are first put in alphabetical order by Name. If merging "BY id", the two SAS data sets must be sorted "BY id" or "BY id name". If either SAS data set is not compatibly sorted for the merge, an error message will appear in the SAS log. The following is the SAS log when merging "BY id" and the error messages contained therein. Real time and CPU times notes have been suppressed.

```
1328  PROC SORT DATA=grades; BY name id;

NOTE: There were 6 observations read from the data set WORK.GRADES.
NOTE: The data set WORK.GRADES has 6 observations and 10 variables.

1329  PROC SORT DATA=finals; BY name id;
1330

NOTE: There were 6 observations read from the data set WORK.FINALS.
NOTE: The data set WORK.FINALS has 6 observations and 5 variables.
1331  DATA all_grades;
1332  MERGE grades finals;
1333  BY  id;
1334

ERROR: BY variables are not properly sorted on data set WORK.GRADES.
Name=Dana ID=567000000 Exam1=97 Exam2=94 Exam3=. Q1=22 Q2=17 Q3=24 Q4=18
Q5=20 PScore=86
PGrade=B FExam=79 FIRST.ID=1 LAST.ID=1 _ERROR_=1 _N_=3
NOTE: The SAS System stopped processing this step because of errors.
NOTE: There were 4 observations read from the data set WORK.GRADES.
NOTE: There were 4 observations read from the data set WORK.FINALS.
WARNING: The data set WORK.ALL_GRADES may be incomplete. When this step
was stopped there were 2 observations and 13 variables.
```

One should get in the habit of reading the SAS log for DATA step operations for the number of observations and number of variables. A researcher knows how much data was recorded and should recognize when these numbers are not correct. In the warning in the last line of the SAS log, only two observations were in SAS data set *all_grades* when the error occurred.

OBJECTIVE 4.11: What happens if these SAS data sets to be merged were concatenated using the SET command instead? (Recall to use SET, the SAS data sets do not have to be sorted first.)

```
DATA all_grades;
SET grades finals;
```

```
PROC PRINT DATA=all_grades NOOBS;
TITLE "Objective 4.11";
TITLE2 "SET instead of MERGE";
RUN;
QUIT;
```

Objective 4.11

SET instead of MERGE

Name	ID	Exam1	Exam2	Exam3	Q1	Q2	Q3	Q4	Q5	PScore	PGrade	FExam
Bill	123000000	85	88	84	20	22	16	.	21	.		.
Carla	456000000	65	78	74	18	20	23	20	24	.		.
Dana	567000000	97	94	.	22	17	24	18	20	.		.
Helen	234000000	96	90	89	16	25	20	18	22	.		.
Lisa	789000000	81	88	92	15	20	22	18	19	.		.
Steven	345000000	80	92	82	19	24	19	20	21	.		.
Bill	123000000	.	.	.	.	.	.	.	.	83	B	88
Carla	456000000	.	.	.	.	.	.	.	.	77	C	92
Dana	567000000	.	.	.	.	.	.	.	.	86	B	79
Helen	234000000	.	.	.	.	.	.	.	.	91	A	84
Lisa	789000000	.	.	.	.	.	.	.	.	79	C	100
Steven	345000000	.	.	.	.	.	.	.	.	94	A	85

The number of observations increases because the variables in the two SAS data sets are not all the same. If one only wishes to expand the number of columns, one should MERGE the SAS data sets and should not use the SET command. If "BY name" were included in the DATA step with the SET statement, the order of the rows would be alphabetized by name, but the number of rows and the missing data would stay the same.

Here are some final remarks about merging SAS data sets.

- More than two SAS data sets can be merged in a single DATA step.
- All SAS data sets to be merged must be compatibly sorted by common key variables before merging. There are some exceptions to this, but in general, sorting the SAS data sets is recommended.
- The BY statement in the DATA step must contain at least one variable that all the SAS data sets have in common. The BY statement must follow the MERGE statement. It cannot appear before the MERGE statement in the DATA step.
- The SAS data sets may have only the BY variables in common or they have more variables in common. Not all common variables have to be specified in the BY statement. Only those variables necessary to successfully merge the SAS data sets are required in the BY statement.
- If the **values** of the common variables are not the same (the variable names are the same however) in the SAS data sets to be merged, the last SAS data set in the MERGE statement will determine the value of the variable. If the difference in the values occurs in the key variables used in the BY statement, then the sorted order

of the data is affected. If, say, in the final exam data, Bill's ID is 123000001, then Bill would mistakenly appear twice in the merged data set.

Name	ID	Exam1	Exam2	Exam3	Q1	Q2	Q3	Q4	Q5	FPScore	FPGrade	FExam
Bill	123000000	85	88	84	20	22	16	.	21	.	.	.
Bill	123000001	.	.	.	.	.	.	.	.	83	B	88

- The MERGE statement increases the number of variables or columns. The number of rows in the resulting SAS data set can increase if one of the SAS data sets to be merged has more lines than the other(s).

4.6 Chapter Summary

In this chapter, some key information about the DATA step has been introduced. Frequently the reader will be referred to this chapter as new material is introduced in later chapters. Adding variables or columns to a SAS data set by creating new variables was introduced. Modifying a SAS data set using IF – THEN statements required the programmer to learn how to write the conditional part of those statements using appropriate syntax. Though the WHERE statement is not limited to usage in the DATA step, the syntax of these conditions is also used in WHERE statements. The SET and MERGE commands modify existing SAS data sets. In a single SAS session, once a SAS data set is created it stays in SAS memory until the session closes thus modifications to the SAS data sets using variable assignment statements or commands, such as SET or MERGE, can be done.

5

Beginning Charts

Very often a chart or a graph is needed to illustrate concepts or findings in a statistical analysis. SAS has procedures that can produce high-resolution bar charts, pie charts, and three-dimensional charts that are presentation quality. Older procedures for graphics are found in the SAS/GRAPH product, and newer procedures are in Statistical Graphics Procedures. This chapter will introduce some very basic concepts in the SAS/GRAPH GCHART procedure. Chapters 16 and 17 will introduce Statistical Graphics and other SAS/GRAPH procedures for plots of data, such as on an XY coordinate system.

5.1 The GCHART Procedure

This introduction to the GCHART procedure provides the reader with some basic instruction for the production of bar charts, histograms, and pie charts. The CHART procedure is an older, low-resolution version of producing charts and is not widely used anymore. The charts produced by the CHART procedure are not presentation quality, but the syntax is quickly learned. The GCHART statements and options given in this chapter are the same for both the CHART and GCHART procedures, and thus are also quickly learned. SAS/GRAPH procedures have many, many options for controlling color, axes, font, line types, and symbols. These graphics options for GCHART and other SAS/GRAPH procedures are covered in Chapter 17. At an introductory level, there are default settings for color, axis, and patterns in the GCHART procedure that are more than adequate. For basic content the introductory syntax to the GCHART procedure is:

```
PROC GCHART DATA=SAS-data-set;
BY variable(s);                                      (optional statement)
WHERE condition;                                     (optional statement)
VBAR variable(s)/<options >;
HBAR variable(s)/<options >;
BLOCK variable(s)/<options >;
PIE variable(s)/< options >;
RUN;
```

The GCHART procedure can be used to generate one or several charts. Each of the GCHART statements VBAR through PIE in the syntax above requests the chart(s) of that type. One of these statements or several can be used in a single GCHART procedure. Available options for each of the statements following PROC GCHART statement are listed in Table 5.1.

BY Statement

A BY statement can be used with PROC GCHART to obtain charts for observations in groups defined by the "BY" variables. PROC SORT (Chapter 2) using the

TABLE 5.1

Summary of Statement Specific Options in the GCHART Procedure

Options Grouped by Function	HBAR	VBAR	BLOCK	PIE
Standard options				
AXIS =	X	X	X	X
FREQ =	X	X	X	X
DISCRETE	X	X	X	X
LEVELS =	X	X	X	X
MIDPOINTS =	X	X	X	X
TYPE =	X	X	X	X
SUMVAR =	X	X	X	X
Separate into groups				
GROUP =	X	X	X	
SUBGROUP =	X	X	X	
Request statistical analysis				
FREQ	X	X		
CFREQ	X	X		
PERCENT	X	X		
CPERCENT				
SUM	X	X		
MEAN	X	X		
NOSTATS	X			
Control chart appearance				
AUTOREF	X	X		
CLIPREF	X	X		
REF =	X	X		
SPACE =	X	X		
NOHEADER			X	X

same or a compatible BY statement must appear prior to the PROC GCHART procedure using a BY statement.

WHERE Statement

The WHERE statement defines a subgroup of the data for which the procedure is to be performed. See Section 4.3 to examine syntax for specifying the condition(s) in the WHERE statement.

VBAR Statement

The VBAR statement lists the variables for which vertical bar charts or histograms are to be produced. The values of the chart variable will appear on the horizontal axis. The default image is a frequency bar chart. The bar chart variables are discrete, and the histogram variables are continuous. One can change the definition of the bar length from frequency to another statistic using VBAR statement options. There is a three-dimensional version of a vertical bar chart, and the statement is:

```
VBAR3D variable(s)/<options >;
```

HBAR Statement

The HBAR statement lists the variables for which horizontal bar charts or histograms are to be produced. The values of this chart variable will appear on the vertical axis. The HBAR statement has many of the same options as the VBAR statement. Additional information in the default image are summary statistics of the mean, frequency, cumulative frequency, and percent for each bar. These statistics can be suppressed with HBAR statement options. There is a three-dimensional version of a horizontal bar chart, and the statement is:

```
HBAR3D variable(s)/<options >;
```

BLOCK Statement

The BLOCK statement lists the variables for which 3-D block charts are needed. Though 3-D vertical bars are produced, this image is very different from the VBAR3D image.

PIE Statement

The PIE statement requests a pie chart for each variable listed. Frequency or percent associated with each slice of the pie can be requested.

A brief description of each of the categories and syntax options within those categories is given here. Standard options are those that control the length and tick marks on the response axis and the number of bars in a bar chart or the number of slices in a pie chart. Bar charts and histograms, by default, measure the frequency or count associated with the value where the bar is centered. Standard options include:

AXIS = *values* where the *values* are the tick marks used in constructing the FREQ, PCT, CFREQ, CPCT, SUM, or MEAN axis. Axis values are not separated with a comma. Syntax examples for this option:

AXIS = 2 4 6 or AXIS = 2 TO 10 BY 2.

FREQ = *variable* Normally, each observation of the chart variable contributes a value of one to the frequency counts. When the specified chart variable has another variable associated with it which has values that represent a count for each observation of the chart variable, the FREQ = option identifies that variable.

SAS has a default number of bars or slices that will be selected for the chart unless one of the following options is included in the chart statement.

DISCRETE is used when the numeric chart variable specified is discrete rather than continuous. If the DISCRETE option is omitted, PROC GCHART assumes that all numeric variables are continuous and automatically chooses the midpoints or intervals for the bars unless one of the MIDPOINTS = or LEVELS = options is used.

LEVELS = *number of midpoints* specifies the number of bars, blocks, or pie chart sections representing each chart variable when the variables given in the chart statement are numeric. SAS will select the midpoints at which the bars are centered. The MIDPOINTS option gives greater control over the number and position of the bars.

MIDPOINTS = *values* of a numeric chart variable, and the *values* identify where the bars are centered and the number of *values* determine the number of bars in a bar chart. These *values* also determine the number of slices in a pie chart. Midpoint values are not separated with a comma. For example:

MIDPOINTS = 100 150 200 250 300 or

MIDPOINTS = 100 TO 300 BY 50.

Bar heights and pie chart sections are frequency counts unless specified otherwise.

TYPE = *type* specifies what the bars in the chart represent. The default TYPE is FREQ. One of the following keywords can be specified for *type*: CFREQ, CPERCENT or CPCT, FREQ, MEAN. PERCENT or PCT, or SUM. For all types except FREQ, a summary variable (SUMVAR =) in the data set must be identified. If SUMVAR = option is specified and TYPE = is omitted, the default is TYPE = SUM.

SUMVAR = *variable* names the numeric variable summarized by the bar height. This option is with the TYPE= option (above) to compute a summary statistic for each bar, such as means, sums, or frequencies.

In addition to controlling the number of bars and the length of the bars, one can choose to place bars in groups, or place groups within each bar.

GROUP = *variable* produces side-by-side charts on the same set of axes, where a chart is associated with each value of the GROUP variable. Data do NOT have to be previously sorted by the group variable when the GROUP option is used.

SUBGROUP = *variable* divides each bar into sections where each section identifies the subgroup variable's contribution to the bar. This is applicable to frequency bar charts or histograms.

Separate from setting the bar length equal to a statistic, one can choose to print a statistic associated with each bar at the end of each bar. These statistics are: FREQ, CFREQ, PERCENT, SUM, and MEAN. When using HBAR, the statistics FREQ, CFREQ, PERCENT, and CPERCENT are printed on the chart by default. NOSTATS suppresses all statistics from being printed on a horizontal chart. This is different from the TYPE = option.

Other options that control the chart appearance are those options that place reference lines on bar charts or histograms, modify the spacing between bars, and suppress default chart headers. Reference lines are produced by the AUTOREF, CLIPREF, and REF = *value* options.

AUTOREF produces reference lines at each value of the tick mark on the axis. These reference lines appear in front of the bars or overlay the bars.

CLIPREF When reference lines are requested using the AUTOREF or REF options, CLIPREF has the effect of producing the reference lines 'behind' the bars. That is, the reference lines are "clipped" so that they appear in the background of the graph. Without CLIPREF, the reference lines are overlaid on the bars.

REF = *value* draws a single reference line on the response axis.

SPACE = *value* identifies the amount of space in between the bars of a chart. SPACE=0 leaves no space between bars as one would request for a histogram. SPACE=0 is the only spacing option utilized in this book.

NOHEADER suppresses the default header line normally printed at the top of a block or pie chart.

To demonstrate the different types of charts and available options, consider the SAS data set *instruction* introduced in Section 3.2.

OBJECTIVE 5.1: Produce the default vertical bar chart for the number of students in each program, that is, a frequency chart with no options selected. Include a title for the chart. It is assumed that the SAS data set *instruction* was previously defined in a DATA step.

```
PROC GCHART DATA=instruction;
VBAR program;
TITLE 'Objective 5.1';
TITLE3 'Vertical Bar Chart for Program';
RUN;
```

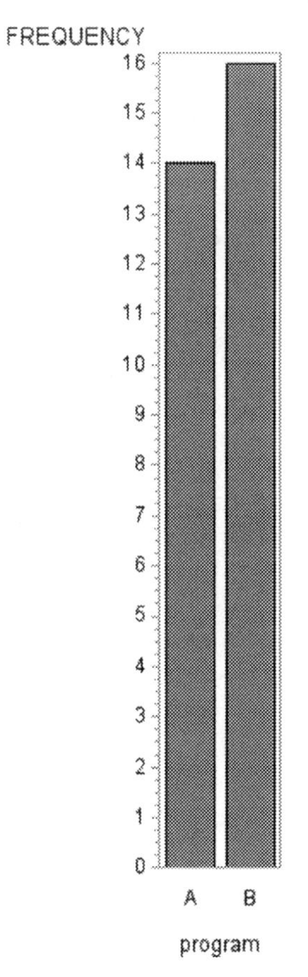

An important option is the DISCRETE option on the various chart statements. In this SAS data set, Program has two values, A and B. Because those values are character or string values, the GCHART correctly produces a bar for A and B.

Suppose the programs were numeric, say 1 and 2, and the INPUT statement identified the variable as numeric. One would also need the DISCRETE option or select the midpoints for the bars. GCHART typically selects 4 or 5 bars for the bar charts (or slices in a pie chart). The DISCRETE option selects a bar for each value of the chart variable. If the programs were 1 and 2 instead of A and B, Figures 5.1 and 5.2 illustrate

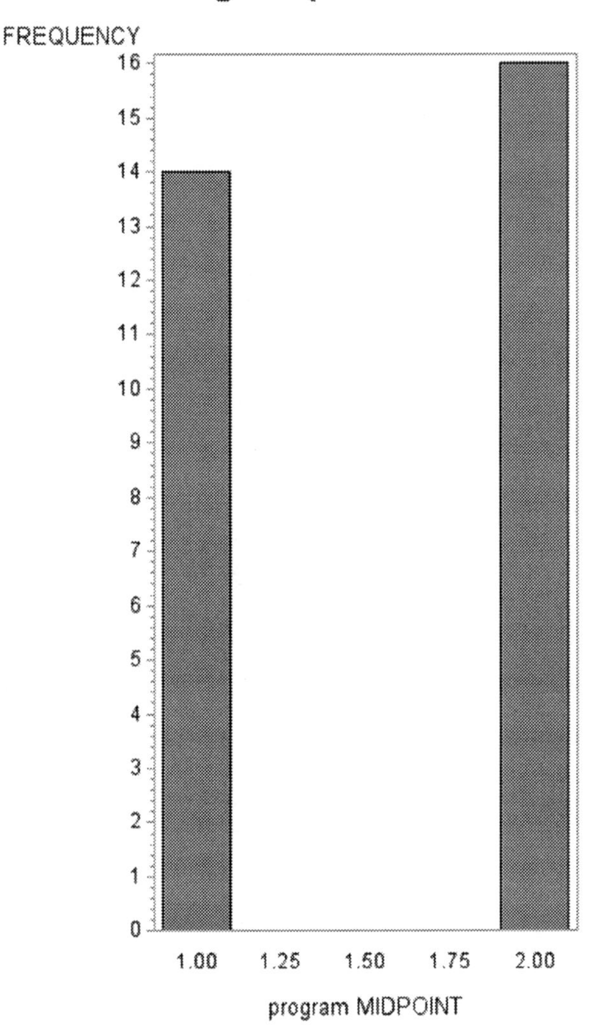

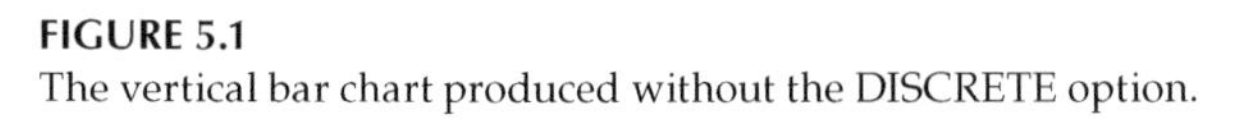
FIGURE 5.1
The vertical bar chart produced without the DISCRETE option.

this point. In Figure 5.1 the values 1.25, 1.50, and 1.75 chosen for midpoints do not correspond to any values of programs. In Figure 5.2, the DISCRETE option identifies only those values of the chart variable found in the data set. One could also have written the VBAR statement as:

```
VBAR program / MIDPOINTS = 1 2 ;
```

to achieve the same result.

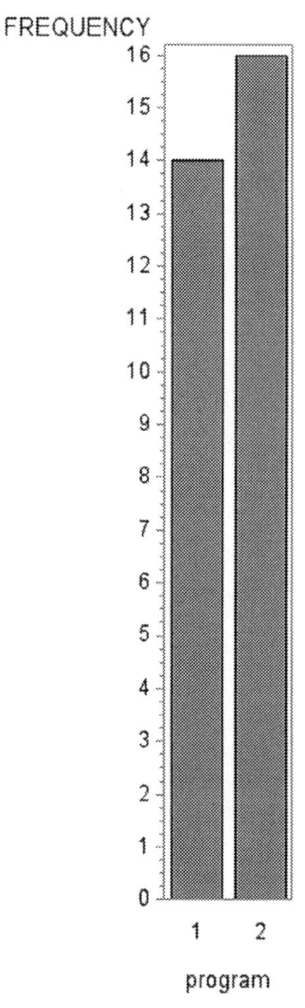

FIGURE 5.2
The DISCRETE option limits the number of bars to the exact numeric values of the chart variable.

Caution:

1. Incorrectly using the DISCRETE option when the chart variable is continuous, would produce too many bars in the chart.
2. If the chart variable is a character or string variable, one must be careful about spelling and capitalization of the data. For example, if there were an error in the data, such as, programs A, B, and b, bars for each of three program values would be produced. This is a reminder to always proofread the data.

OBJECTIVE 5.2: Produce two vertical histograms (bar charts with no spaces between bars) for the number of students in each program. Include a title.

```
PROC SORT DATA=instruction; BY program;
PROC GCHART DATA=instruction; BY program;
VBAR score / SPACE=0;
TITLE 'Objective 5.2';
TITLE3 'Histogram of the Scores for Each Program';
RUN;
```

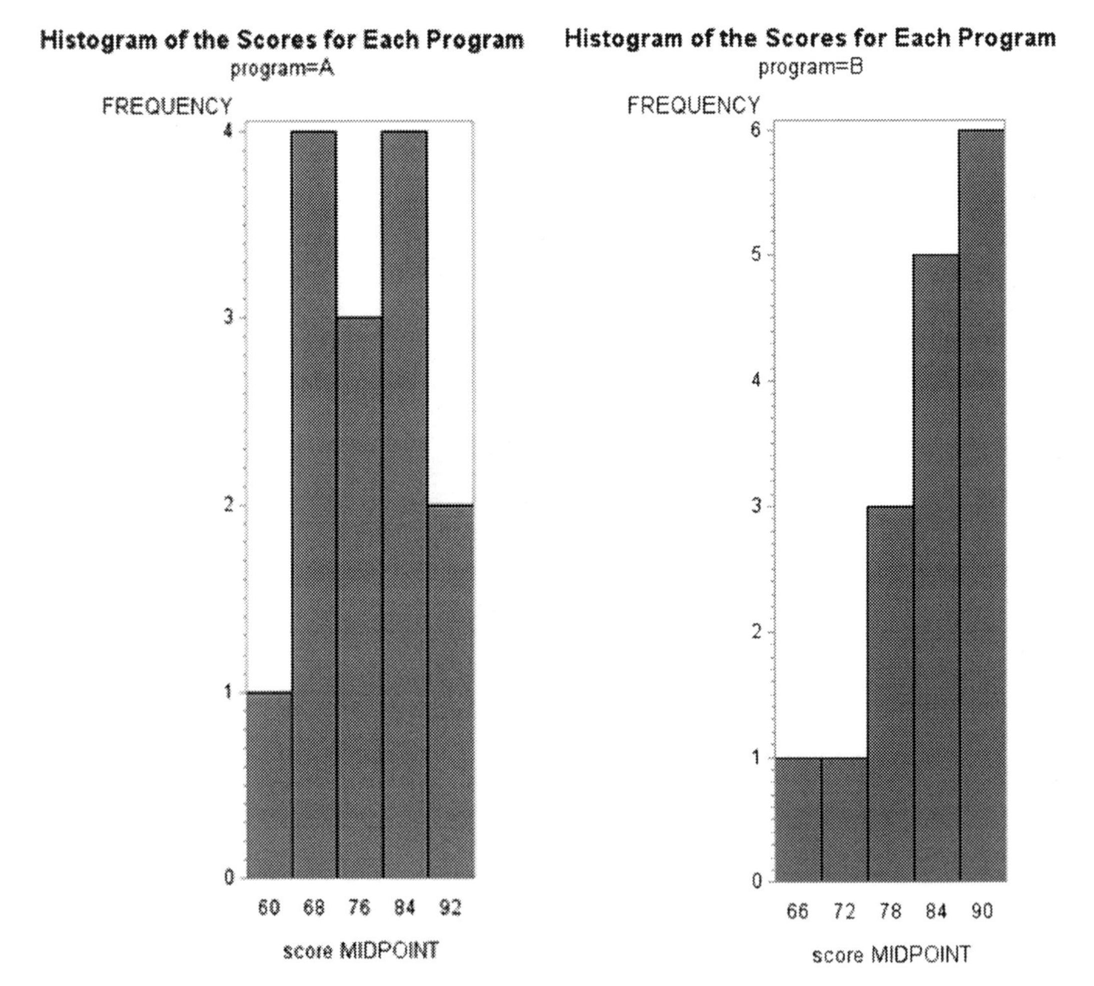

Note:

1. The SPACE = 0 option removes the spaces between the bars. The histograms in Objective 5.2 can be compared with the images in Objective 5.1 where spaces between bars are present.
2. When using a BY statement, if the data have not previously been sorted by the program variable, a sort must be done.

3. The vertical axes for frequency and the horizontal axes for the midpoints may not be the same for the two images. That makes it more difficult to compare images if one or both axes are not the same.

OBJECTIVE 5.3: Redo the charts in Objective 5.2. This time however, control the vertical axis for the frequency and the horizontal axis for the number and position of the bars.

```
PROC GCHART DATA=instruction; BY program;
VBAR score / SPACE=0  AXIS= 1 TO 6   MIDPOINTS = 60 65 70 75 80 85 90;
TITLE 'Objective 5.3';
TITLE2 'Histogram for the Scores for Each Program';
RUN;
```

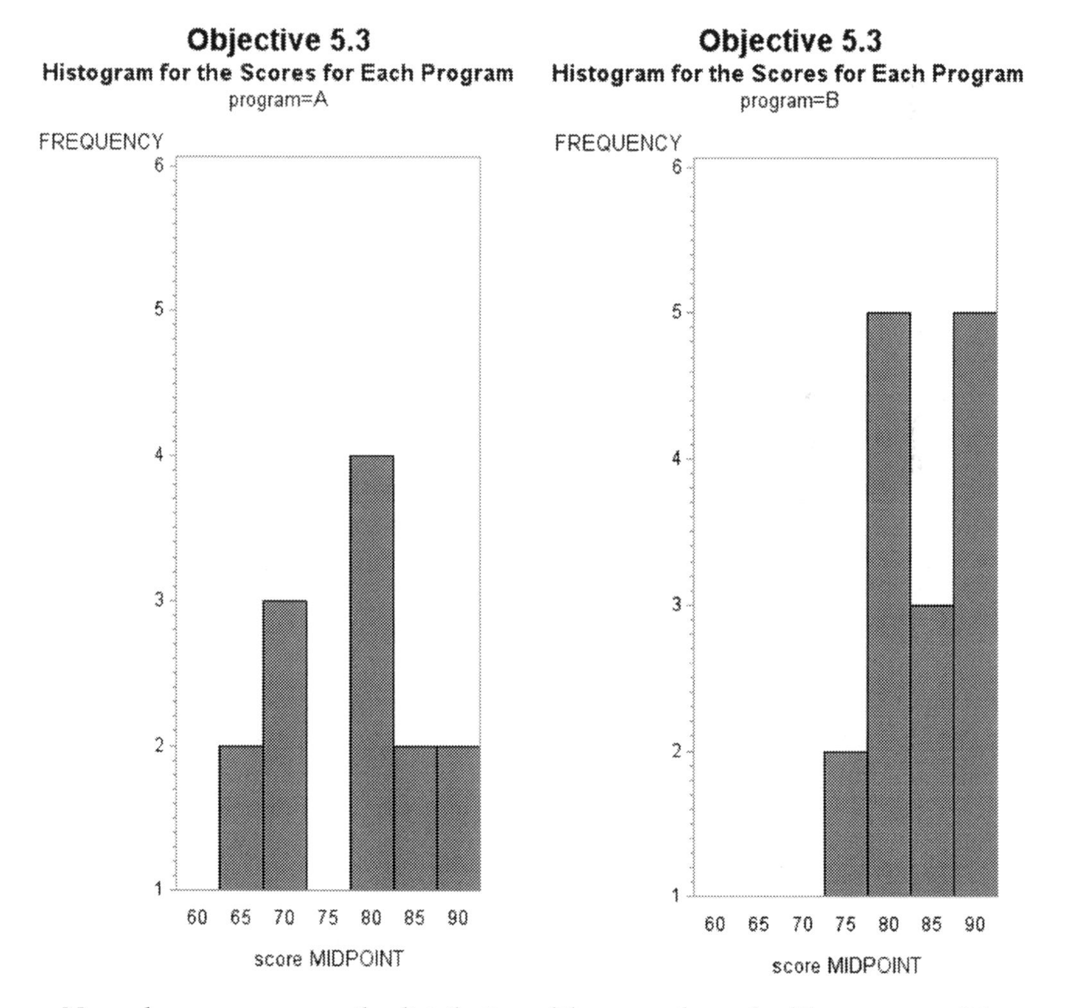

Now when one compares the distribution of the scores for each of the programs, it is more easily seen that the scores in Program B tend to be larger than those for Program A. The data had previously been sorted by program, so it was not necessary to sort the data again.

It is left to the reader to run this procedure again using the LEVELS = option, say, LEVELS = 5, instead of the MIDPOINTS option. While the number of bars is controlled, there is no control of the midpoint values at which the bars are centered.

OBJECTIVE 5.4: Modify Objective 5.3. This time place both programs' histograms on the same set of axes. That is, only one chart containing the same information as Objective 5.3 is to be produced.

```
PROC GCHART DATA=instruction;
VBAR score / SPACE=0  GROUP = program  AXIS= 1 TO 6
             MIDPOINTS = 60 65 70 75 80 85 90;
TITLE 'Objective 5.4';
TITLE2 'Histogram for the Scores for Each Program';
RUN;
```

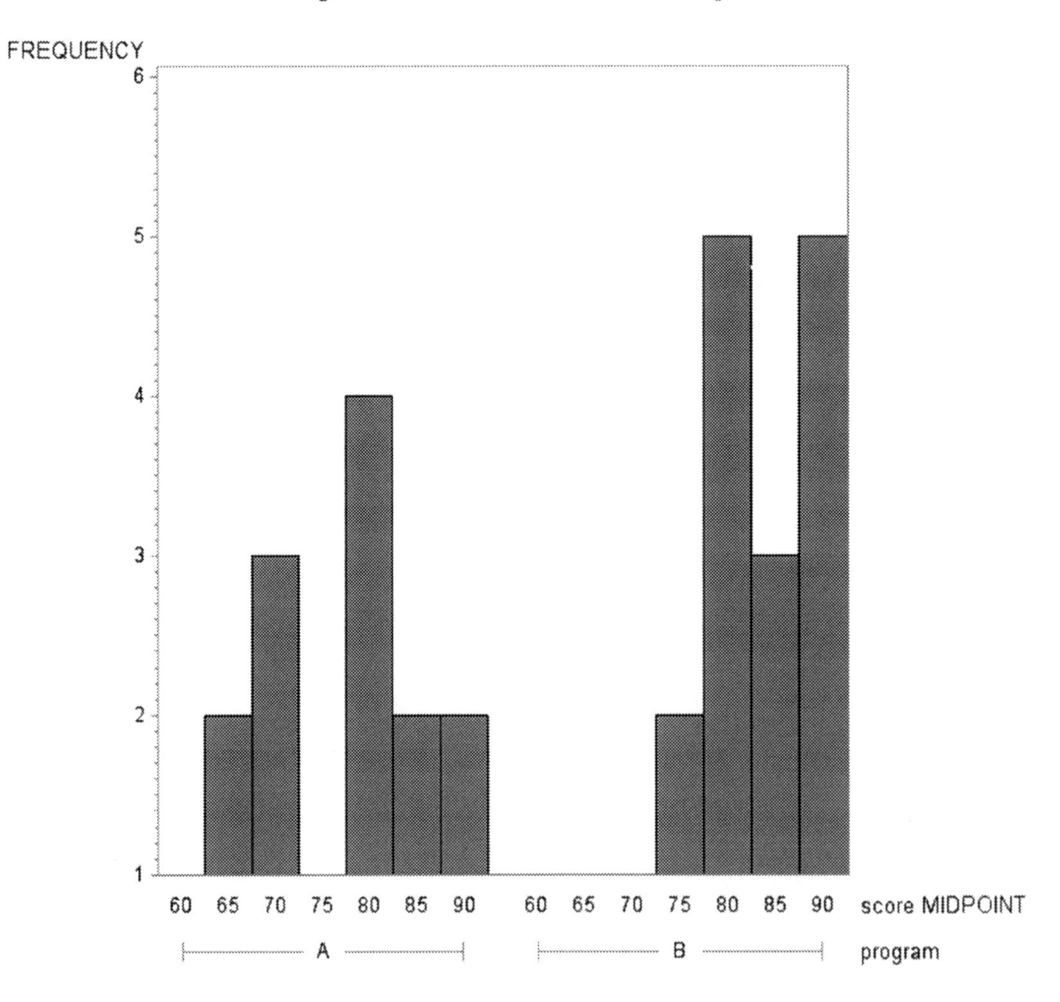

Note:

- Charts for both programs appear on the same set of axes.
- For the GROUP = option, the data did not have to be previously sorted by group.

- A grouped chart may not work well for large numbers of levels of the GROUP variable.
- Without the MIDPOINTS and AXIS options, the GCHART procedure will select the same five midpoints for each level of the GROUP variable, and the vertical axis will be determined by the tallest bars in the chart.
- Without the MIDPOINTS one could select the LEVELS = 5 option (or values other than 5). The selected midpoints will be the same for each level of the GROUP variable, but the midpoints GCHART selects may not be values one would prefer to use.

OBJECTIVE 5.5: Produce the default horizontal bar chart for the score variable. Include a single line of title for the chart.

```
PROC GCHART DATA=instruction;
HBAR score ;
TITLE 'Objective 5.5 - Default Horizontal Bar Chart for Score';
RUN;
```

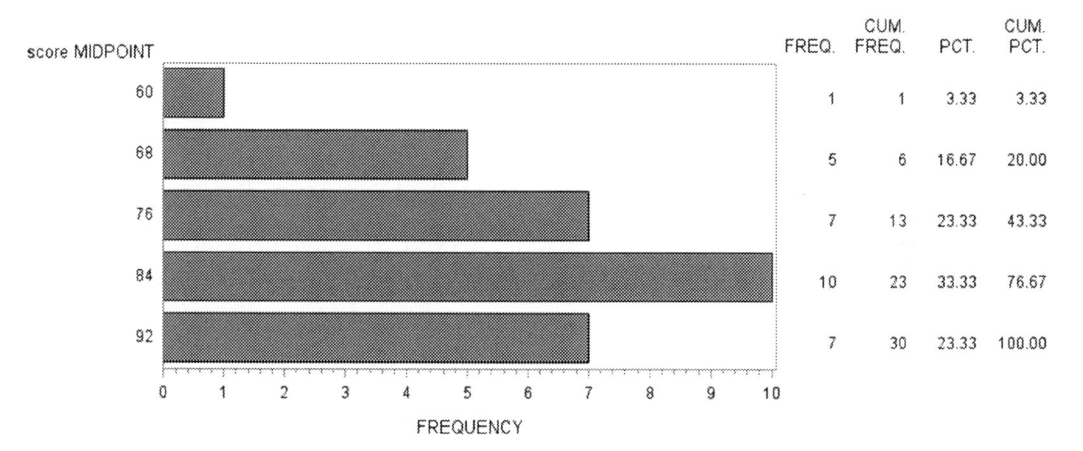

The default statistics that HBAR (and HBAR3D) computes appear on the right side of the chart. Since score is a continuous variable, a histogram (using the SPACE=0 option) is more appropriate.

OBJECTIVE 5.6: Produce a horizontal histogram for the variable score suppressing all frequency statistics. Use four bars in the histogram. Let the procedure select the midpoints for the bars. Include reference lines at each tick mark in the background and a title.

```
PROC GCHART DATA=instruction;
HBAR score / SPACE=0 LEVELS = 4 NOSTATS AUTOREF CLIPREF;
TITLE 'Objective 5.6 - Histogram';
TITLE3 'Number of Levels Specified - All Frequency Information is
Suppressed';
TITLE4 'Reference lines are included in the background.';
RUN;
```

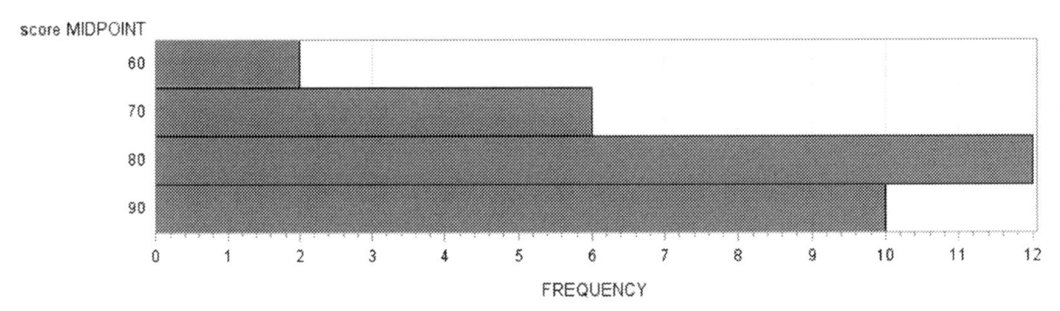

Note:

- NOSTATS suppressed all four of the frequency statistics. If one or more (but not all four values) statistics are to be used, one must specify which statistics are wanted on the chart, such as FREQ PERCENT.
- LEVELS = 4 requested four bars in the chart, but not the midpoints for the bars. The MIDPOINTS = *values* option must be used for more control of the bars instead of the LEVELS = option.
- AUTOREF draws lines at each tick mark on the frequency axis. CLIPREF places them in the background of the chart. Without the CLIPREF option, the reference lines are overlaid across the bars.
- Lines 1, 3, and 4 of TITLE were used. Skipping TITLE2 separates the first line of title from the other lines.

OBJECTIVE 5.7: Produce a horizontal bar chart where each bar length is the mean score for each of the programs. Print the MEAN on the chart. Include title(s).

```
PROC GCHART DATA=instruction;
HBAR program / TYPE=MEAN  SUMVAR=score  MEAN;
TITLE 'Objective 5.7';
TITLE3 'The mean of each program';
RUN;
```

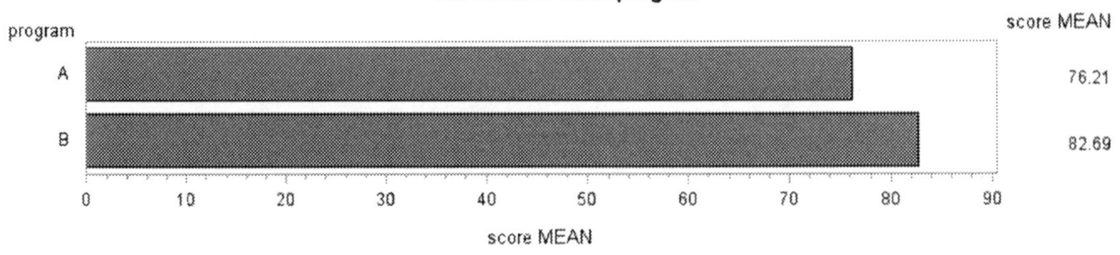

A vertical bar chart could similarly be done. Due to the low number of bars there is a large blank space between the title and the bar chart.

OBJECTIVE 5.8: Redo Objective 5.7 using a three-dimensional vertical bar chart and also a block chart.

```
PROC GCHART DATA=instruction;
VBAR3D program / TYPE=MEAN SUMVAR=score MEAN;
TITLE 'Objective 5.8: Mean of each program in a 3-D vertical bar chart';
RUN;
BLOCK program / TYPE=MEAN SUMVAR=score NOHEADING;
TITLE 'Objective 5.8: Mean of each program in a block chart';
RUN;
```

Objective 5.8: Mean of each program in a 3-D vertical bar chart

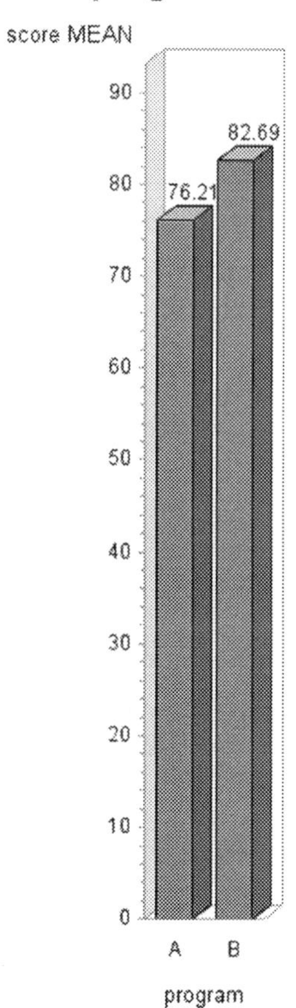

Objective 5.8: Mean of each program in a block chart

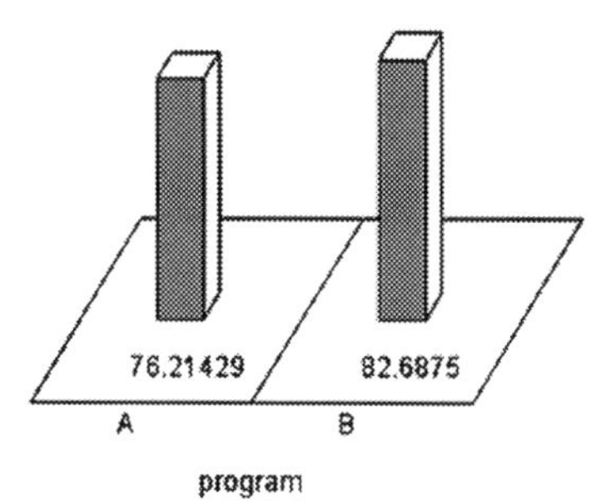

More than one type of chart can be produced in a single GCHART procedure. The titles on the images produced in the same procedure block of code can be changed by placing a RUN statement after each chart and TITLE statement pair. Without the first RUN statement, only the second title will be produced, and it will be on both charts. The NOHEADING option suppresses a default heading for the block chart, "BLOCK CHART OF MEAN" which would appear in the space above the block chart and below any lines of title selected for the image.

OBJECTIVE 5.9: Produce a pie chart for the percent of students in each of the programs.

```
PROC GCHART DATA=instruction;
PIE program / TYPE=PERCENT ;
TITLE 'Objective 5.9: Pie Chart';
TITLE2 'for the percent of subjects in each of the programs.';
RUN;
```

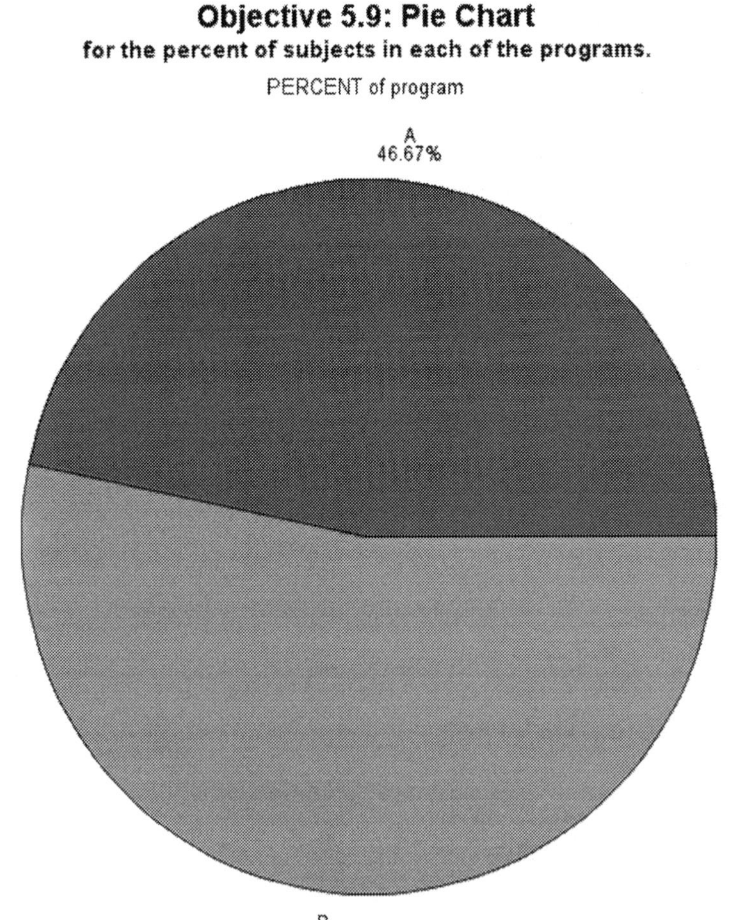

- The most obvious thing to notice when obtaining pie charts in grayscale can be the lack of a prominent distinction between the slices. When printing this grayscale image, the chart will likely appear as a dark gray circle. Clearly, color options will improve the appearance of the pie charts. This will be done in Chapter 17.
- The default setting for the pie chart is for the frequency count (TYPE=FREQ) of each level of the chart variable. Here the percentage of the data in each program (TYPE=PERCENT) was requested.
- If the chart variable is numeric, one of the DISCRETE, MIDPOINTS, or LEVELS options may be needed to obtain slices for the specific values of the chart variable since the GCHART procedure selects four slices by default when there are many more than four numeric levels possible.
- Including the NOHEADING option on the PIE statement will produce an image without the "PERCENT of Program" text above the pie chart.

All statements for these objectives could be done in a single program. Beginning SAS programmers may wonder what the program would look like. It would look like this.

```
DM 'LOG; CLEAR; ODSRESULTS; CLEAR; ';
TITLE;

DATA instructions;
INPUT Program $ Score @@;
DATALINES;
A 71 A 82 A 88 A 64 A 59 A 78 A 72
A 81 A 83 A 66 A 83 A 91 A 79 A 70
B 65 B 88 B 92 B 76 B 87 B 89 B 85
B 90 B 81 B 91 B 78 B 81 B 86 B 82
B 73 B 79
;
RUN;

PROC GCHART DATA=instruction;
VBAR program;
TITLE 'Objective 5.1';
TITLE3 'Default Vertical Bar Chart for Program';
RUN;

PROC SORT DATA=instruction; BY program;
PROC GCHART DATA=instruction; BY program;
VBAR score / SPACE=0;
TITLE 'Objective 5.2';
TITLE3 'Histogram of the Scores for Each Program';
RUN;

PROC GCHART DATA=instruction; BY program;
VBAR score / SPACE=0  AXIS= 1 TO 6   MIDPOINTS = 60 65 70 75 80 85 90;
TITLE 'Histogram for the Scores for Each Program';
TITLE2 'Objective 5.3';
RUN;
```

```
PROC GCHART DATA=instruction;
VBAR score / SPACE=0  GROUP = program  AXIS= 1 TO 6   MIDPOINTS = 60 65
70 75 80 85 90;
TITLE 'Histogram for the Scores for Each Program';
TITLE2 'Objective 5.4';
RUN;

PROC GCHART DATA=instruction;
HBAR score;
TITLE 'Default Horizontal Bar Chart for Score';
TITLE 'Objective 5.5 - Default Horizontal Bar Chart for Score';
RUN;

PROC GCHART DATA=instruction;
HBAR score / LEVELS = 4 NOSTATS AUTOREF CLIPREF ;
TITLE 'Objective 5.6: Horizontal Bar Chart';
TITLE3 'Number of Levels Specified - All Frequency Information is
Suppressed';
TITLE4 'Reference lines are included in the background.';
RUN;

PROC GCHART DATA=instruction;
HBAR program / TYPE=MEAN SUMVAR=score  MEAN;
TITLE 'Objective 5.7';
RUN;

PROC GCHART DATA=instruction;
VBAR3D program / TYPE=MEAN SUMVAR=score MEAN;
TITLE 'Objective 5.8: Mean of each program in a 3-D vertical bar chart';
RUN;
BLOCK program / TYPE=MEAN SUMVAR=score NOHEADING ;
TITLE 'Objective 5.8: Mean of each program in a block chart';
RUN;

PROC GCHART DATA=instruction;
PIE program / TYPE=PERCENT;
TITLE 'Objective 5.9: Pie Chart ';
TITLE2 'for the percent of subjects in each of the programs.';
RUN;

QUIT;
```

Included in the SAS code is the display manager command at the top of the program that resets or clears the Log and Results Viewer windows of content produced earlier in the SAS session. Resetting the titles at the top of the program using TITLE; clears all active lines of titles created in the current SAS session. If one uses footnotes, clearing the lines of footnotes is also recommended.

Blank lines have been used to separate blocks of code but this is not necessary. A block of code refers to statements that are a part of one procedure or a single DATA step. The blank line is the author's choice. Some programmers indent lines of code and/or use blank lines to separate blocks of code. A RUN statement can be placed at the end of each procedure block of code or only once in the line before the QUIT statement. Beginning programmers will develop their own style as they learn more SAS programming. Methods of annotating programs are strongly recommended and are covered in Section 20.2.

5.2 Saving Graphics Images

For any of the procedures covered so far, the contents of the Results Viewer window can be printed or saved as an HTML file. Additionally, individual images in the Results Viewer can be saved in a graphics format. By right clicking on a graphics image in the Results Viewer, **Save picture as** can be selected from the pop-up menu. In the next dialog window, the folder and the filename are specified. One can select either **PNG (*.png)** (pronounced as "ping") or **Bitmap (*.bmp)** file type when graphics image is named. Saved images can then be imported into presentation or report software.

5.3 Chapter Summary

In this chapter, the GCHART procedure was introduced. The GCHART procedure produces horizontal or vertical bar charts or histograms, block charts, and pie charts. GCHART is one of the legacy procedures in SAS/GRAPH. More recent additions to SAS procedures now include ODS Graphics. In this chapter, color selection and other visual enhancements to the charts were not covered, but the basic code to produce graphics was introduced. More SAS/GRAPH procedures and visual enhancements to graphs are presented in Chapter 17.

6

One and Two Population Hypothesis Tests about the Means

Recall in Chapter 3 both the UNIVARIATE and MEANS procedures are able to produce tests about a population mean assuming the population is normally distributed. Another one of the basic test procedures in a first statistics course is the comparison of two population means. These population means can also be tested for equality using a t-test. Recall that in the two-sample t-tests, the procedures were different when the two samples were independent versus when the samples were paired or dependent. For these test methods the populations sampled were normally distributed. Other techniques are appropriate when the data are not normally distributed.

6.1 The TTEST Procedure

The TTEST procedure has the capability to test a single population mean or compare two population means based on either dependent or independent samples.

The syntax of the TTEST procedure is:

PROC TTEST DATA=*SAS-data-set* <options> ;
CLASS *variable* ;
PAIRED *pair-lists* ;
VAR *variables* ;
RUN;

No statement can be used more than once in a single TTEST procedure. The statements can appear in any order after the PROC TTEST statement. BY and WHERE statements are again optional statements. When using a BY statement, remember that the SAS data set must be compatibly sorted prior to this procedure. The TTEST procedure can also produce Output Delivery System (ODS) Graphics that support or illustrate the results.

Options on the PROC TTEST statement include:

ALPHA=*p* where $0 < p < 1$; this value is used to determine the $(1 - p)100\%$ level of confidence used in the confidence interval calculations. The default setting for all confidence intervals is 95% (ALPHA=0.05) unless otherwise specified using the ALPHA option.

CI=EQUAL An equal tailed confidence interval for σ is computed. This is a default setting.

CI=NONE No confidence interval for σ is printed. CI's for the means and the difference between the population means are among the default output and will still be computed.

H0 =*value* requests a test of the mean versus the specified value. By default, H0=0. H0 is "H zero", not "H oh".

PLOTS = (*list*) ODS Graphics plots requested. Available options for ODS Graphics are: ALL, NONE, HISTOGRAM, BOXPLOT, INTERVAL, QQ, PROFILES, AGREEMENT, SUMMARY. When requesting more than one ODS Graphic option, enclose the list in parenthesis and modify PLOTS with (ONLY), such as PLOTS(ONLY) = (QQ PROFILES). If one does not wish to include ODS Graphics in the results, the PLOTS=NONE option should be included.

SIDES = 2 | L | U (or SIDED or SIDE)

SIDES = 2 specifies the two-sided test and confidence interval for the mean.

SIDES = L specifies the *lower* one-sided tests and the confidence interval from negative infinity to the upper confidence bound for the mean.

SIDES = U specifies the *upper* one-sided tests and the confidence interval from the lower confidence bound for the mean up to positive infinity.

There are more options available for the PROC TTEST statement than those listed here, but this list is a healthy introduction to the procedure. For other options see SAS Help and Documentation.

CLASS Statement

A CLASS statement identifying the single grouping variable must accompany the PROC TTEST statement when conducting an independent samples t-test. The grouping or class variable must have exactly two levels, either numeric or character. The means of the variables listed in the VAR statement are the responses to be compared. The CLASS statement in this procedure functions much differently than the CLASS statement in the UNIVARIATE and MEANS procedures.

VAR Statement

The VAR statement names the continuous variable means which are to be tested. If there is no CLASS statement, then a single sample t-test for the population mean is performed. If a CLASS statement is used, then an independent samples t-test for the population means is performed for each response variable in the VAR statement. If a CLASS statement is used and the VAR statement is omitted, all numeric variables in the input data set (except a numeric variable in the CLASS statement) will be tested using an independent samples t-test for the population means.

PAIRED Statement

The means of the variables in the PAIRED statement are to be compared using a paired or dependent samples t-test. Differences are computed using the variable on the left minus the variable on the right in the PAIRED statement. Allowable syntax for this statement is summarized in Table 6.1.

TABLE 6.1

Syntax for the PAIRED Statement of the TTEST Procedure

Overview of PAIRED Statements	These Comparisons are Produced
PAIRED A*B ;	A – B
PAIRED A*B C*D ;	A – B and C – D
PAIRED (A B) * (C D);	A – C, A – D, B – C, B – D
PAIRED (A B) * (B C);	A – B, A – C, B – C

The CLASS statement and the VAR statement **cannot** be used in the same TTEST procedure with the PAIRED statement. Therefore, if the analysis requires an independent samples test and a dependent samples t-test for different response variables, two separate blocks of TTEST code must be used.

6.2 One Population Test and Confidence Interval for the Mean

Recall the notation from Chapter 3 when there is a single population.

	Population Parameters	Sample Statistics
Mean	μ	$\bar{y}$
Variance	σ^2	s^2
Standard Deviation	σ	s
		sample size = n

Chapter 3 included only the two-sided test for the mean using the UNIVARIATE and MEANS procedures. In addition to the two-sided test where H_1: $\mu \neq \mu_0$ can be tested, one can also test the one-sided alternatives, H_1: $\mu > \mu_0$ (upper test), and H_1: $\mu < \mu_0$ (lower test). These tests and confidence limits are summarized in Table 6.2. For these tests an observed significance level or p-value is calculated by the TTEST procedure, and again, one would reject H_0 for small p-values ($p \leq \alpha$).

To investigate the one population test method, the objectives will utilize the SAS data set *instruction* introduced in Chapter 3, Objective 3.1. That is,

```
DATA instruction;
INPUT program $ score @@;
DATALINES;
A 71 A 82 A 88 A 64 A 59 A 78 A 72
A 81 A 83 A 66 A 83 A 91 A 79 A 70
B 65 B 88 B 92 B 76 B 87 B 89 B 85
B 90 B 81 B 91 B 78 B 81 B 86 B 82
B 73 B 79
;
```

TABLE 6.2

Summary of Hypothesis Testing and Confidence Limits for a Single Population Mean

Hypotheses	Test Statistic	Reject H_0 if	
H_0: $\mu = \mu_0$ H_1: $\mu \neq \mu_0$ (μ_0 is specified)	$t = \dfrac{\bar{y} - \mu_0}{s/\sqrt{n}}$	$\lvert t \rvert \geq t_{\alpha/2,\,df}$	**(1 – α)100% Confidence Interval for μ** $\bar{y} \pm t_{\alpha/2,df} \dfrac{s}{\sqrt{n}}$
H_0: $\mu = \mu_0$ H_1: $\mu > \mu_0$	$t = \dfrac{\bar{y} - \mu_0}{s/\sqrt{n}}$	$t \geq t_{\alpha,\,df}$	**(1 – α)100% Confidence Lower Bound for μ** $\bar{y} - t_{\alpha,df} \dfrac{s}{\sqrt{n}}$
H_0: $\mu = \mu_0$ H_1: $\mu < \mu_0$	$t = \dfrac{\bar{y} - \mu_0}{s/\sqrt{n}}$	$t \leq -t_{\alpha,\,df}$	**(1 – α)100% Confidence Upper Bound for μ** $\bar{y} + t_{\alpha,df} \dfrac{s}{\sqrt{n}}$

where $t_{p,df}$ is the critical t-value that determines a right tail area of p and df = n – 1 in this application.

OBJECTIVE 6.1: Test whether or not the mean score is 75 (for the combined programs) versus the two-sided alternative.

```
PROC TTEST DATA=instruction H0=75;
VAR score;
TITLE 'Objective 6.1';
RUN;
```

This is the minimal code one would use to test H_0: $\mu = 75$ versus H_1: $\mu \neq 75$. A word of caution: omitting the option H0=75 will produce a two-sided test for H_0: $\mu = 0$ versus H_1: $\mu \neq 0$.

Objective 6.1

The TTEST Procedure

Variable: Score

N	Mean	Std Dev	Std Err	Minimum	Maximum
30	79.6667	8.8448	1.6148	59.0000	92.0000

The TTEST procedure generates a few summary statistics for each variable in the VAR statement.

Mean	95% CL Mean		Std Dev	95% CL Std Dev	
79.6667	76.3640	82.9694	8.8448	7.0440	11.8902

A 95% confidence interval for the mean (76.3640, 82.9694) and a 95% confidence interval for the standard deviation (7.044, 11.8902) are included in the default output of the procedure. A 95% confidence is the default. To change this, use the ALPHA=p option in the PROC TTEST statement and enter a value for p, typically 0.10 or smaller.

DF	t Value	Pr > |t|
29	2.89	0.0072

The test statistic for comparing the population mean to 75 is $t = 2.89$ with $p = 0.0072$ though the final table with this test statistic does not identify $\mu_0 = 75$. One would conclude that the mean score of students is different from 75 ($\alpha = 0.05$, $t_{29} = 2.89$, $p = 0.0072$).

Included with the test output is a histogram of the data with the normal curve superimposed, shown in Figure 6.1. A box plot with the mean of 79.6667 marked by a ◊ is included in the lower margin of the ODS Graphic. The rectangular shaded area beneath the histogram and in the background of the boxplot identifies the 95% confidence interval for the mean. This ODS Graph is identified by its ODS Graphic name Summary Panel. Figure 6.2 is a Q-Q plot of the data used as a diagnostic tool to examine normality. Data values that lie on or close to the 45-degree reference line are approximately normally distributed. (Chambers, Cleveland, Kleiner, and Tukey, 1983). These two ODS Graphs are among the default output. In the syntax in Section 6.1 these are identified in the PLOTS option of the PROC TTEST statement as SUMMARY and QQ.

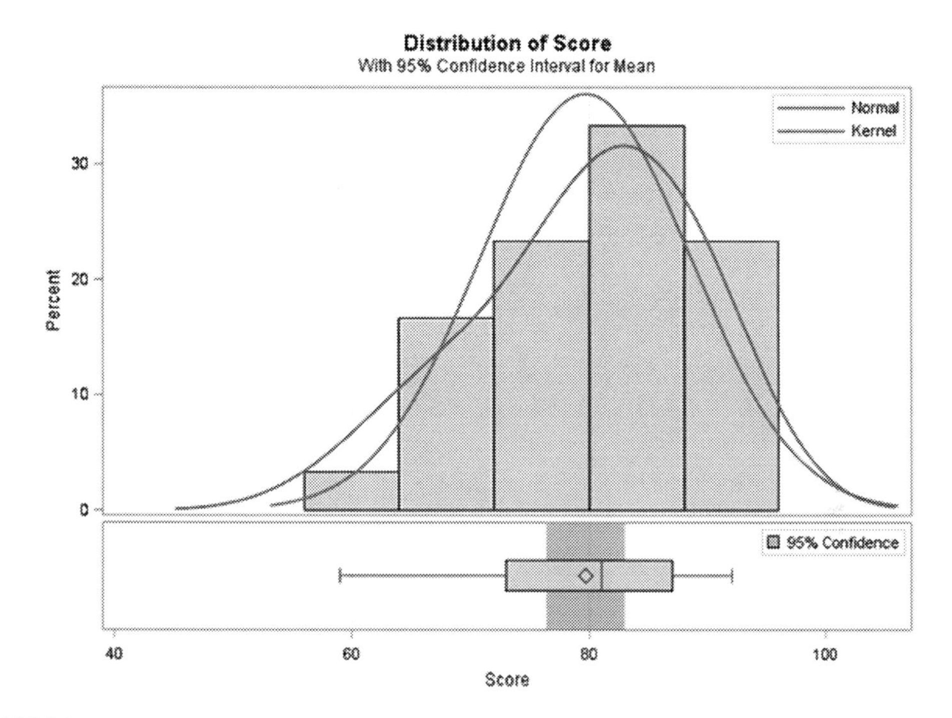

FIGURE 6.1
ODS graph for the distribution of score.

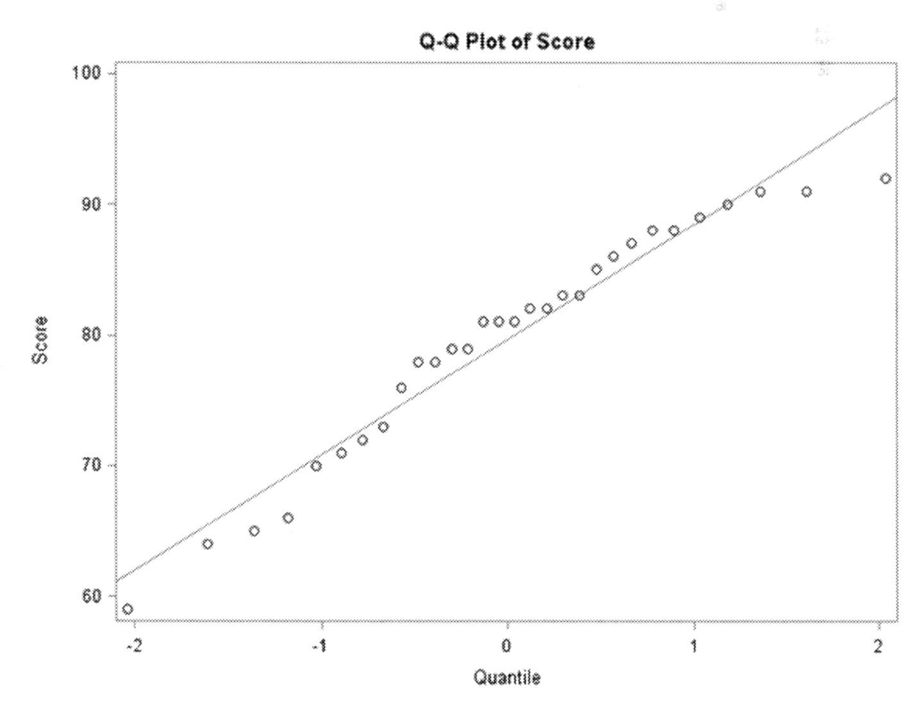

FIGURE 6.2
ODS graph of the Q-Q plot.

OBJECTIVE 6.2: Produce 98% confidence intervals for the mean and standard deviation of the population of scores, test whether the mean score differs from 75 at $\alpha = 0.02$, and suppress all ODS Graphs.

```
PROC TTEST DATA= instruction H0=75 ALPHA=0.02 PLOTS=NONE;
VAR score;
TITLE 'Objective 6.2';
RUN;
```

Changing the value of ALPHA ***will only affect the confidence interval calculations*** in the printed output.

Objective 6.2

The TTEST Procedure

Variable: Score

N	Mean	Std Dev	Std Err	Minimum	Maximum
30	79.6667	8.8448	1.6148	59.0000	92.0000

The summary statistics have not changed, of course.

Mean	98% CL Mean		Std Dev	98% CL Std Dev	
79.6667	75.6909	83.6424	8.8448	6.7639	12.6148

Note the changes in confidence level in the column headers in the computed endpoints of the confidence intervals compared to Objective 6.1.

DF	t Value	Pr > |t|
29	2.89	0.0072

The test statistic is $t = 2.89$ with $p = 0.0072$. Though α changed from 0.05 to 0.02, the test statistic "t Value" and observed significance level "Pr > |t|" do not change. These values are functions of the data; not of α.

No graphs were included in the output because of the PLOTS=NONE option. If one wanted only the ODS Graph Summary Panel, then the PLOTS option should be changed to PLOTS(ONLY)=SUMMARY. PLOTS(ONLY)=QQ would produce only the Q-Q plot.

OBJECTIVE 6.3: Produce one-sided t-tests for each program. Specifically, does the mean score for each program exceed 75? Suppress all ODS Graphics. Do this in a single TTEST procedure.

```
PROC SORT DATA= instruction; BY program;
PROC TTEST DATA= instruction PLOTS=NONE H0=75 ALPHA=0.02 SIDES=U;
BY program;
VAR score;
TITLE 'Objective 6.3';
RUN;
```

Data will need to be SORTed by program if it has not previously been done.

SIDES=U will produce the upper test of the mean, that is, H_1: $\mu > 75$. The TTEST procedure is executed BY program to get the test for each program.

Objective 6.3

The TTEST Procedure

Variable: Score

Program=A

N	Mean	Std Dev	Std Err	Minimum	Maximum
14	76.2143	9.4069	2.5141	59.0000	91.0000

Mean	98% CL Mean		Std Dev	98% CL Std Dev	
76.2143	70.4781	Infty	9.4069	6.4457	16.7363

DF	t Value	Pr > t
13	0.48	0.3186

Objective 6.3

The TTEST Procedure

Variable: Score

Program=B

N	Mean	Std Dev	Std Err	Minimum	Maximum
16	82.6875	7.3277	1.8319	65.0000	92.0000

Mean	98% CL Mean		Std Dev	98% CL Std Dev	
82.6875	78.5683	Infty	7.3277	5.1323	12.4106

DF	t Value	Pr > t
15	4.20	0.0004

Note the CL for the mean has "Infty", ∞, as the right endpoint. For an upper or right-side test, only a confidence lower bound is computed. See Table 6.2.

For Program A the 98% confidence lower bound for the mean score is 70.4781. Since this value is *below 75*, we cannot conclude that the mean score for Program A is greater than 75. Note the df corresponds to n – 1 = 13 for Program A. The observed significance level is now "Pr > t". For Program A one would conclude that the mean score does not exceed 75 ($\alpha = 0.02$, $t_{13} = 0.48$, $p = 0.3186$). This is in agreement with the confidence lower bound computed at the same α.

For Program B the 98% confidence lower bound for the mean score is 78.5683. Since this value is *above 75*, we can conclude that the mean score for Program B is greater than 75. Using the t-test results for Program B one would conclude that the mean score does exceed 75 ($\alpha = 0.02$, $t_{15} = 4.20$, $p = 0.0004$). This also is in agreement with the confidence lower bound computed for Program B.

SIDES=L on the PROC TTEST statement would have produced one-sided results if the alternative was H_1: $\mu < 75$. "Pr < t" would identify the observed significance level for this one-sided test, and a confidence upper bound for μ would be computed. The confidence lower bound would be "-Infty", -∞.

CAUTION: A BY statement must be used to conduct the one population tests for each program. The TTEST procedure does have a CLASS statement, but the usage of the CLASS statement is strictly for comparing two population means based on independent samples. That is, using the SAS statement: CLASS program; instead of BY program; in the TTEST procedure will compare the programs to each other using an independent samples t-test, and the H0=75 option will be ignored. The two samples topic is covered later in this chapter.

6.3 Overview: t-tests Produced by TTEST, UNIVARIATE, and MEANS Procedures

At this point, one may realize that three different procedures can test the population mean using a t-test. How do these procedures compare side by side? Consider the two-sided test in Objective 6.1, H_0: $\mu = 75$ versus H_1: $\mu \neq 75$ with $\alpha = 0.05$.

The TTEST Procedure

```
PROC TTEST DATA= instruction SIDES=2  ALPHA=0.05 H0=75;
VAR score;
RUN;
```

- SIDES=2 is optional since the two-sided test is the default.
- One-sided options for tests and confidence limits are available.
- ALPHA=0.05 is the default, but other values can be chosen for the confidence limits.
- More than one variable can be analyzed in the VAR statement, but the mean of each variable would be compared to 75. There is presently no option for specifying multiple values for μ_0 when multiple variables are specified in the VAR statement.
- ODS Graphics illustrating the distribution of the data and the confidence interval for the mean are available.

The TTEST Procedure

Variable: Score

N	Mean	Std Dev	Std Err	Minimum	Maximum
30	79.6667	8.8448	1.6148	59.0000	92.0000

Mean	95% CL Mean		Std Dev	95% CL Std Dev	
79.6667	76.3640	82.9694	8.8448	7.0440	11.8902

DF	t Value	Pr > |t|
29	2.89	0.0072

< Summary Panel and Q-Q Plot not shown>

- CI=NONE in the PROC TTEST statement would suppress the CI for σ, (7.0440, 11.8902).

- The absolute value bars around t in "Pr > |t|" identify this test information as a two-sided test. Unfortunately, the table containing the df, t Value, and Pr > |t| is not labeled with the hypothesized value of 75. One would have to be careful when interpreting this result.

The UNIVARIATE Procedure

```
PROC UNIVARIATE DATA= instruction ALPHA=0.05 CIBASIC MU0=75;
VAR score;
RUN;
```

- A two-sided test is the only option. One-sided test options are NOT available.
- Confidence intervals must be requested with the CIBASIC option. ALPHA=0.05 is the default, but other values can be chosen for the confidence level.
- More than one variable can be analyzed in the VAR statement. Options for including more than one value in the MU0= option were shown in Objective 3.3.
- No default ODS Graphics produced by this code.
- One could include a HISTOGRAM statement for all or some of the variables in the VAR statement.

The UNIVARIATE Procedure

Variable: Score

Basic Confidence Limits Assuming Normality

Parameter	Estimate	95% Confidence Limits	
Mean	79.66667	76.36398	82.96936
Std Deviation	8.84477	7.04403	11.89015
Variance	78.22989	49.61840	141.37574

Tests for Location: Mu0=75

Test	Statistic		p Value	
Student's t	t	2.889888	**Pr > |t|**	0.0072
Sign	M	7	**Pr >= |M|**	0.0161
Signed Rank	S	126.5	**Pr >= |S|**	0.0069

- The CIBASIC option produces confidence intervals for the mean, standard deviation, and variance at the specified value of α.
- MU0=75 option results are clearly identified in the Tests for Location table.
- UNIVARIATE computes three test statistics for the means in the Tests for Location table. The absolute value indicated in these p-value expressions identifies that these are two-sided tests.

The MEANS Procedure

```
DATA instruction;
INPUT Program $ Score @@;
Score_75 = Score - 75;
DATALINES;
```

```
<data not shown>
.
.
.
PROC MEANS DATA= instruction MEAN STD STDERR N ALPHA=0.05 CLM T PRT;
VAR score score_75;
RUN;
```

- The MEANS procedure only tests whether a mean is different from zero, a two-sided test. Thus, in a DATA step one must create a new variable by subtracting the μ_0 value in the null hypothesis from every score in the SAS data set, and then test whether that mean difference differs from zero.
- In the PROC MEANS statement, PRT will only compute the observed significance level for a two-sided test comparing the mean of each variable in the VAR statement to zero.
- CLM produces the confidence interval which supports the two-sided test of the mean at the specified α value.
- No ODS Graphics are available for this procedure at this time.

The MEANS Procedure

Variable	Mean	Std Dev	Std Error	N	Lower 95% CL for Mean	Upper 95% CL for Mean	t Value	Pr > |t|
score	79.6666667	8.8447660	1.6148259	30	76.3639768	82.9693566	49.33	<.0001
Score_75	4.6666667	8.8447660	1.6148259	30	1.3639768	7.9693566	2.89	0.0072

- The confidence interval for Score agrees with the TTEST and UNIVARIATE procedures, but the t-statistic (49.33) and observed significance level (<.0001) do not. This value of t and the observed significance level are for the test of H_0: $\mu = 0$ versus H_1: $\mu \neq 0$.
- The t = 2.89 and its 0.0072 observed significance level for Score_75 agree with the other two test procedures. The confidence interval for Score_75 is not meaningful.
- In the MEANS procedure, one can request LCLM or UCLM to get either the lower confidence limit for the mean or the upper confidence limit for the mean, respectively, for the given α. But the observed significance level or p-value for either one-sided test is not available at this time.

6.4 Two Populations Tests and Confidence Intervals for the Difference between Means

For two sample tests comparing two population means, the notation is modified to identify the appropriate sample or population. See Table 6.3.

Tests of hypotheses and confidence interval methods depend on whether or not the samples are dependent (or paired) samples or are independent samples. When samples are dependent, one computes $d = x - y$ for each of the n pairs of observations, x and y. The mean of these differences is $\bar{d}$, and the standard deviation of these differences is s_d. After obtaining these statistics, the test statistic and confidence interval are like those for single

TABLE 6.3

Notation for Two Population Parameters and Two Sample Statistics

	Population 1	Population 2
Mean	μ_1	μ_2
Variance	σ_1^2	σ_2^2
	Sample 1	**Sample 2**
Mean	$\bar{y}_1$	$\bar{y}_2$
Variance	s_1^2	s_2^2
Sample Size	n_1	n_2

population procedures. Only the two-sided alternative is shown in the top one-third of Table 6.4. There are also one-sided alternatives as in the one population case in Table 6.2. In the second part of Table 6.4 are the methods for two independent samples.

For the independent samples t-test, there are two procedures: one when the population variances are equal and another when the population variances are not equal. There is a test for the equality of two population variances to determine which method of comparing the population means is appropriate. The F-test for the equality of variances is shown in

TABLE 6.4

Summary of Hypothesis Testing and Confidence Intervals for Two Normal Population Means

Hypotheses	Test Statistic	Reject H_0 if	$(1-\alpha)100\%$ CI
Dependent Samples			
$H_0: \mu_1 - \mu_2 = 0$ $H_1: \mu_1 - \mu_2 \neq 0$	$t = \dfrac{\bar{d}}{\sqrt{s_d / n}}$	$\lvert t \rvert \geq t_{\alpha/2, df}$	$\bar{d} \pm t_{\alpha/2, df}\sqrt{s_d/n}$ where $df = n - 1$
Independent Samples			
$H_0: \mu_1 - \mu_2 = 0$ $H_1: \mu_1 - \mu_2 \neq 0$ if $\sigma_1^2 = \sigma_2^2$	$t = \dfrac{\bar{y}_1 - \bar{y}_2}{\sqrt{s_p^2\left(\dfrac{1}{n_1} + \dfrac{1}{n_2}\right)}}$	$\lvert t \rvert \geq t_{\alpha/2, df}$	$(\bar{y}_1 - \bar{y}_2) \pm t_{\alpha/2, df}\sqrt{s_p^2\left(\dfrac{1}{n_1} + \dfrac{1}{n_2}\right)}$
	where $s_p^2 = \dfrac{(n_1 - 1)s_1^2 + (n_2 - 1)s_2^2}{n_1 + n_2 - 2}$ is the pooled variance estimate and $df = n_1 + n_2 - 2$		
$H_0: \mu_1 - \mu_2 = 0$ $H_1: \mu_1 - \mu_2 \neq 0$ if $\sigma_1^2 \neq \sigma_2^2$	$t = \dfrac{\bar{y}_1 - \bar{y}_2}{\sqrt{\left(\dfrac{s_1^2}{n_1} + \dfrac{s_2^2}{n_2}\right)}}$	$\lvert t \rvert \geq t_{\alpha/2, df^*}$	$(\bar{y}_1 - \bar{y}_2) \pm t_{\alpha/2, df^*}\sqrt{\left(\dfrac{s_1^2}{n_1} + \dfrac{s_2^2}{n_2}\right)}$
			where $df^* = \dfrac{\left(\dfrac{s_1^2}{n_1} + \dfrac{s_2^2}{n_2}\right)^2}{\left[\dfrac{\left(\dfrac{s_1^2}{n_1}\right)^2}{(n_1 - 1)} + \dfrac{\left(\dfrac{s_2^2}{n_2}\right)^2}{(n_2 - 1)}\right]}$

TABLE 6.5

Summary of Hypothesis Testing for Two Normal Population Variances

Hypotheses	Test Statistic	Reject H_0 if
H_0: $\sigma_1^2 = \sigma_2^2$ H_1: $\sigma_1^2 \neq \sigma_2^2$	$F = \frac{s_1^2}{s_2^2}$	$F \geq F_{\alpha/2, n_1-1, n_2-1}$ $\alpha/2$ is the right-side area under the F-distribution, $n_1 - 1$ is the numerator df, and $n_2 - 1$ is the denominator df.

Table 6.5. Although this is a two-sided test, the test statistic is usually calculated with the larger observed sample variance in the numerator and only the right side of the rejection region is then specified.

6.4.1 Dependent or Paired Samples

When samples are dependent, the analysis must take into account the possible outside influence of pairing observations. Two examples of dependent samples are: 1. Samples where a subject has measurements taken before and after a treatment is applied. 2. A panelist influences a response measurement in each of the samples, such as a taste test panel comparing two products.

Paired Data Example Subjects are measured for improvement after receiving a treatment. Responses for each subject are in the table below. Higher scores are preferred or indicate improvement.

Subject	Before	After
1	138	324
2	284	520
3	234	318
4	132	220
5	183	232

For paired data, the responses for the same subject appear on the same line in the DATA step. In this case,

```
DATA improvement;
INPUT  subject  before  after ;
DATALINES;
1 138 324
2 284 520
3 234 318
4 132 220
5 183 232
;
RUN;
```

OBJECTIVE 6.4: Compute a 99% confidence interval for the true mean difference between the before and after means. Examine the default ODS Graphics.

```
PROC TTEST DATA=improvement CI=NONE ALPHA=0.01;
PAIRED before*after ;
TITLE 'Objective 6.4';
RUN;
```

The SAS statement: PAIRED before*after; resulted in before – after differences in the output.

For differences of after – before, use PAIRED after*before;

Objective 6.4

The TTEST Procedure

Difference: before - after

N	Mean	Std Dev	Std Err	Minimum	Maximum
5	–128.6	78.7452	35.2159	–236.0	–49.0000

The values in the first table are: n, $\bar{d}$, s_d, $\frac{s_d}{\sqrt{n}}$, minimum difference, and maximum difference for the five pairs.

Mean	99% CL Mean	
–128.6	–290.7	33.5374

The observed average difference is –128.6, and the 99% CI for the mean difference is (–290.7, 33.537). Since zero is a likely value for the mean difference, it cannot be concluded that any improvement occurred.

DF	t Value	Pr > |t|
4	–3.65	0.0217

There is not a significant difference between the before and after treatment means ($\alpha = 0.01$, $t_4 = -3.65$, $p = 0.0217$). By default, the TTEST procedure computes the test statistic and p-value for the two-sided tests of all pairs identified in the PAIRED statement. One-sided alternatives can be tested. However, one must correctly order the variables in the PAIRED statement to conform to the side of the test needed.

To enable or disable graphics, one needs to use the ODS Graph Names. For the paired t-test using the TTEST procedure, there is a default set of ODS graphs produced.

ODS Graph Names: HISTOGRAM (with overlaid normal density), BOXPLOT, INTERVAL (overlaid confidence interval band) are arranged in a single panel in Figure 6.3. The collection of all of these graphics in Figure 6.3 is identified as the Summary Panel and by the ODS Graph Name SUMMARY.

In Figure 6.4 Paired Profiles Graph, the first variable "before" is connected to the second variable "after" for each subject or pair. Note the means for each of the paired variables are also plotted and are connected by the bold line. Though not a formal test, the tendency for each of the segments to have a positive slope supports the inquiry, "Are the after measures larger than before?" The Paired Profiles ODS Graph Name is PROFILES. As the name of the graph suggests this plot is applicable to paired data. For large samples this plot would become very congested.

In the Agreement Plot in Figure 6.5, the second response "after" (in the PAIRED statement) is plotted on the vertical axis and the first response "before" is plotted on the horizontal axis. The observed pairs of observations are plotted using the smaller circles. The farther from the 45-degree reference line, the more likely a significant difference is detected. The ordered pairs above the reference line illustrates the tendency of the variable on the vertical axis to be larger than the variable on the horizontal axis. The mean of each variable is also plotted

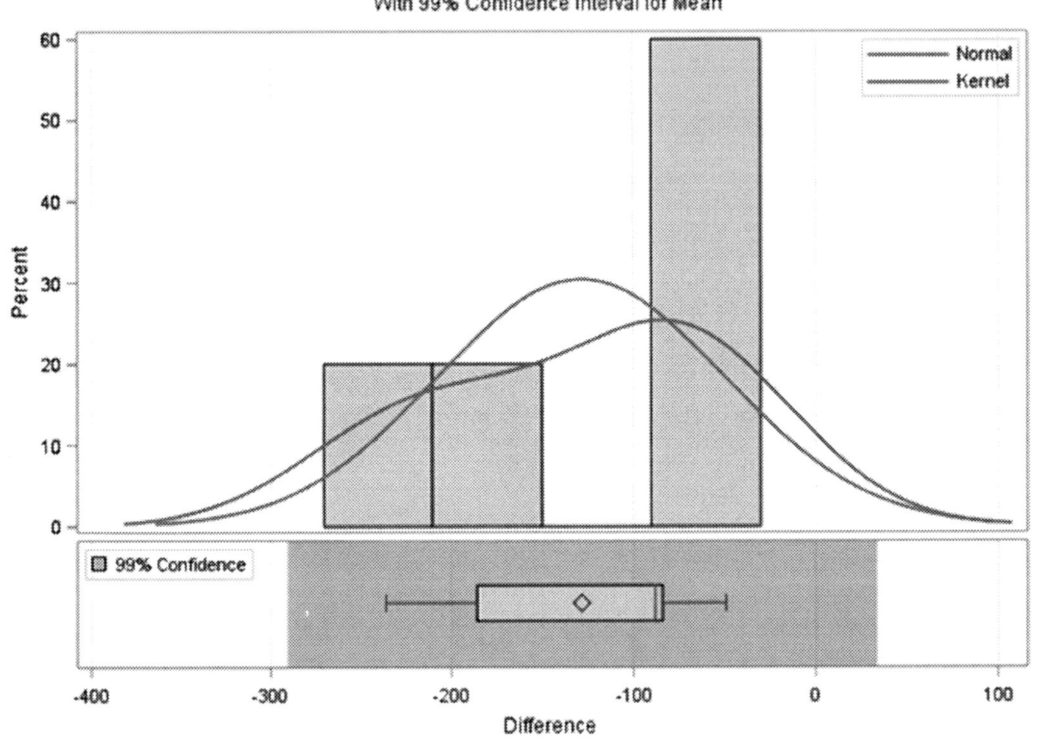

FIGURE 6.3
Summary panel for paired data.

as an ordered pair indicated by the larger, bold circle. The ODS Graph Name for this plot is AGREEMENT. An agreement plot is applicable for paired data only.

Figure 6.6 is a normal Q-Q plot. The observed differences are the plotted small circles. When the data are normally distributed the points lie "close" to the 45-degree reference line. The ODS Graph Name is QQ. A Q-Q plot can be used for single sample data or paired data.

For paired data the defaults plots produced by the TTEST procedure are:

PLOTS=(HISTOGRAM BOXPLOT INTERVAL QQ PROFILES AGREEMENT)

OBJECTIVE 6.5: Rerun the Objective 6.4 program with two different PLOTS options on the PROC TTEST statement. Specifically,

```
i.     PROC TTEST DATA=improvement ALPHA=0.01
                       PLOTS(ONLY)=(HISTOGRAM BOXPLOT);
       PAIRED before*after;
       TITLE 'Objective 6.5';

ii.    PROC TTEST DATA=improvement CI=NONE ALPHA=0.01
                       PLOTS(ONLY)=(BOXPLOT INTERVAL) SIDED=L;
       PAIRED before*after;
       TITLE 'Objective 6.5';
```

The output for these two options is not included in this text but are left as an exercise for the reader.

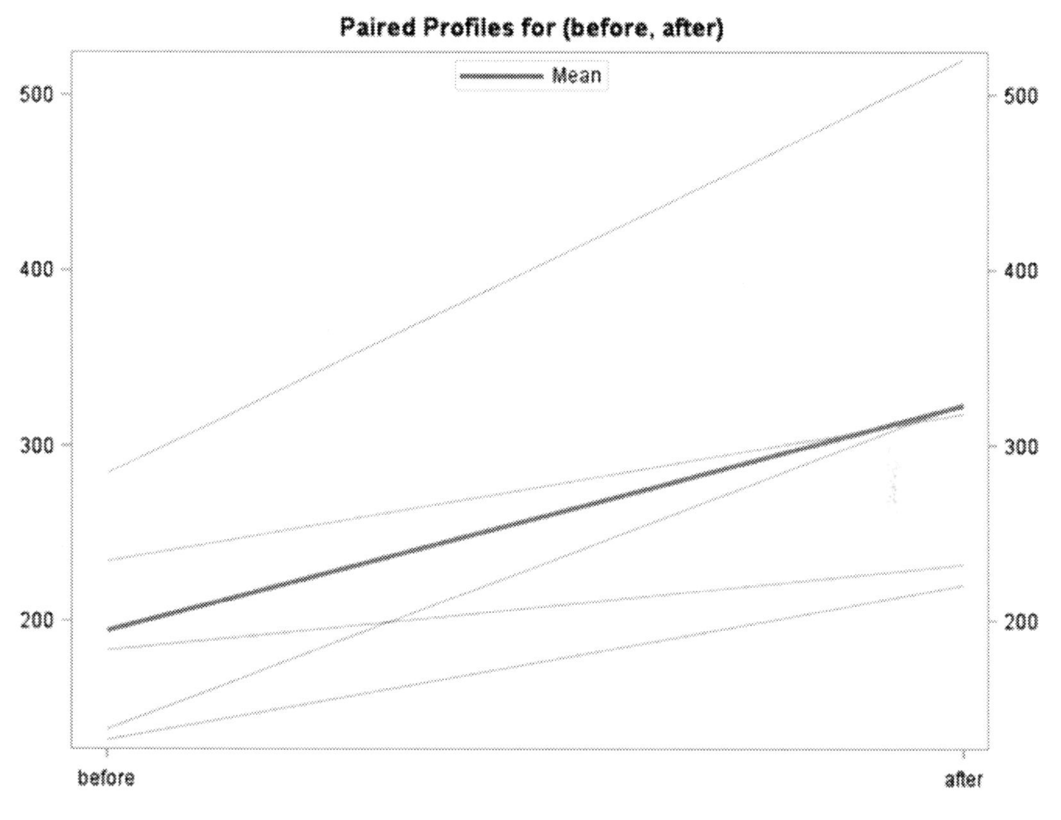

FIGURE 6.4
Paired profiles graph.

PLOTS(ONLY) will restrict the plots to those requested. When ODS Graphics are enabled, all default plots are printed. If "(ONLY)" is omitted in the above two options, all four of the plots overviewed in Objective 6.4 will appear in the HTML output. PLOTS=NONE will suppress all ODS Graphics in this procedure.

When one reverses the order of the variables in the PAIRED statement,

```
PAIRED after*before ;
```

Difference: after - before					
N	**Mean**	**Std Dev**	**Std Err**	**Minimum**	**Maximum**
5	128.6	78.7452	35.2159	49.0000	236.0

Mean	**99% CL Mean**	
128.6	-33.5374	290.7

DF	**t Value**	**Pr > |t|**
4	3.65	0.0217

The mean and t values in the tables are the opposite sign of the first scenario in Objective 6.4. Minimum and Maximum are also affected. The conclusion for the two-sided hypothesis test is the same. For a one-sided test, one has to specify the correct side **and** the correct

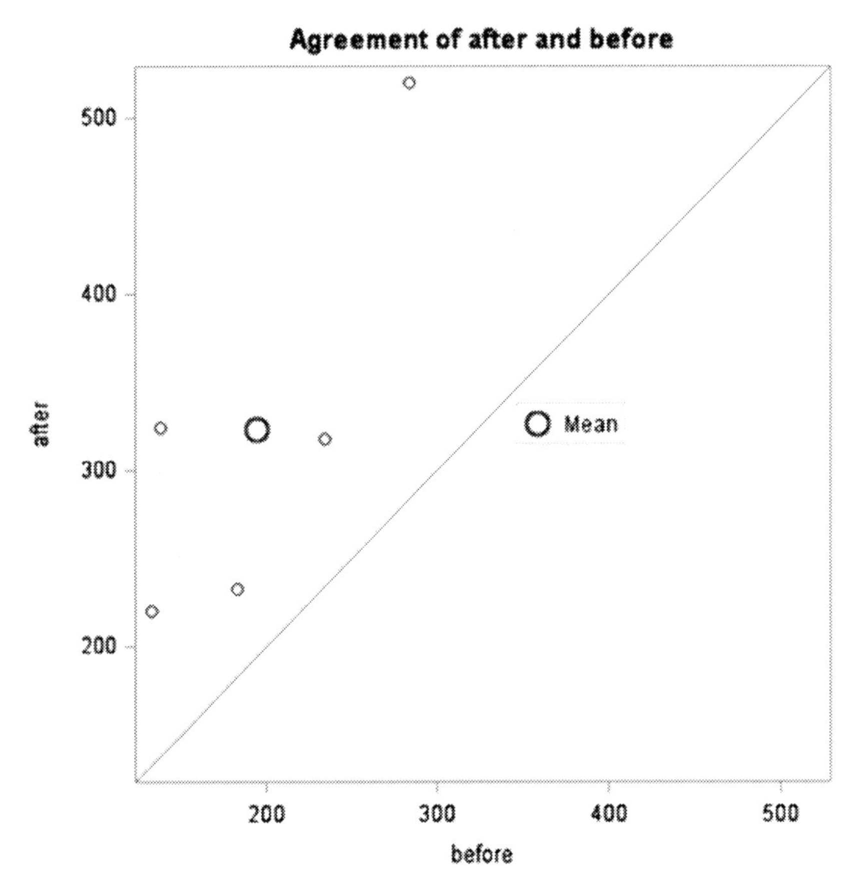

FIGURE 6.5
Agreement plot for paired data.

PAIRED statement. In this example, if one wanted to test whether or not there was significant improvement after the treatment, that is, the "after" mean is larger, there are two options for the essential code.

Option 1

```
PROC TTEST DATA=two SIDES=U ALPHA=0.01;
PAIRED after * before;
RUN;
```

Option 2

```
PROC TTEST DATA=improvement SIDES=L ALPHA=0.01;
PAIRED before * after;
RUN;
```

6.4.2 Independent Samples

To investigate the two population test method with independent samples, these objectives will utilize the instructional program data introduced and used in Objectives 6.1 – 6.3 in this chapter.

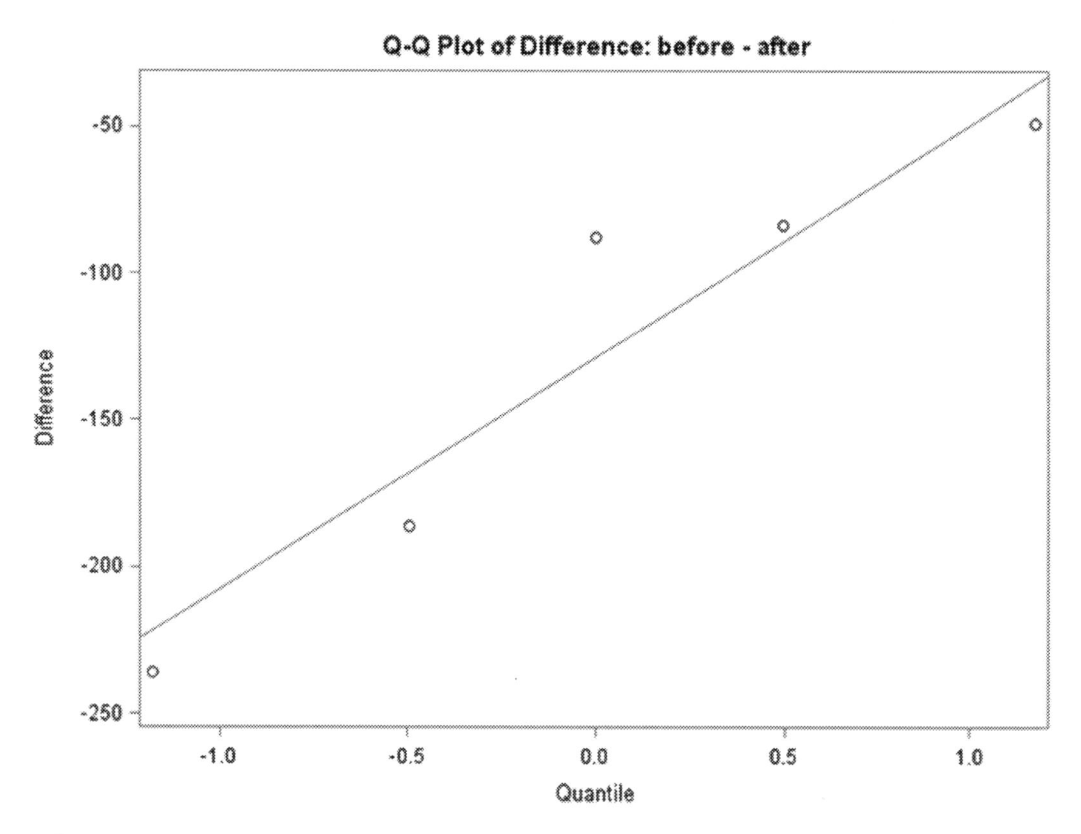

FIGURE 6.6
Q-Q plot for the paired data example.

```
DATA instruction;
INPUT Program $ Score @@;
DATALINES;
A 71 A 82 A 88 A 64 A 59 A 78 A 72
A 81 A 83 A 66 A 83 A 91 A 79 A 70
B 65 B 88 B 92 B 76 B 87 B 89 B 85
B 90 B 81 B 91 B 78 B 81 B 86 B 82
B 73 B 79
;
```

OBJECTIVE 6.6: Compare the means of the two program populations. Assume that the two populations of scores are normally distributed. Is there evidence that one program yields higher results than the other? Compute a 95% confidence interval for the mean. Identify whether the equal variance or unequal variances approach to the analysis is more appropriate.

```
PROC TTEST DATA=instruction;
CLASS program;
VAR score;
TITLE 'Objective 6.6';
TITLE3 't-test for the Difference Between Two Independent Means';
RUN;
```

This is the minimal code needed to obtain test results. PLOTS, SIDES, and other options can be used in the PROC TTEST statement.

Objective 6.6

t-test for the Difference Between Two Independent Means

The TTEST Procedure

Variable: Score

Program	N	Mean	Std Dev	Std Err	Minimum	Maximum
A	14	76.2143	9.4069	2.5141	59.0000	91.0000
B	16	82.6875	7.3277	1.8319	65.0000	92.0000
Diff (1-2)		-6.4732	8.3576	3.0586		

Program	Method	Mean	95% CL Mean		Std Dev	95% CL Std Dev	
A		76.2143	70.7829	81.6456	9.4069	6.8195	15.1549
B		82.6875	78.7828	86.5922	7.3277	5.4130	11.3411
Diff (1-2)	Pooled	-6.4732	-12.7384	-0.2080	8.3576	6.6324	11.3033
Diff (1-2)	Satterthwaite	-6.4732	-12.8867	-0.0597			

The CLASS statement names the variable which identifies the two populations to be compared. The order of the values of the class variable are ascending or alphabetical. Note the difference between the means is A – B.

The value of 8.3576 in the first and second tables is s_p, the pooled estimate of the standard deviation (Table 6.4). The confidence intervals for the difference between means are computed for both the equal variance (Pooled) and unequal variances (Satterthwaite) conditions (Satterthwaite, 1946).

Method	Variances	DF	t Value	Pr > |t|
Pooled	Equal	28	-2.12	0.0433
Satterthwaite	Unequal	24.487	-2.08	0.0481

The TTEST procedure computes the test statistics, "t Value", for both the equal and unequal variances conditions. The value 28 is the $df = n_1 + n_2 - 2$ and 24.487 is the value of df* which are defined for this application in Table 6.4. The absolute value bars around t in "Pr > |t|" identify that a two-sided test has been conducted.

Equality of Variances

Method	Num DF	Den DF	F Value	Pr > F
Folded F	13	15	1.65	0.3527

The final table **Equality of Variances** is labeling the test of the variances. The reader must interpret the test of equal variances before interpreting the test of the means in the previous table.

In Figure 6.7, the HISTOGRAM for each value of the CLASS variable are default ODS Graphs. The population label is located in the upper left corner of each histogram. The BOXPLOT for each value of the CLASS variable appears below the histograms. The labels

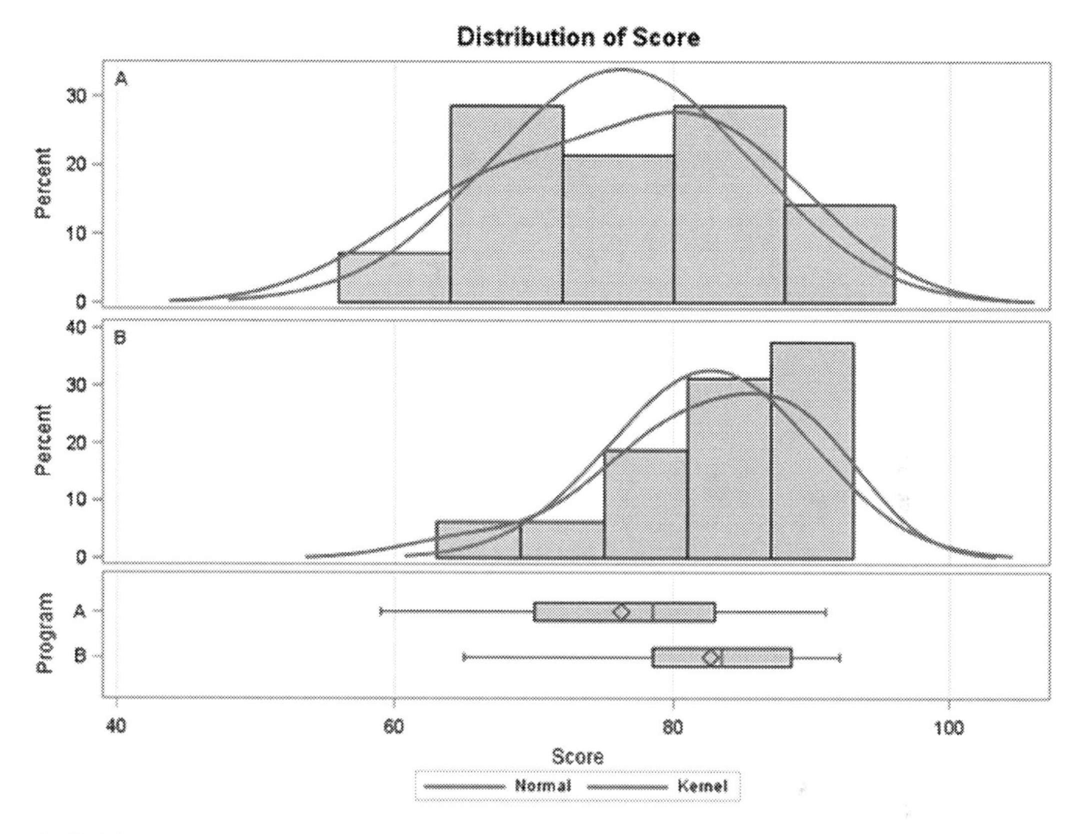

FIGURE 6.7
Distribution of score for two independent samples.

for the two populations appear on the left axis of the boxplots. The horizontal axis is applicable to both histograms and boxplots.

The Q-Q plots for each sample are also default ODS Graphs (Figure 6.8). The populations are identified in the upper left corner of each Q-Q plot. The ODS Graphs that are the default for TTEST for independent samples are HISTOGRAM, BOXPLOT, and QQ.

Overall remarks or conclusions:

The mean for program A is 76.214 and the mean for program B is 82.688. The difference between means (A – B) is –6.473. The standard deviations are 9.4069 for A and 7.3277 for B. The t-tests are for H_0: $\mu_A = \mu_B$ versus H_0: $\mu_A \neq \mu_B$. The test statistic $t_{28} = -2.12$ is calculated under the equal variance assumption, and $t_{24.5} = -2.08$ is calculated under the unequal variance assumption. SAS does not determine which is appropriate. The reader must do that.

Equality of Variances: This is the F-test for testing H_0: $\sigma^2_1 = \sigma^2_2$ versus H_1: $\sigma^2_1 \neq \sigma^2_2$. Here $F_{13,15} = 1.65$. "Folded F" in the last table implies that the larger sample variance was placed in the numerator in the computation of the F-statistic.

Since a 95% CI was specified, then $\alpha = 0.05$ will be used for all tests. Here, the equal variances condition has been met ($\alpha = 0.05$, $F_{13,\,15} = 1.65$, $p = 0.3527$). The appropriate test to compare the means is the pooled variance t-test. There is a difference in mean score between the two programs ($\alpha = 0.05$, $t_{28} = -2.12$, $p = 0.0433$). Since the test is significant and the observed sample B mean 82.6875 is larger than the sample A mean 76.2143, one can conclude that Program B has a larger population mean.

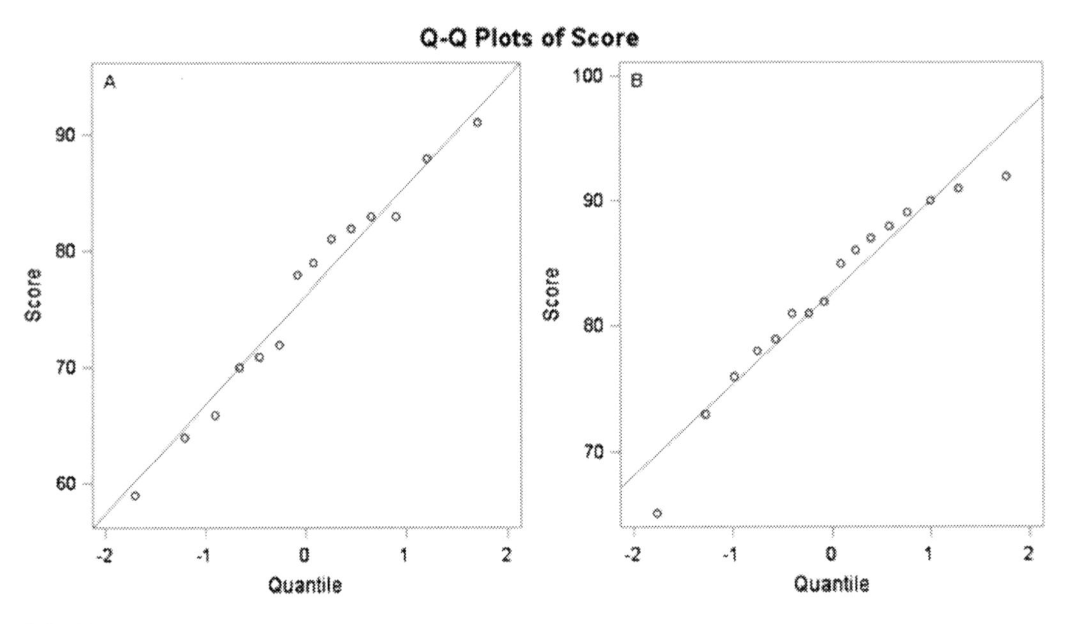

FIGURE 6.8
Q-Q plots for each independent sample.

OBJECTIVE 6.7: Rerun Objective 6.6 with the SIDED = L option and observe the output. Repeat using the SIDED = U option. How are the differences computed? Does this make a difference in the results of the t-test?

SIDED=L

Program	Method	Mean	95% CL Mean		Std Dev	95% CL Std Dev	
A		76.2143	70.7829	81.6456	9.4069	6.8195	15.1549
B		82.6875	78.7828	86.5922	7.3277	5.4130	11.3411
Diff (1-2)	Pooled	–6.4732	–Infty	–1.2702	8.3576	6.6324	11.3033
Diff (1-2)	Satterthwaite	–6.4732	–Infty	–1.1554			

Method	Variances	DF	t Value	Pr < t
Pooled	Equal	28	–2.12	0.0217
Satterthwaite	Unequal	24.487	–2.08	0.0240

For the left-side test, one observes that A – B is still the difference being computed. The negative infinity left endpoint for the confidence intervals for the mean difference is consistent with a left-sided test as are the observed significance levels "Pr < t" are 0.0217 and 0.0240 (half of the values from the two-sided tests in Objective 6.6). The equal variance test information has been suppressed by the author since it does not change.

SIDED=U

Program	Method	Mean	95% CL Mean		Std Dev	95% CL Std Dev	
A		76.2143	70.7829	81.6456	9.4069	6.8195	15.1549
B		82.6875	78.7828	86.5922	7.3277	5.4130	11.3411
Diff (1-2)	Pooled	–6.4732	–11.6763	Infty	8.3576	6.6324	11.3033
Diff (1-2)	Satterthwaite	–6.4732	–11.7911	Infty			

Method	Variances	DF	t Value	Pr > t
Pooled	Equal	28	−2.12	0.9783
Satterthwaite	Unequal	24.487	−2.08	0.9760

For the right-side test, A – B is still the difference being computed. Now confidence lower bounds are computed, positive infinity is the right endpoint for the confidence intervals for the mean difference, and the observed significance level "Pr > t" is 0.9783 or 0.9760 (1 minus the p-value from the left-side tests). These observed significance values are quite large consistent with the lack of evidence supporting $\mu_A > \mu_B$. The equal variance information is again suppressed by the author since it does not change.

Reminder: The CLASS statement can have only one variable, and the variable must have exactly two levels. If the class variable has only one level, you will get an error message in the Log window:

```
ERROR: The CLASS variable does not have two levels.
```

Data are case sensitive. So, if, for example, the values of the Program variable are A, B, and b (where b is a typographical error), or there are more than two programs, say A, B, and C, the TTEST procedure will not run. An error message will appear in the Log window. In either case, the error message is:

```
ERROR: The CLASS variable has more than two levels.
```

Trying to compare more than two populations, say A, B, and C, will require a different procedure. (More on this in Chapter 7.)

6.5 Chapter Summary

In this chapter, the TTEST procedure was overviewed. The TTEST procedure is applicable when the populations the samples are drawn from are normal. The TTEST procedure can be used for confidence intervals and t-tests about either one population mean, or for the comparison of two population means. In the case of two populations, the samples can be either dependent (paired samples) or independent samples. One-sided and two-sided tests of population means are possible in this procedure. ODS Graphics are available to support any of these three types of tests. While the TTEST procedure does not test for normality of the data, ODS Graphics (histograms and Q-Q plots) may assist in the determination of the normality of the data.

7

One-Way ANOVA Methods, Non-Parametric Methods and Ranking Data

This chapter presents the SAS programming code for comparing t population means where $t \geq 2$. When the populations are normally distributed and have equal variance, Analysis of Variance (ANOVA or AOV) methods are used to test the t population means for equality. When the populations are not normally distributed, a non-parametric test called the Kruskal-Wallis rank sum test may be used. The Kruskal-Wallis test and some other non-parametric methods utilize the ranks of the observations rather than the value of the observations themselves. For this reason, a SAS procedure for ranking data is also presented in this chapter.

7.1 ANOVA: Hypothesis Testing

When comparisons among two or more normal population means are to be made using t independent samples, an ANOVA can be done to analyze the sample means and draw inferences about the populations. Very often the populations to be compared are referred to in statistics methods literature as "treatments" or "treatment groups". There are t independent samples of size n_i obtained where $i = 1, 2, \ldots, t$. So, for each of the t treatment samples, the sample means, $\bar{y}_i$, and sample variances, s_i^2, are measured. The overall mean from the t samples is given by $\bar{y}$.

The hypotheses and test statistic for the ANOVA F-test are given in Table 7.1

MSTrt is the mean square for the differences among treatment means or "mean square between groups", and MSE is the mean square due to error or "mean square within group". The "between" and "within" jargon is popular in some elementary statistical methods books. The phrasing for the null and alternative hypotheses could also be, H_0: There is no treatment effect versus H_1: There is a treatment effect.

NOTE: When t=2, the independent t-test statistic, t_{N-t}, yields the same conclusion or result as the ANOVA F-test. Observe that the denominator of the F-statistic in Table 7.1 is a function of each of the t sample variances. These sample variances are pooled together into one estimate of the variance, MSE, which these t populations have in common. Algebraically, it can be shown that when t=2 groups, $t_{N-t}^2 = F_{1,N-t}$, and $s_p^2 = MSE$.

One can also compute confidence interval estimates for the population or treatment means using the pooled variance estimate, MSE.

$$(1-\alpha)100\% \text{ Confidence Interval for } \mu_i$$

$$\bar{y}_i \pm t_{\alpha/2,df} \frac{\sqrt{MSE}}{\sqrt{n}}$$

where $df = N - t$ in this application.

TABLE 7.1

Summary of Hypothesis Testing in ANOVA

Hypothesis	Test Statistic	Reject H_0 if
H_0: $\mu_1 = \mu_2 = \ldots = \mu_t$ H_1: at least one μ_i is different	$F_{t-1,N-t} = \dfrac{\frac{1}{t-1}\sum_{i=1}^{t} n_i(\bar{y}_i - \bar{y}_{\cdot})^2}{\frac{1}{N-t}\sum_{i=1}^{t}(n_i - 1)s_i^2} = \dfrac{MSTrt}{MSE}$	$F \geq F_{\alpha,t-1,N-t}$

where N is the grand sample size; $N = \sum_{i=1}^{t} n_i$ and $F_{\alpha,ndf,ddf}$ is the critical F-value determining a right-side probability of α, ndf is the numerator df, and ddf is the denominator df for the application.

If the ANOVA F-test is significant, there are several applicable post-hoc multiple comparison procedures available to contrast the population means. These post-hoc analyses are not detailed here, but a few options will be listed with the syntax below. Additionally, there are alternative methods when there is not a common variance for the t populations. Mixed models methods modeling multiple variances may be used for the analysis, or some nonparametric method may be used.

7.2 The GLM Procedure

The GLM (General Linear Model) procedure is one of the most common SAS procedures used to perform an analysis of variance. The GLM procedure has many, many facets and hence, many more syntax statements and options available. Only a brief introduction to the procedure for ANOVA methods is presented here. The syntax of the procedure is as follows:

PROC GLM DATA=*SAS-data-set <options>*;
CLASS *group;*
MODEL *response = group;*
MEANS *group/<options>;*

BY and WHERE statements can be added to this procedure as needed.

PROC GLM statement options include:

```
PLOTS= NONE | DIAGNOSTICS RESIDUALS BOXPLOT
```

ODS Graphics can be requested in the PROC GLM statement. If ODS Graphics are enabled (Section 19.1), the BOXPLOT is the default image produced. One can suppress all graphics by including PLOTS=NONE as the option. If a list of ODS Graphics are to be included, one should enclose the list in parenthesis, such as, PLOTS = (DIAGNOSTICS RESIDUALS).

DIAGNOSTICS and RESIDUALS are by default assembled as a 3 × 3 panel of graphs. One can produce each graph individually by "unpacking" them. The

UNPACK option is a global plot option. It appears to the left of the equal sign and is enclosed in parenthesis such as, PLOTS(UNPACK) = (DIAGNOSTICS RESIDUALS).

CLASS statement

This statement names the variable that identifies the populations to be compared. Class variables can be either numeric or character and must have two or more different values. At most one CLASS statement can be used in a single GLM procedure, and it must appear before the MODEL statement.

MODEL statement

The MODEL statement specifies the response(s) to be analyzed and the class variable(s). The order is important in that response variables are always listed on the left side of the equal sign, and the class variables always appear on the right. One can list several response variables in a single MODEL statement, such as

```
response1 response2 ... responsen = group ;
```

For each response variable, an ANOVA table (F-test) will be produced. Only one MODEL statement can be used in a single GLM procedure, and it must appear after the CLASS statement.

When ODS Graphics are enabled, a BOXPLOT with the ANOVA F-test results inset will be generated unless it is suppressed using PLOTS=NONE in the PROC GLM statement.

The MODEL statement does have several options, none of which are needed to conduct a basic ANOVA; thus, they are not presented here.

MEANS statement

This statement will produce the sample means and standard deviations for each of the t levels of the class variable listed in this statement. An additional BOXPLOT will be produced by default when using a MEANS statement.

There are several options for the MEANS statement. For the one-way ANOVA, these are the most helpful:

CLM will produce a (1 – p)100% confidence interval for each of the t population means. However, to produce individual confidence intervals, one must also include the LSD (**L**east **S**ignificant **D**ifference) option.

ALPHA=*p* where 0 < p < 1; this value is used to determine the (1 – p)100% level of confidence used in the confidence interval calculations. The default setting for all confidence intervals is 95% (ALPHA=0.05) unless otherwise specified using the ALPHA option.

There are many other options for the MEANS statement that identify multiple comparisons procedures at the specified value of ALPHA. Here are a few:

LSD – Fisher's least significant difference

TUKEY – Tukey's honest significant difference

BON – Bonferroni adjusted pairwise comparisons

SCHEFFE – Scheffe adjusted pairwise comparisons

CLDIFF or LINES for illustrating the pairwise comparison conclusions

These multiple comparison methods and the additional statements for more general contrasts are not overviewed in this book as they are beyond the scope of a first statistics course. Readers familiar with multiple comparison procedures should consult SAS Help and Documentation for more options on the MEANS Statement, and they should also become familiar with the LSMEANS statement which also conducts post-hoc analyses.

ANOVA Example Data – In meat science, treatments that improve the quality of cooked meat are investigated. In this experiment, there is a Control group, denoted by C, and two experimental treatments, denoted T1 and T2, applied to cuts of beef prior to commercial packaging. Similar steaks are independently assigned to each of the three treatment groups. After cooking the steaks by the same method to the same internal temperature (a measure of doneness), pH and cook yield are measured on each steak. There are five independently treated steaks in each treatment group. The data from this experiment appear in Table 7.2.

TABLE 7.2

Data From a Meat Science Experiment

Treatment Group					
C		T1		T2	
pH	Cook Yield	pH	Cook Yield	pH	Cook Yield
6.14	20.9	5.98	22.4	6.18	23.4
5.98	22.1	6.32	23.8	6.22	21.8
6.30	21.8	5.89	23.0	6.03	22.6
6.25	20.3	6.08	24.5	2.97	24.8
6.07	21.2	6.11	22.8	5.93	25.1

OBJECTIVE 7.1: Create the SAS data set *meat* from the data in Table 7.2. Compare the response means for three treatment groups by conducting an ANOVA. Use $\alpha = 0.05$ in your conclusions. Compute the sample means for each of the three treatment groups. Reset the log and Results Viewer first.

```
DM 'LOG; CLEAR; ODSRESULTS; CLEAR; ';

DATA meat;
INPUT Group $ pH CookYield;
DATALINES;
C  6.14  20.9
C  5.98  22.1
C  6.30  21.8
C  6.25  20.3
C  6.07  21.2
T1 5.98  22.4
T1 6.32  23.8
T1 5.89  23.0
T1 6.08  24.5
```

```
T1 6.11  22.8
T2 6.18  23.4
T2 6.22  20.8
T2 6.03  22.6
T2 5.97  24.8
T2 5.93  25.1
;
PROC GLM DATA=meat;
CLASS group;
MODEL ph cookyield = group;
MEANS group;
TITLE 'Objective 7.1 - ANOVA';
RUN;
QUIT;
```

In the previous code, each line after DATALINES must contain the complete record for each observation. One could have included more than one record per line by using the double trailing @ control for the DATA step (Objective 2.4), such as:

```
DATA meat ;
INPUT Group $ pH CookYield @@;
DATALINES;
C  6.14  20.9  C  5.98  22.1
C  6.30  21.8  C  6.25  20.3
C  6.07  21.2  T1 5.98  22.4
T1 6.32  23.8  T1 5.89  23.0
T1 6.08  24.5  T1 6.11  22.8
T2 6.18  23.4  T2 6.22  20.8
T2 6.03  22.6  T2 5.97  24.8
T2 5.93  25.1
;
```

However, experiments typically have many more response columns than this example. The style without the double trailing @ control is recommended when there are several columns of variables. In Chapters 8 and 14, some methods for reading in data from external files are presented. Reading data from external files certainly may be more efficient, and most often do not have more than one observation per row or line of data.

Objective 7.1 – ANOVA

The GLM Procedure

Class Level Information

Class	**Levels**	**Values**
Group	3	C T1 T2

Number of Observations Read	15
Number of Observations Used	15

The CLASS statement of the GLM procedure identifies the populations being compared and the number of observations in the data set in the **Class Level Information** table. This is another opportunity for the programmer to proofread the data by checking the number of **Levels** of the class variable, the **Values** of the class variable, and the **Number of Observations**.

Objective 7.1 – ANOVA

The GLM Procedure

Dependent Variable: pH

Source	DF	Sum of Squares	Mean Square	F Value	Pr > F
Model	2	0.02001333	0.01000667	0.50	0.6162
Error	12	0.23812000	0.01984333		
Corrected Total	14	0.25813333			

R-Square	Coeff Var	Root MSE	pH Mean
0.077531	2.310547	0.140866	6.096667

Source	DF	Type I SS	Mean Square	F Value	Pr > F
Group	2	0.02001333	0.01000667	0.50	0.6162

Source	DF	Type III SS	Mean Square	F Value	Pr > F
Group	2	0.02001333	0.01000667	0.50	0.6162

The MODEL statement produces the ANOVA table, a table of summary information (R^2= coefficient of determination, Coefficient of Variation = $CV = \frac{\sqrt{MSE}}{\bar{y}.}100\%$, $\sqrt{MSE}$, and $\bar{y}.$), and two additional tables. The response variable being analyzed appears in the line above the ANOVA table: "Dependent Variable: pH". In this ANOVA table, the test of whether or not the three meat treatments result in the same mean pH of the prepared steak is computed.

Conclusion: There is no difference among the three treatment methods in the mean pH of prepared steaks ($\alpha = 0.05$, $F_{2,12} = 0.50$, $p = 0.6162$).

In this introductory ANOVA example, the value of F (0.50) is the same in the ANOVA table and in both the Type I and Type III analyses below the ANOVA. In analyses where there is more than one class variable, the reader should become accustomed to reading the Type I and/or Type III analysis depending upon the application. For now, by the simplicity of the one-way analysis, the Type I, Type III, and ANOVA table F-tests are the same.

The BOXPLOT for each dependent variable is the default ODS graph accompanying the MODEL statement. The inclusion of the ANOVA F-statistic and observed significance level in the corner of the graph in Figure 7.1 is repeated. The three sample pH means are marked by the diamonds, ◊, and are all quite similar as the ANOVA F-statistic determined.

Because two response variables were included in the MODEL statement, a second ANOVA table, summary statistics table, Type I and III tests, and boxplot are produced for the second response variable, CookYield.

Objective 7.1 – ANOVA

The GLM Procedure

Dependent Variable: CookYield

Source	DF	Sum of Squares	Mean Square	F Value	Pr > F
Model	2	14.14933333	7.07466667	4.96	0.0270
Error	12	17.12400000	1.42700000		
Corrected Total	14	31.27333333			

R-Square	Coeff Var	Root MSE	CookYield Mean
0.452441	5.277928	1.194571	22.63333

Source	DF	Type I SS	Mean Square	F Value	Pr > F
Group	2	14.14933333	7.07466667	4.96	0.0270

Source	DF	Type III SS	Mean Square	F Value	Pr > F
Group	2	0.02001333	0.01000667	0.50	0.6162

Conclusion: There is a difference among the three treatments in the mean Cook Yield ($\alpha = 0.05$, $F_{2,12} = 7.96$, $p = 0.0270$).

The BOXPLOT in Figure 7.2 illustrates the differences among the treatment groups. The control mean appears lower than the means for the other treatments. **A graph never replaces a formal test of significance**, but supports the conclusions drawn from an analysis. One needs to know about multiple comparison methods to draw more conclusions about which of these treatment groups differs from another.

The MEANS statement requests sample means for each level of the class variable for each response variable and those appear in a table after a second set of boxplots for each dependent variable. In this release of SAS there is some redundancy in ODS Graphs in the GLM procedure when the MEANS statement is included in the GLM procedure code. The BOXPLOT for each response variable appears again; this time without the inset of the ANOVA F-statistic and its observed significance level. See Figures 7.3 and 7.4

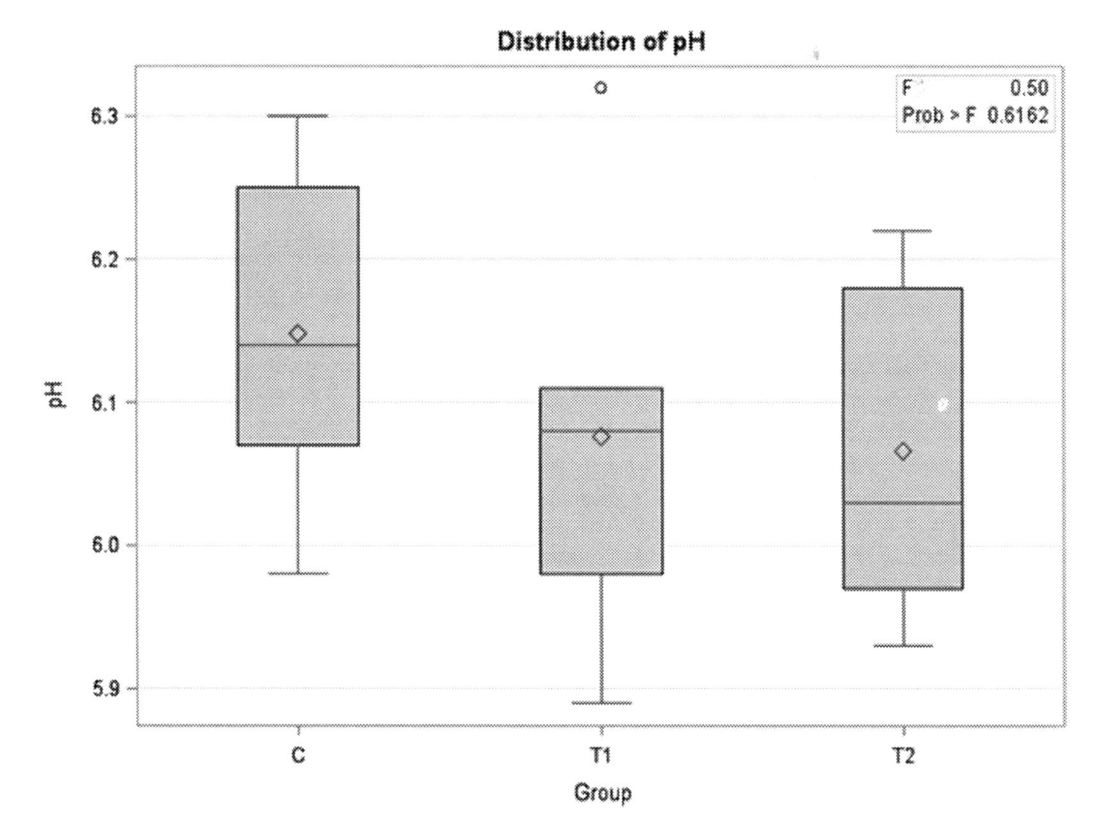

FIGURE 7.1
The default BOXPLOT for the MODEL statement in the pH analysis.

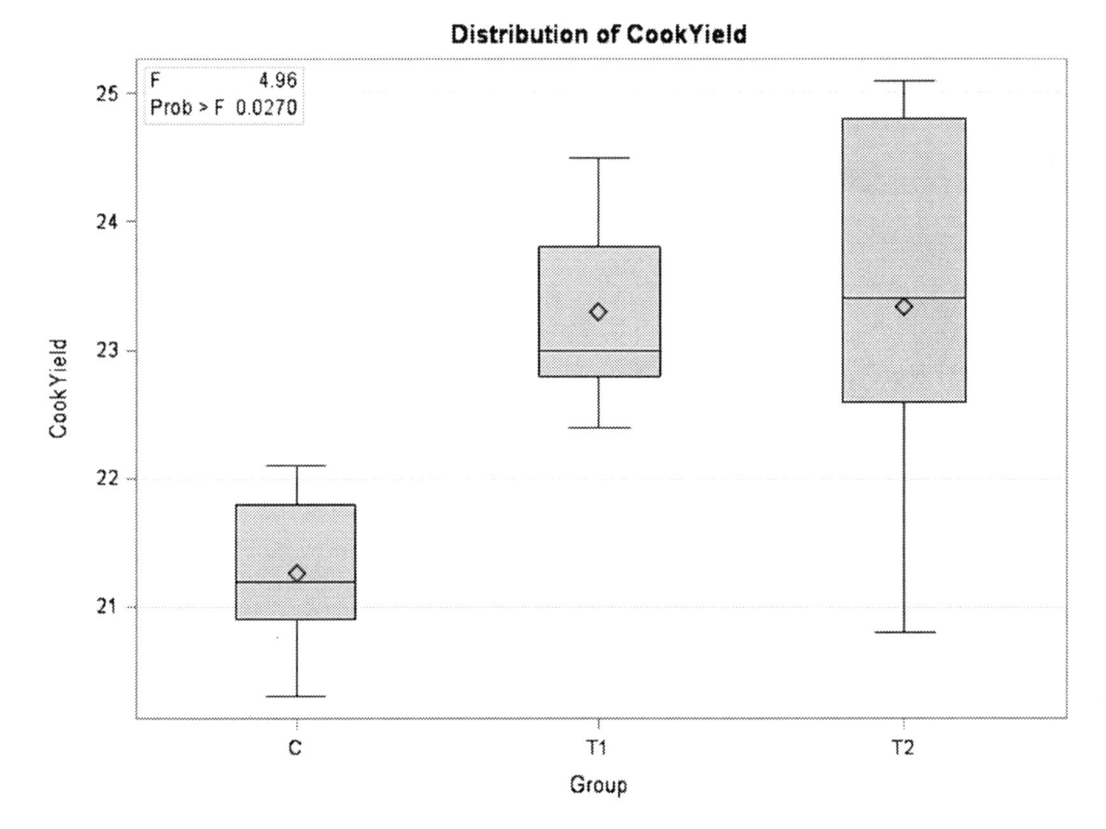

FIGURE 7.2
The default BOXPLOT for the MODEL statement in the Cook Yield analysis.

		pH		CookYield	
Level of Group	**N**	**Mean**	**Std Dev**	**Mean**	**Std Dev**
C	5	6.14800000	0.13026895	21.2600000	0.71624018
T1	5	6.07600000	0.16164777	23.3000000	0.84261498
T2	5	6.06600000	0.12817956	23.3400000	1.74871381

The MEANS statement produces a table with columns for the levels of the class variable Group, sample sizes, sample means, and standard deviations for each level. Though there were equal sample sizes in this example, that is not a requirement for this procedure. Note standard errors for each mean are not computed. The LSMEANS statement can compute the standard errors of the means using the STDERR option. That is,

```
LSMEANS group/STDERR;
```

See the LSMEANS statement for the GLM procedure in SAS Help and Documentation to see more of the options available for the LSMEANS statement.

Naturally, if only one ANOVA is needed, then only specify a single dependent variable in the MODEL statement, such as: MODEL ph = group; or MODEL cookyield = group;

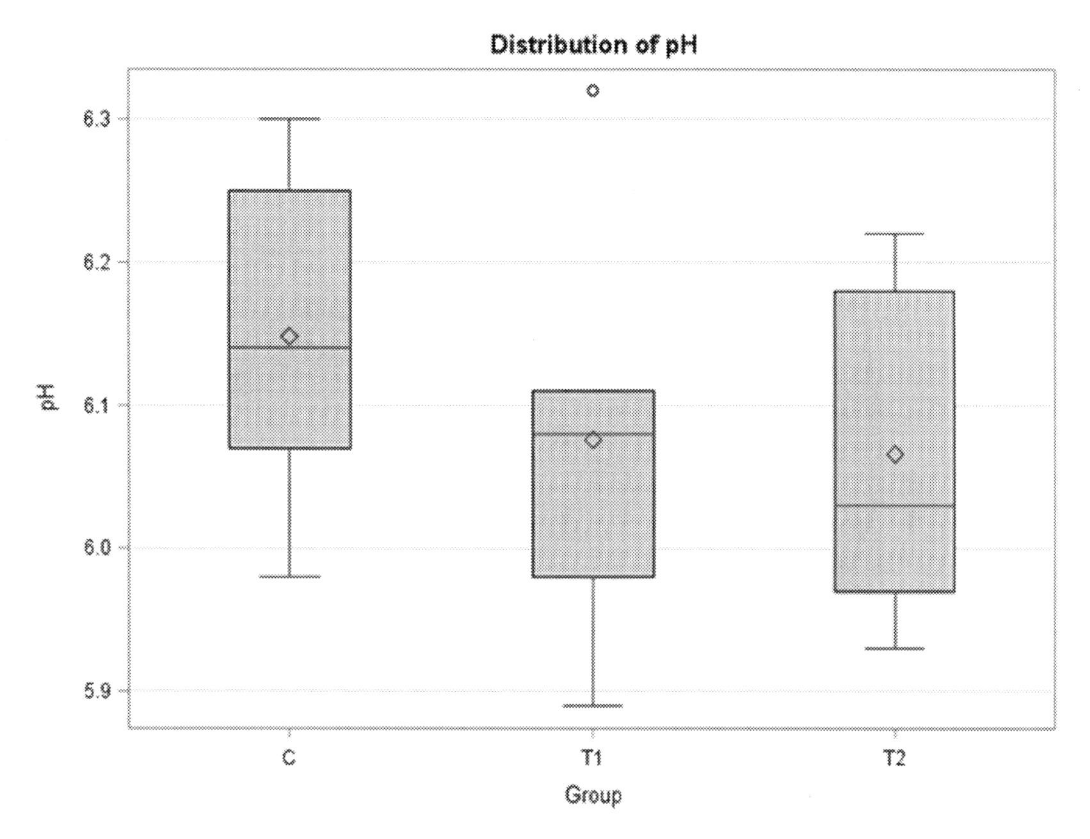

FIGURE 7.3
The BOXPLOT produced by the MEANS statement for the pH response.

OBJECTIVE 7.2: Modify Objective 7.1 by suppressing all of the boxplots. Simply including the PLOTS=NONE option on the PROC GLM statement will do this. No output is shown here since the MODEL results are the same as the output from Objective 7.1.

OBJECTIVE 7.3: Compute the ANOVA for only the cook yield response variable. Include 95% confidence intervals for the group means. Include only the ODS Graphs for the residual analysis to investigate the normality option. Include these residual graphs in a panel plot.

```
PROC GLM DATA=meat PLOTS(ONLY)=(RESIDUALS DIAGNOSTICS);
CLASS group;
MODEL cookyield = group;
MEANS group / CLM LSD;
TITLE "Objective 7.3 - ANOVA, CI's & Residuals";
RUN;
```

Because more than one plot is being requested, the list of ODS Graphs must be enclosed in parenthesis. The default BOXPLOT from the MODEL statement will be suppressed by the inclusion of the ONLY option. Note the usage of double quotation marks (") in the TITLE statement in order to use the apostrophe (or single quote mark ') in the text of the title (Section 2.4).

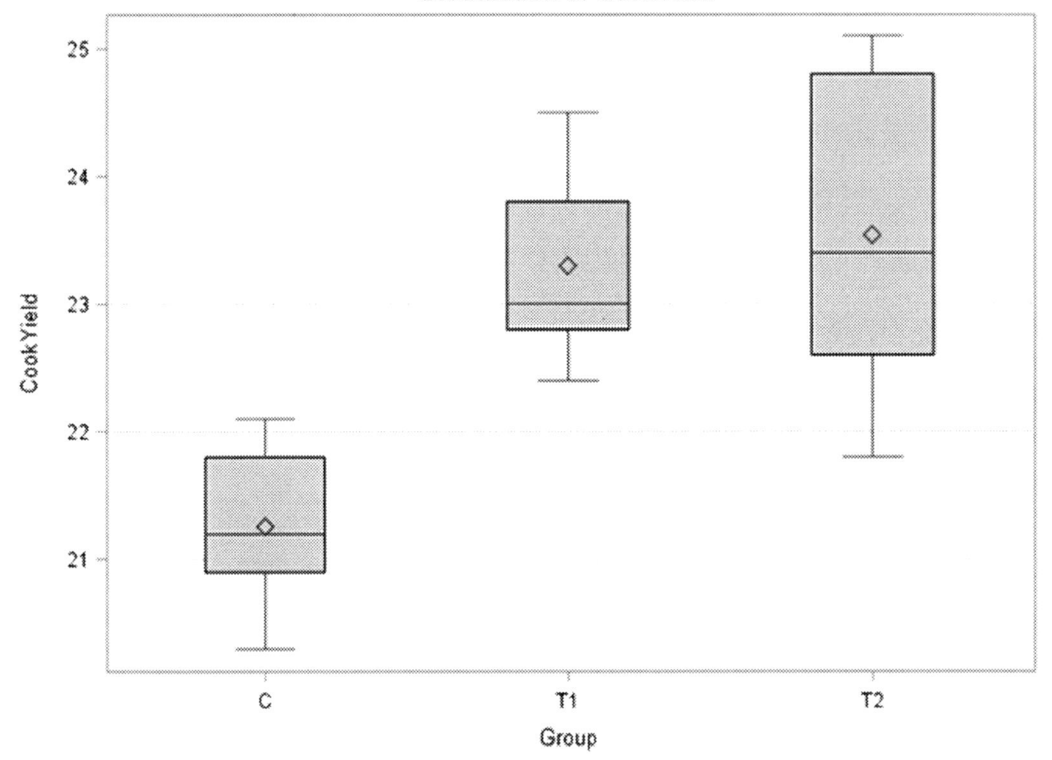

FIGURE 7.4
The BOXPLOT produced by the MEANS statement for the Cook Yield response.

The ANOVA results are the same as in Objective 7.1 and are not repeated here. Only the panel graph, **Fit Diagnostics for CookYield**, (Figure 7.5) and means with 95% confidence intervals are shown here. In the panel graph the first column of the panel are the residual graphs. In the first of these, a random scatter of the residuals supports the assumptions of the ANOVA. In the second row is a Q-Q plot of the residuals. If the residuals lie along the diagonal reference line, that supports the normality assumption. And lastly, the histogram of the residuals is in the third row. The residuals can be examined for normality rather than the original data since the sample sizes were small, and the effects of the treatments have been "removed" from the residuals allowing a larger sample size for the normality investigation. The remaining six panels in the second and third columns are the DIAGNOSTICS requested. A background in regression analysis is needed to fully understand and interpret the diagnostic graphs. The last image (row 3, column 3) is not an ODS Graph but a table of summary information.

To obtain individual (larger) images of the eight graphs in the Figure 7.5 panel, the UNPACK option can be used. Thus, the PLOTS option can be modified as:

```
PLOTS(UNPACK) = (RESIDUALS DIAGNOSTICS)
```

These larger images for the residual analysis and any default images associated with other statements in the GLM procedure will appear in the HTML output. To suppress all

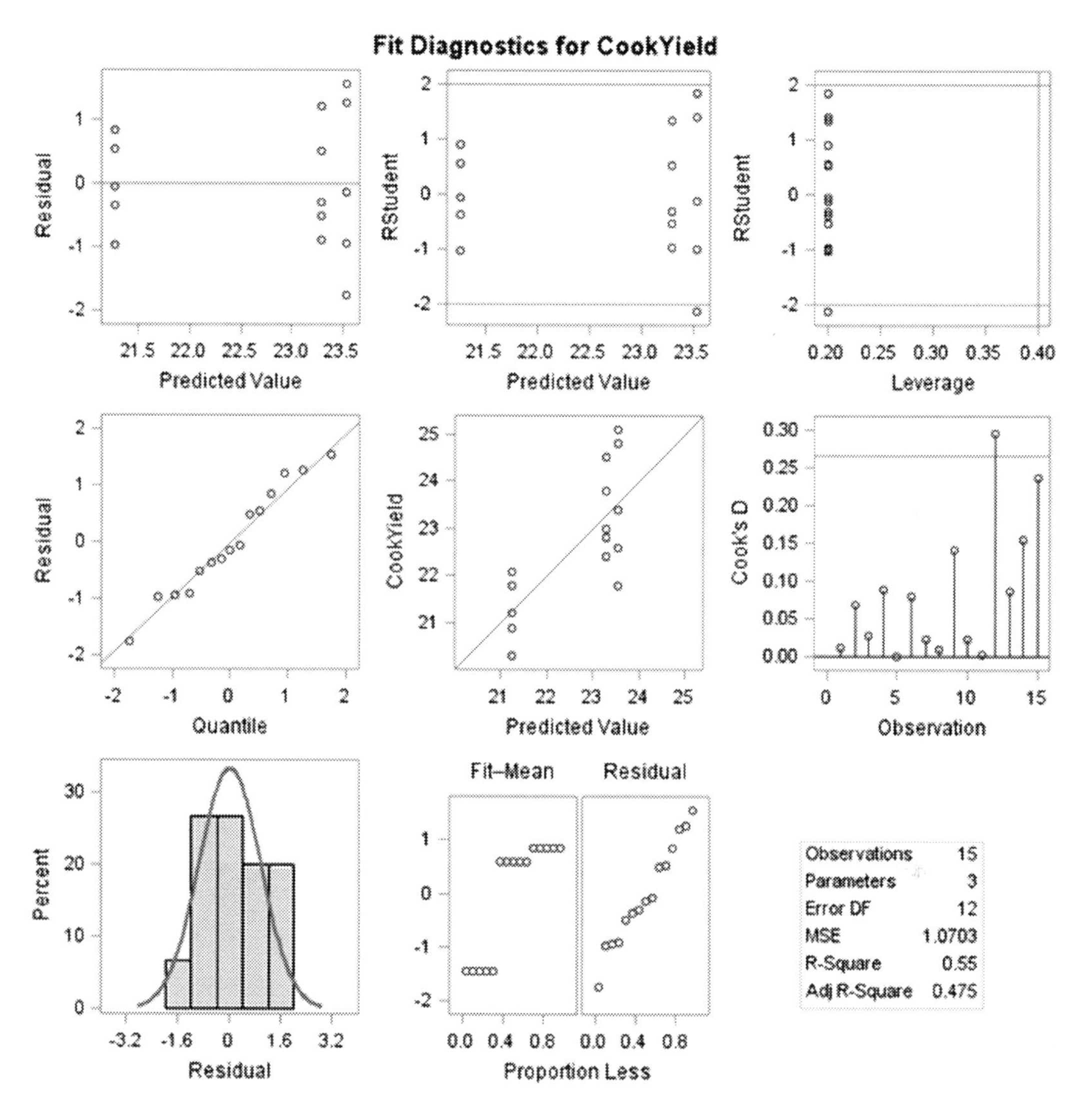

FIGURE 7.5
Panel plot of residuals and diagnostics.

default ODS graphics generated by the GLM procedure, the PLOTS options can include the ONLY modifier option:

```
PLOTS(ONLY)=(RESIDUALS DIAGNOSTICS)
```

The UNPACK option can be used with the ONLY option, such as:

```
PLOTS(UNPACK ONLY)=(RESIDUALS DIAGNOSTICS)
```

This would produce the larger images for the residual analysis and no other ODS Graphics.

In this present SAS version, both RESIDUALS and DIAGNOSTICS must be selected. Just one of these types of graphics cannot be selected. There are many other ODS Graphs available in the GLM procedure that are not presented here. The ODS Graphs presented here are those that are most useful when computing a one-way analysis of variance.

Objective 7.3 – ANOVA, CI's & Residuals

The GLM Procedure
t Confidence Intervals for CookYield

Alpha	0.05
Error Degrees of Freedom	12
Error Mean Square	1.427
Critical Value of t	2.17881
Half Width of Confidence Interval	1.163984

Group	N	Mean	95% Confidence Limits	
T2	5	23.3400	22.1760	24.5040
T1	5	23.3000	22.1360	24.4640
C	5	21.2600	20.0960	22.4240

Two tables of output result for the **t Confidence Intervals for CookYield**. The first table identifies the value of α, and hence, the confidence level. This value can be changed using the ALPHA = p option on the MEANS statement. The remainder of the first table, cites values used in the computation of the confidence interval, $\bar{y}_i \pm t_{\alpha/2,df}\frac{\sqrt{MSE}}{\sqrt{n}}$. The "Half Width of the Confidence Interval" is computed by $t_{\alpha/2,df}\frac{\sqrt{MSE}}{\sqrt{n}}$ for this choice of α and is 1.163984 in this analysis. Selecting T1 as an example, the sample mean is 23.3, and the 95% confidence interval for the T1 group mean is (22.1360, 24.4640).

The LSMEANS statement can equivalently produce confidence intervals for a single class variable with the following statement in the GLM procedure:

```
LSMEANS group/CL ALPHA=0.05;
```

No multiple comparison option needs to be included in the LSMEANS statement to produce these intervals. The LSMEANS statement is computationally different from the MEANS statement, in general. For the simple case of the one-way ANOVA, the MEANS and LSMEANS are computationally equivalent.

7.3 Non-Parametric Tests

When testing the equality of two or more population means, the ANOVA method can be used when the populations are normally distributed. What if the populations are not normally distributed? Since ANOVA procedures no longer apply, a class of tests called nonparametric tests could be used. Nonparametric methods do not assume a particular type of distribution. Many nonparametric methods rank the data low to high and base comparisons on functions

TABLE 7.3

Rank Sum Test Methods for Comparing Two or More Groups

Hypotheses	Test Statistic	Reject H_0 if	Comment
H_0: $\mu_1 = \mu_2 = \ldots = \mu_t$ H_1: at least one μ_i is different	$\chi^2 = \frac{12}{N(N+1)} \sum_{i=1}^{t} \frac{R_i^2}{n_i} - 3(N+1)$ where $N = \sum_{i=1}^{t} n_i$	$\chi^2 \geq \chi^2_{\alpha,t-1}$	This is referred to as the **Kruskal-Wallis** test for $t \geq 2$ groups.
H_0: $\mu_1 = \mu_2$ H_1: $\mu_1 \neq \mu_2$	$\mu = \frac{n_1(n_1+n_2+1)}{2}$ and $\sigma = \sqrt{\frac{n_1 n_2 (n_1+n_2+1)}{12}}$ $z = \frac{R_1 - \mu}{\sigma}$	$\lvert z \rvert \geq z_{\alpha/2}$	This is referred to as the **Wilcoxon Rank Sum Test** or the Mann-Whitney test for $t = 2$. This is algebraically equivalent to the Kruskal-Wallis Test when $t = 2$.

where $\chi^2_{\alpha,df}$ is the critical χ^2 determined by right side probability α, and $df = t - 1$ in this application. Also, z_p is the critical standard normal value determined by right side probability p.

of these ranks. Two nonparametric procedures based on ranks of data in independent samples are outlined here. They are often referred to as "rank sum" tests. (Conover, 1999).

Steps for a non-parametric rank sum test:

1. Combine the samples.
2. Rank the observations in the combined sample from smallest to largest. Assign the average rank to scores that are tied.
3. Sum the ranks from each of the t samples. Call this rank sum, R_i for $i = 1, 2, \ldots, t$.
4. Compute the test statistic(s) given in Table 7.3.

7.4 The NPAR1WAY Procedure

The NPAR1WAY procedure performs an analysis on ranks, and it can compute several statistics certain based on rank scores of a response variable. Only one class variable can be specified, hence the "1WAY" part of the procedure name. The NPAR1WAY procedure is used for independent samples only. Introductory statistics courses generally include the Wilcoxon or Rank Sum type tests.

The syntax of the NPAR1WAY procedure is:

```
PROC NPAR1WAY DATA=SAS-data-set <options>;
CLASS variable;
VAR variables;
RUN;
```

PROC NPAR1WAY, CLASS, and VAR are required statements. BY and WHERE statements are, again, optional statements that can also be used.

The options for the PROC NPAR1WAY statement are:

ANOVA performs an analysis of variance on the raw data.

WILCOXON performs an analysis of the ranks of the data. For two levels, this is the same as the Wilcoxon rank-sum test. For any number of levels, this is a Kruskal-Wallis test. For the two sample cases, the procedure uses a continuity correction.

PLOTS = *list* Each of the test methods has an associated plot. Select the plot type that supports the test procedure selected. For the ANOVA, the PLOTS option is PLOTS = ANOVABOXPLOT. For a Wilcoxon test the PLOTS option should read PLOTS=WILCOXONBOXPLOT. If all tests are performed, one would select PLOTS=ALL, and plots can be suppressed by PLOTS=NONE.

The following non-parametric methods (and their associated ODS Graphs) are also available in this procedure but not covered in this text.

EDF (EDFPLOT) – empirical distribution function

MEDIAN (MEDIANPLOT)

SAVAGE (SAVAGEBOXPLOT)

VW (VWBOXPLOT) – Van der Waerden analysis

If no option is specified in the PROC NPAR1WAY statement, *all* of the available test methods and supporting ODS Graphs are produced.

CLASS Statement

The CLASS statement names only one classification variable. Each value or level of this variable identifies groups in the data, and these values can be character or numeric. Class variables must have two or more levels.

VAR Statement

The VAR statement identifies the response variable(s) to be analyzed. If the VAR statement is omitted, the NPAR1WAY procedure analyzes all numeric variables in the data set (except for the CLASS variable if it is numeric).

OBJECTIVE 7.4: Using the SAS data set *meat* created in Objective 7.1, compare the distribution of pH for only the Control and T1 using non-parametric rank sum methods. Use $\alpha = 0.05$ in the conclusions.

```
PROC NPAR1WAY DATA=meat WILCOXON ;
WHERE Group ne "T2" ; ** T2 group is excluded from the analysis;
CLASS Group ;
VAR pH;
TITLE 'Objective 7.4';
RUN;
```

If no options are specified on the PROC NPAR1WAY statement, several pages of output will be produced as the procedure will run several tests. Here, only the non-parametric Wilcoxon rank sum tests are overviewed.

Objective 7.4

The NPAR1WAY Procedure

Wilcoxon Scores (Rank Sums) for Variable pH Classified by Variable Group					
Group	**N**	**Sum of Scores**	**Expected Under H0**	**Std Dev Under H0**	**Mean Score**
C	5	30.50	27.50	4.772607	6.10
T1	5	24.50	27.50	4.772607	4.90

Average scores were used for ties.

The **Wilcoxon Scores (Rank Sums)** are $R_C = 30.50$ and $R_{T1} = 24.50$. Below the table of Wilcoxon scores, it is noted that average scores or ranks were used for tied values of pH. When none of the response variables are tied, there is no such footer added to this table. The tables of NPAR1WAY output should be examined carefully. Only in the title of the first table does the response variable pH appear.

Wilcoxon Two-Sample Test	
Statistic	30.5000
Normal Approximation	
Z	0.5238
One-Sided Pr > Z	0.3002
Two-Sided Pr > |Z|	0.6004
t Approximation	
One-Sided Pr > Z	0.3065
Two-Sided Pr > |Z|	0.6131

Z includes a continuity correction of 0.5.

The Wilcoxon Two Sample Test is applicable when t=2, and the z-score method is the second method overviewed in Table 7.3. SAS indicates that it employs a continuity correction; that is, $R_1 - 0.5$ is used in the computation of z in Table 7.3 rather than R_1.

Kruskal-Wallis Test	
Chi-Square	0.3951
DF	1
Pr > Chi-Square	0.5296

When t=2, one obtains both the Wilcoxon Rank Sum results and the Kruskal-Wallis results. If t > 2, only the Kruskal-Wallis results are produced. (See Objective 7.5).

Conclusion: There is no difference in pH distribution between the Control and T1 group ($\alpha = 0.05$, $z = 0.5238$, $p = 0.6004$, or $\chi_1^2 = 0.3951$, $p = 0.5296$). Only the two-sided test is interpreted since that is all that was covered in the brief methods review in Table 7.3.

In Figure 7.6, the **Distribution of Wilcoxon Scores** for pH is the WILCOXONBOXPLOT (ODS Graph name) produced by default when the WILCOXON option is selected. The values on the vertical axis are ranks (ranging from 1 to $n_1 + n_2$). The means and medians illustrated in the boxplots are the means and medians of the ranks. To suppress this plot, one could use the option PLOTS=NONE on the PROC NPAR1WAY statement.

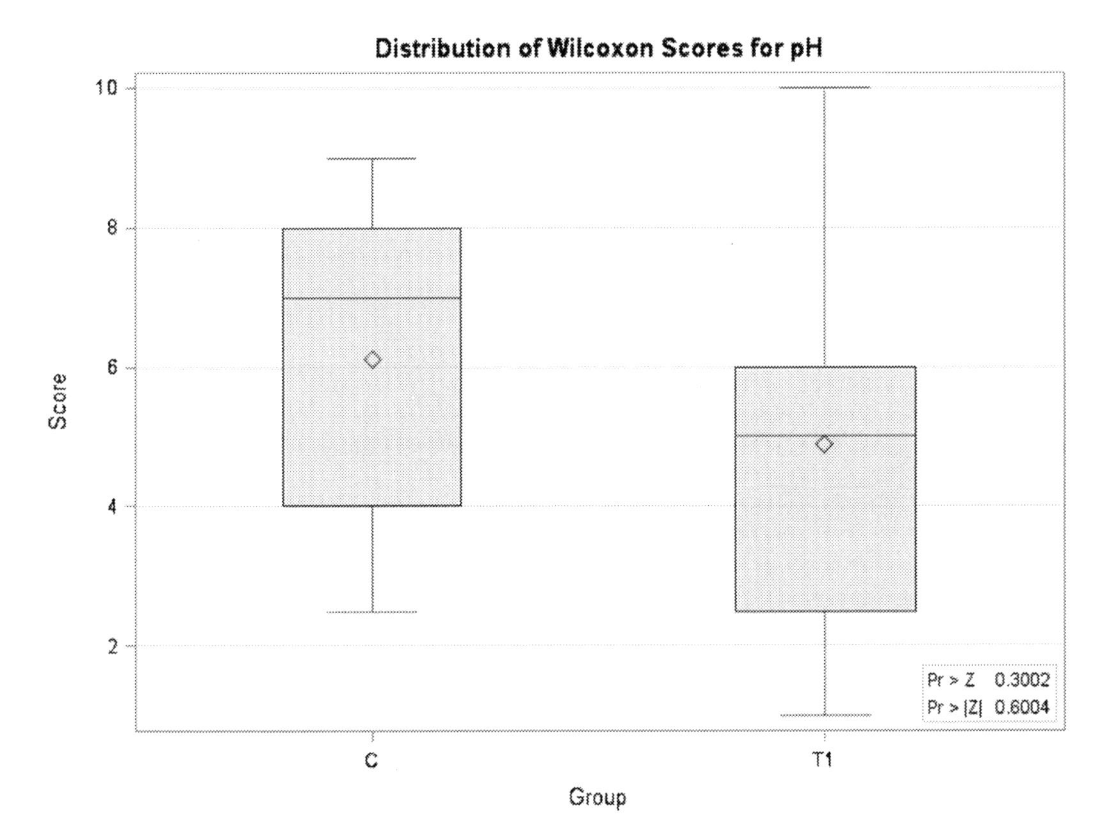

FIGURE 7.6
The default WILCOXONBOXPLOT for the t = 2 non-parametric test.

OBJECTIVE 7.5: Use the NPAR1WAY procedure to conduct an analysis of variance and Kruskal-Wallis test for all three treatment groups for the CookYield response variable only. Suppress all plots. Interpret the tests at $\alpha = 0.05$.

```
PROC NPAR1WAY DATA=meat WILCOXON ANOVA PLOTS=NONE;
CLASS Group ;
VAR CookYield ;
TITLE 'Objective 7.5';
RUN;
```

Objective 7.5

The NPAR1WAY Procedure

Analysis of Variance for Variable CookYield Classified by Variable Group		
Group	**N**	**Mean**
C	5	21.260
T1	5	23.300
T2	5	23.340

Source	DF	Sum of Squares	Mean Square	F Value	Pr > F
Among	2	14.149333	7.074667	4.9577	0.0270
Within	12	17.124000	1.427000		

The CLASS statement in the NPAR1WAY procedure identifies the classification variable Group and its levels, and the sample size and sample mean for each level in the **Analysis of Variance** (first) table. The response variable analyzed is also identified in this same table title. From the ANOVA it can be concluded that there is a difference in mean cook yield among the three treatment groups ($\alpha = 0.05$, $F_{2,12} = 4.9577$, $p = 0.0270$) which is the same as the result in Objective 7.1.

There is no option in NPAR1WAY to check the residuals as there is in the GLM procedure. Also, if one is interested in post-hoc multiple comparisons of the means, the GLM procedure is recommended if the normality condition is satisfied.

Objective 7.5

The NPAR1WAY Procedure

Wilcoxon Scores (Rank Sums) for Variable CookYield Classified by Variable Group

Group	N	Sum of Scores	Expected Under H0	Std Dev Under H0	Mean Score
C	5	19.0	40.0	8.164966	3.80
T1	5	51.0	40.0	8.164966	10.20
T2	5	50.0	40.0	8.164966	10.00

Kruskal-Wallis Test

Chi-Square	6.6200
DF	2
Pr > Chi-Square	0.0365

The response variable is in the title of the **Wilcoxon Scores** table. The Kruskal-Wallis test agrees with the ANOVA; that is, there is a difference in cook yield among the three treatment groups ($\alpha = 0.05$, $\chi_2^2 = 6.6200$, $p = 0.0365$). Since $t = 3$, observe that no table of Wilcoxon z-tests is produced as it was in Objective 7.4.

When samples are dependent or paired, the UNIVARIATE procedure should be used. The UNIVARIATE procedure performs a sign test for paired samples, and the Wilcoxon Signed Rank test for dependent samples. See Conover (1999) for details on this method.

7.5 Ranking Items in a Data Set Using the RANK Procedure

Because the non-parametric methods can be based upon the ranks of data, it would be good to look at a procedure that ranks observations in a data set next. While the SORT procedure can arrange the data in order, it does not assign a place or rank to the sorted observations. The RANK procedure can be used to assign ranks to items in a data set in either ascending or descending order. The data do **NOT** have to be sorted before running the RANK procedure. The RANK procedure by itself does not generate printed output. A SAS data set is created when the RANK procedure is run. The RANK procedure does not change the order of the observations in the data set. ODS Graphics are not available for the RANK procedure.

The syntax of the RANK procedure is as follows:

```
PROC RANK DATA=SAS-data-set <options>;
VAR variable(s);
RANKS new-variable(s);
RUN;
```

Options on the PROC RANK statement include the following:

OUT=*newsetname* creates a SAS data set containing the original data set items and the newly created ranks

DESCENDING reverses ranking from largest to smallest;
SAS will rank observations in ascending order by default, and no option is needed to select ascending as the sorting order.

TIES = MEAN | HIGH | LOW This option controls how tied observations are to be ranked. Pick only one: MEAN, HIGH, or LOW for this option. For non-parametric tests, average ranks (TIES=MEAN) are usually used in the analysis.

VAR statement
The VAR statement specifies one or more numeric variables in the SAS data set to be ranked. All specified variables will be ranked in ascending order. If the DESCENDING option is used in the PROC RANK statement, then all specified variables will be ranked in descending order.

RANKS statement
For each variable in the VAR statement, a new variable must be created. The values of these variables are the ranks assigned. There must be the same number of variables in the RANKS statement as there is in the VAR statement. The order of the variables listed in the RANK statement is important. The first variable identifies the ranks of the first variable listed in the VAR statement, and so on.

There are other options available, but this is a brief introduction to this procedure. See SAS Help and Documentation for more information.

For the ranking objectives, a small data set in Table 7.4 will be used. There are no identical observations for X, but for Y and Z tied observations are included.

OBJECTIVE 7.6: Create the SAS data set *six* for the data in Table 7.4. Rank each of the response variables X, Y, and Z in ascending order. Print the original data and the associated ranks.

TABLE 7.4

Data for the RANK Procedure Demonstrations

X	Y	Z
89	25	41
47	33	37
73	27	37
66	25	29
50	42	37

```
DM 'LOG; CLEAR; ODSRESULTS; CLEAR;';
DATA six;
INPUT X Y Z ;
DATALINES;
89      25      41
47      33      37
73      27      37
66      25      29
50      42      37
;
```

```
PROC RANK DATA=six OUT=new6;
VAR x y z;
RANKS RX RY RZ;
PROC PRINT DATA=new6;
TITLE 'Objective 7.6';
RUN;
QUIT;
```

Programming Note: RX, RY, and RZ were used as variable names in the RANKS statement simply so they could be recognized as the rank associated with a given variable. X, Y, and Z (the variables ranked) do not have to be a part of the rank variable names, although (in the author's opinion) it makes it easier to recognize which ranks belong to which variables.

Objective 7.6

Obs	X	Y	Z	RX	RY	RZ
1	89	25	41	5	1.5	5
2	47	33	37	1	4.0	3
3	73	27	37	4	3.0	3
4	66	25	29	3	1.5	1
5	50	42	37	2	5.0	3

By default, the average rank is used when there are ties. For X there are no ties, and each value of X has a unique rank. For Y there are two observations of 25, and 25 is the lowest value observed for Y. Since there are two values of 25, they are in the first and second ordered positions. The average of 1 and 2 is 1.5, RY is observed to be 1.5 for these two scores. The next score would be the third score, and if there is no tie there, RY will be 3.

For Z 37 is the value of the second, third, and fourth ordered scores. Therefore, 37 should receive the average of ranks 2, 3, and 4 which is 3. The next ordered score would be in position 5 and receives rank 5. When there are no ties, the ranks will take values 1, 2, 3, … ., n where n is the sample size. When there are ties at the lowest ordered position and/or the largest ordered position, ranks of 1 and/or n may not be observed.

One may find it helpful to reorder the columns of variables printed so that the ranks are next to the original variable. That is,

```
PROC  PRINT DATA=new6;
VAR X RX Y RX Z RZ;
TITLE 'Objective 7.6';
RUN;
```

OBJECTIVE 7.7: Run the above program with each of the following options on the PROC RANK statement: TIES=HIGH, TIES=HIGH DESCENDING, TIES=LOW and TIES=LOW DESCENDING options and note the effects of each of the options. This is left as an exercise for the reader.

OBJECTIVE 7.8: Rank the X observations in ascending order and the Y observations in descending order in a new SAS data set named *eight*. Z will not be ranked. Print the data set *eight* in ascending order according to X along with the associated ranks. Use only one PRINT procedure.

There are two options presented for Objective 7.8. Either is OK.

Option 1: Track the SAS data sets by name through this program. Note that the SAS data sets must be sorted before the data MERGE (Section 4.2 MERGE Statement).

```
DATA six;
INPUT X Y Z;
DATALINES;
89      25      41
47      33      37
73      27      37
66      25      29
50      42      37
;
PROC RANK DATA=six OUT=sixA;
VAR X;
RANKS RX;

PROC RANK DATA=six OUT=sixB DESCENDING;
VAR Y;
RANKS RY;

PROC SORT DATA=sixA; BY X Y;
PROC SORT DATA=sixB; BY X Y;

/*
Sorting the data sets by X (or RX) before merging is all that is
necessary since X has all unique values. If there were duplicate values
for X one would want to include more than one variable in the BY
statements of the SORT procedures.
See Section 20.2 for information on annotating programs with block
comments such as this.
*/

DATA eight;
MERGE sixA sixB; BY X Y;

PROC PRINT DATA=eight;
TITLE 'Objective 7.8 - Option 1';
RUN;
QUIT;
```

Option 2: In this option, notice that X, Y, Z, and the ranks for X are in new6. NEW6 is then read into a second RANK procedure, and the ranks of Y are added to the variables in NEW6 to form the SAS data set eight. No additional DATA step for a MERGE is necessary.

```
DATA six;
INPUT  X  Y  Z ;
DATALINES;
89      25      41
47      33      37
73      27      37
66      25      29
50      42      37
;
```

```
PROC RANK DATA=six OUT=new6;
VAR X ;
RANKS RX ;

PROC RANK DATA=new6 OUT=eight DESCENDING;
VAR Y;
RANKS RY;

PROC SORT DATA=eight ; BY X;              *or sort by RX;
PROC PRINT DATA=eight;
VAR X RX Y RY Z;    *optional statement to specify the order of columns;
TITLE 'Objective 7.8 - Option 2';
RUN;
QUIT;
```

The reader is encouraged to investigate each of these options.

7.6 Chapter Summary

In this chapter, SAS procedures for comparing the means of two or more populations were overviewed. ANOVA methods are performed by the GLM procedure where it is assumed that the populations are normally distributed. There is an ANOVA procedure, but one rarely sees it used since the GLM procedure has more capabilities. Hence, the ANOVA procedure was not covered here. The GLM procedure has a great many more types of statistical analyses that it can perform, but they are beyond the scope of this book. When one cannot assume normality, non-parametric methods are applicable. There are many non-parametric tests available in the NPAR1WAY procedure, but the (Wilcoxon) rank sum methods were presented here. The GLM procedure can accommodate more than one class variable in an analysis, but the NPAR1WAY procedure (as its name suggests) can accommodate only one class variable. Since the non-parametric methods presented here utilized the ranks of the observations, the RANK procedure was included in this chapter to demonstrate how to rank observations in either ascending or descending order.

8

Data Step Information 3 – Reading Data Files and Labeling Variables

Data measurements do not have to be included as part of the SAS program. Data may be saved in external files and read into a SAS program. There are several ways to read in a data file. Here, one of the older, but still very reliable, methods of reading in data, is the INFILE statement in the DATA step.

8.1 The INFILE Statement

When the data is stored in a *text* file, the file name commonly has the file extension *.txt, *.dat, or *prn. Other extensions may also be in use. These types of files may be read into a SAS session using an INFILE statement. When an INFILE statement is used in a SAS program, it must follow the DATA statement and come before the INPUT statement. When an INFILE statement is used, no DATALINES (or CARDS) statement is needed. Data transformations or variable assignment statements can still be a part of this DATA step. As before, these statements always follow the INPUT statement, where the INPUT statement names the variables in the external file and identifies which, if any, of the variables are character rather than numeric.

The syntax of the FILENAME and INFILE statements:

FILENAME *fileref 'location and name of file';*
DATA *SAS-data-set;*
INFILE *fileref <options>;*
INPUT *variable1 ... variablen;*
data transformations and/or variable assignments - if any

The FILENAME statement is a stand-alone statement. It should appear before the DATA step is submitted, but it does not have to appear in the line immediately before the DATA step. The purpose of the FILENAME statement is to give the location (directory information – drive:\folder) and name of the data file and assign it a temporary or convenience name, *fileref*, to be used in the SAS session. The drive:\folder\filename in the FILENAME statement is enclosed in quotation marks (either single or double), and the *fileref* must follow SAS variable naming guidelines (Section 2.2) and additionally is limited to eight characters.

The FILENAME statement is not necessary if the complete file path and filename are specified in the INFILE statement. If the complete file path is specified in the INFILE statement, it must be enclosed in quotation marks (either single or double).

When using an INFILE statement, one must know the variable names and the order in which they occur in the external data file. These eternal text files can be data only, or they may contain column headers or other information at the top or bottom of the file. Only one INFILE statement is permitted within a DATA step.

A few of the options that can be used in the INFILE statement are:

DELIMITER = *delimiters* or DLM = *delimiters*

Delimiters are the symbols or characters used to separate pieces of information in an external data file. Up to this point, when entering a row of data, variable values have typically been separated by at least one space. When no spaces separate the data columns, column pointer controls in the INPUT statement should be used. See Section 2.2. A delimiter in an external data file can be a comma (,), a tab, a space, a letter, or a combination of letters or symbols. When the delimiter is a space, it is not necessary to use a DELIMITER option.

For example, consider a comma delimited file (typically with a .csv extension) called temp.csv stored in the drive or folder identified in this illustration as a:\

75,6,120
219,14,101

To read in a file such as this:

```
DATA one;
INFILE 'a:\temp.csv' DLM = ',' ;
INPUT x y z;
RUN;
```

Note that in this example a FILENAME statement is not used, and the complete file path and file name are specified in quotation marks in the INFILE statement.

When a file is tab-delimited, then DLM='09'X is the appropriate option. In other file types DLM='8888', DLM='##', and DLM='%20' are other possible examples of delimiters separating pieces of information within rows of an external data file.

DSD

The DSD option requests that two adjacent delimiters indicate a missing observation. For example, if first row in the external data file in the above example is 75,,120 and the delimiter is a comma, a missing observation (.) will be recorded for Y in line 1.

FIRSTOBS = *record-number*

This indicates that reading the external file begins at the record or line number specified, rather than beginning with the first record or line. This allows one to store information other than the data "at the top" of an external data file. For example,

```
INFILE datafile FIRSTOBS = 10;
```

specifies that reading the data begins at line 10 of the external file. In the first nine lines one could record the variable names as column headers or other information about the experiment that yielded this data. It is the recommendation of the author that a minimum of one line of column headers should be included in the external files, if possible. If not, a second external file that is an explanation of the contents of the external data file should be created.

LRECL = *record-length*

Specifies the record length in bytes. The default is 256. Values between 1 and 1,048,576 (1 megabyte) can be specified. This is useful for external data files that are "wide", that is, the files contain a large number of columns.

MISSOVER

This prevents a SAS program from going to a new line if it does not find values in the current line for all the INPUT statement variables. When an INPUT statement reaches the end of the current record, instead of "wrapping around" to a new line, the values that are expected but not found in the current line are set to missing.

OBS = *record-number*

The record or line number of the last record that is to be read from an input file can be specified. For example,

```
INFILE datafile OBS = 30;
```

indicates that the last record from the external file to be read is the 30th line item in the file. This option is useful if there is extraneous information "below" the record number specified, or if one does not wish to read all lines of the external file into the SAS data set.

STOPOVER

The DATA step stops processing when an INPUT statement reaches the end of the current record without finding values for all variables in the INPUT statement.

OBJECTIVE 8.1: Create a data file that is a space-delimited text file. Read the data file into a SAS session using an INFILE statement.

This objective will be examined using three different external file types so that options on the INFILE statement can be examined. The data file is created in the SAS Editor in this objective, but this is not required. Likely, other software and/or instruments, such as, sensors or data loggers, create text files containing data. One will have to become familiar with the format of the data files created by those applications.

In the SAS Editor, enter the following data exactly as it appears below. Insert at least one space between the entries in a single row. In the illustration below more than one space was used, but this is to emphasize that at least one space is necessary. Do not use the TAB key. TAB has a different effect than space. TAB is covered in the next objective.

```
2017     7     35
2018     8     54
2019    11    101
```

Save this file as *tornado1.dat* by selecting **File – Save as – Save as type – Data files and enter file name in the blank**. Note the drive:\folder in which the file is saved. Modify the above text so that it appears exactly as it appears below. Again, no TABS, use the space bar only.

```
YEAR    NUM     DAMAGE
2017      7      35
2018      8      54
2019     11     101
```

Save this file as *tornado2.dat* in the same drive:\folder. Modify the text one more time so that it appears exactly as it appears below. Where a space in the records occurs, it is exactly one space.

```
2017 7 35
2018 8 54
201911101
```

Save this file as *tornado3.dat*. Clear the text in the Editor window.

Enter the following SAS program in the Editor. Specify the drive:\folder in which the three files have been stored rather than **a:** in the program below. Note that SAS data set *three* requires one to use column pointer controls in the INPUT statement (Section 2.2) in conjunction with the INFILE statement.

```
FILENAME  t1  'a:\tornado1.dat' ;
FILENAME  t2  'a:\tornado2.dat' ;
FILENAME  t3  'a:\tornado3.dat' ;

DATA one;
INFILE  t1;
INPUT  YEAR  NUMBER  DAMAGE;
PROC PRINT DATA=ONE;
TITLE 'Objective 8.1';

DATA two;
INFILE t2  FIRSTOBS=2 ;
INPUT Year Number Damage;
PROC PRINT DATA=two;

DATA three;
INFILE  t3;
INPUT Year 1-4  Number 5-6  Damage 7-9;
PROC PRINT DATA=three;
RUN;
QUIT;
```

The printed output of these three files should be identical with the exception of the column headings (variable names). In the Results Viewer, one will notice that the uppercase and lowercase letters appear exactly as entered in the INPUT statement. When reading in any data file using the INFILE statement, the Log should always be checked for the appropriate number of observations in the SAS data set. For the SAS data set *two* the following information is observed in the log window.

```
6     DATA two;
7     INFILE t2  FIRSTOBS=2 ;
8     INPUT Year Number Damage;

NOTE: The infile T2 is:
      Filename=a:\tornado2.dat,
      RECFM=V,LRECL=32767,File Size (bytes)=79,

NOTE: 3 records were read from the infile T2.
      The minimum record length was 17.
      The maximum record length was 17.
NOTE: The data set WORK.TWO has 3 observations and 3 variables.
NOTE: DATA statement used (Total process time):
      real time           0.52 seconds
      cpu time            0.01 seconds
```

In the second NOTE, the number of records (or lines) from the external file that were read are indicated, and in the third NOTE the number of observations (rows) and the number of variables (columns) in SAS data set *two* are indicated.

As stated above, the FILENAME statement is not necessary if the full location and filename are specified in the INFILE statement. To read in the data file *tornado1.dat,* the following commands can also be used.

```
DATA one;
INFILE 'a:\tornado1.dat';
INPUT Year Number Damage;
RUN;
```

Text files of data do not have to be created in the SAS Editor. They can be created using other software. They must be saved as a text file though. Depending upon the software used to create these files, the file extensions could be quite different. A couple of common extensions are .txt and .prn. Many spreadsheets have the option of saving the file as a tab delimited file. One can import these into SAS also. The advantage of tab delimited files is evident when that file is opened and viewed. The columns, and hence the variables, are aligned and more easily viewed. Using the Windows platform, the tab delimiter is identified by its hexadecimal code, '09'X. For files that are tab delimited, here is the basic SAS code to infile that data:

```
FILENAME adata 'a:\temp.dat';
DATA one;
INFILE adata DLM = '09'X  DSD;
INPUT x y z;
RUN;
```

NOTE: DSD indicates that two adjacent tabs will imply that a value is missing.

OBJECTIVE 8.2: In the Editor, open tornado3.dat. Insert tabs between variable values – two tabs on each line. Save this tab delimited file as tornado4.dat. Clear the Editor window. Write the SAS code to read and print the contents of this data table. This time, do not use a FILENAME statement.

```
DATA four;
INFILE 'a:\tornado4.dat' DLM='09'X  DSD;
INPUT Year Number Damage;

PROC PRINT DATA=four ;
TITLE 'Objective 8.2';
RUN;
```

The printed output from these two objectives is not included since it is repetitive. It is left to the reader to investigate each of these simple cases.

Text files containing data are a convenience when data files are particularly large, and the inclusion of all the lines of data in the SAS program would make the DATA step quite lengthy. Data files are also convenient when they are to be distributed to several people on a portable storage device, on Internet or cloud sites, or by e-mail.

There are many more options, and thus, more capability of the INFILE statement, but this is a good starting point for beginning SAS programmers. Other SAS procedures exist that read external data files. The IMPORT procedure is a popular choice which will be covered in this book in Chapter 14.

8.2 The LABEL Statement

In an INPUT statement, variable names are specified. Variable names can be chosen to be lengthy as long as there are no spaces in the variable name, but SAS programmers often prefer shorter variable names that are less cumbersome. Rather than using a lengthy variable name to identify or define the variable, a LABEL statement can be used. A LABEL statement is used in a DATA step after the INPUT statement. Thus, the LABEL statement can be used in conjunction with an INFILE statement or when a DATALINES (or CARDS) statement is used. It can also be used in a DATA step after the SET or MERGE commands.

```
DATA one;
INPUT x $ y z;
LABEL x = 'variable x definition' y = 'variable y definition'
      z = 'variable z definition';
DATALINES;
```

In a LABEL statement, the variable definition can be up to 40 characters long including blank spaces. The text of the label is enclosed in single or double quotation marks. Several variables can be specified in a single LABEL statement. Use a semicolon only after the last variable label definition is given. The order of the variables and labels in the LABEL statement does not matter. That is,

```
LABEL       y = 'variable y definition'
            x = 'variable x definition'
            z = 'variable z definition';
```

is equivalent to the LABEL statement in the above DATA step. One may choose to label all or only some of the variables in a SAS data set. One still uses the variable name in the programming code, but the label enhances the appearance of the output.

The LABEL statement only applies to the variable name. It does not define the values the variable can take. For that the FORMAT procedure is needed, and that procedure is covered in Chapter 18.

Labels allow the user to apply a simple naming convention for the variable names in the program and have variables well-defined in the output. If the LABEL statement is included in the DATA step, SAS will use the defined labels in the output for most procedures. One procedure that does not automatically use the labels is the PRINT procedure. To request the PRINT procedure to use the labels, the LABEL option on the PROC PRINT statement must be included. (See Section 2.3 for the PRINT Procedure.)

OBJECTIVE 8.3: Continuing with SAS data set *four* containing the tornado data. Label the Number and Damage variables as the number of tornadoes and the damage assessment measured in ten thousand dollar units. Examine the PRINT procedure without and with labels, and also examine the usage of labels in the output of the MEANS procedure.

```
DATA four;
INFILE 'a:\tornado4.dat' DLM='09'X DSD;
INPUT Year Number Damage;
LABEL Number="Number of Tornadoes"
            Damage="Damage Assessment, x$10,000";
```

```
PROC PRINT DATA=four;
TITLE 'Objective 8.3';
TITLE2 'PRINT with no LABEL option';

PROC PRINT DATA=four LABEL;
TITLE2 'PRINT with LABEL option';

PROC MEANS DATA=four;
VAR Number Damage;
TITLE2;
RUN;
QUIT;
```

Objective 8.3

PRINT with no LABEL option

Obs	Year	Number	Damage
1	2017	7	35
2	2018	8	54
3	2019	11	101

Objective 8.3

PRINT with LABEL option

Obs	Year	Number of Tornadoes	Damage Assessment, x$10,000
1	2017	7	35
2	2018	8	54
3	2019	11	101

Objective 8.3

The MEANS Procedure

Variable	Label	N	Mean	Std Dev	Minimum	Maximum
Number	Number of Tornadoes	3	8.6666667	2.0816660	7.0000000	11.0000000
Damage	Damage Assessment, x$10,000	3	63.3333333	33.9754814	35.0000000	101.0000000

The first PRINT procedure produces output exactly as seen when the procedure was introduced in Chapter 2, and the column headings will appear exactly as the variable names were initialized in the DATA step. The second PRINT procedure has the detail of the labels in the column headings. NOOBS or other options can still be used when the LABEL option is included. The output for the MEANS procedure includes both the variable names and the labels. It was not necessary to specify that the labels be used in the results of the MEANS procedure.

8.3 ViewTable and Table Editor

Up to this point, nearly all of this book's objectives where SAS data sets were created, the contents of those data sets were examined using the PRINT procedure and observing the data in the Results Viewer. One can certainly proofread data using the PRINT procedure.

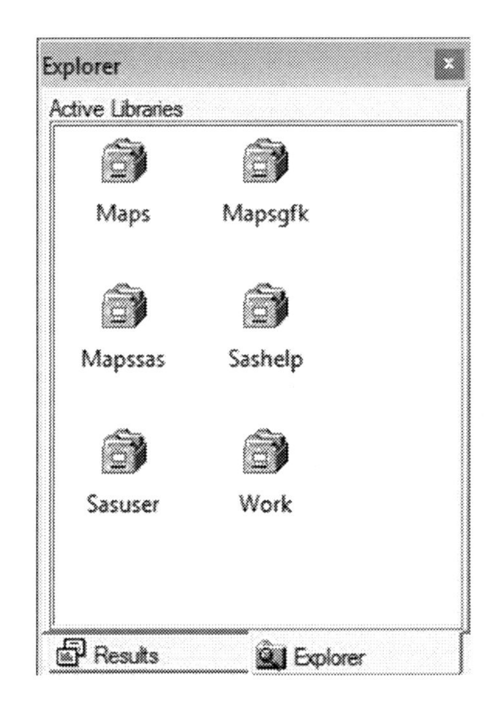

FIGURE 8.1
Active libraries in the SAS Explorer window.

For large data sets, this can make the output unnecessarily lengthy. PRINT procedure commands can be removed or disabled using commenting syntax (Section 20.2) once corrections are made to the SAS data set.

There is also an "on screen" point-and-click approach to examining the contents of a SAS data set. To do this the Explorer tab in the lower left of the SAS screen is selected. From the Explorer window, **Work** is selected from the Active Libraries. See Figure 8.1. The four data sets created in Objectives 8.1 and 8.3 are contained in the **Work** library. These SAS data sets are listed by name in alphabetical order. See the left side of Figure 8.2. All of the

Explorer
Contents of 'Work'
Four One
Three Two

VIEWTABLE: Work.Two

	Year	Number	Damage
1	2017	7	35
2	2018	8	54
3	2019	11	101

FIGURE 8.2
Using ViewTable to review the data in a SAS data set.

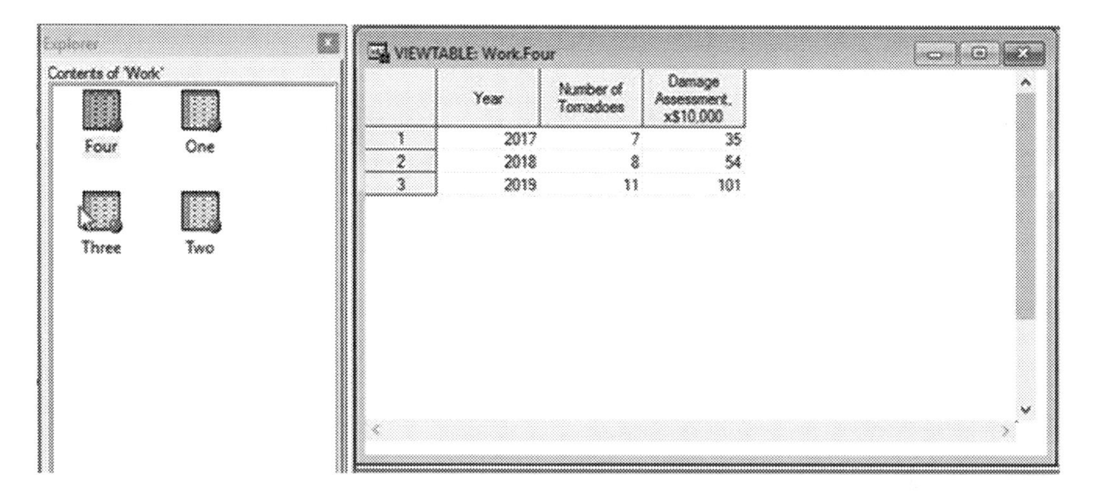

FIGURE 8.3
ViewTable of SAS data set four with labels.

SAS data sets created in the same SAS session are kept in temporary storage in the **Work** library. In Chapter 12, creating libraries other than **Work** and permanent SAS data sets are overviewed.

One can open the SAS data set **Two** by clicking on it in the Explorer window, and a new window with the heading ViewTable will open as shown in Figure 8.2. (Any of the SAS data sets could have been selected.) The name of the SAS data set also appears in this heading. The SAS data set is identified by its two-level SAS name "**Work.Two**". **Two** is the name assigned in the previous objectives, and **Work** identifies the library in which the SAS data set **two** is located. One can review the contents of a SAS data set in the ViewTable window rather than writing and submitting PRINT procedure code.

In ViewTable the column headers are the variable names unless the variables have been defined using a LABEL statement. In Figure 8.3 the SAS data set *Work.Four* is in ViewTable, and the column headers are the labels given in Objective 8.3.

To view the variable names rather than the column labels, while the ViewTable window is active, select **View – Column Names** from the pull down menu, shown in Figure 8.4. If column labels have been created, those column labels are the default setting for the ViewTable display.

Once one confirms the contents of the SAS data set examined in ViewTable, it is recommended that the ViewTable window be closed for the following reasons: 1. The SAS display can become very congested with many open windows. 2. If one wishes to modify a SAS data set in the Enhanced Editor and that SAS data set is open in ViewTable, an error will occur. Suppose *Work.Four* is open in ViewTable (as in Figure 8.4), and additional modifications to *Work.Four* are submitted from the Enhanced Editor. Those modifications could be sorting, adding variables, or deleting variables. In the Log window, one would observe:

```
ERROR: You cannot open WORK.FOUR.DATA for output access with member-level
control because WORK.FOUR.DATA is in use by you in resource environment
ViewTable Window.
```

This error easily corrected by closing ViewTable for *Work.Four* and resubmitting the revised program from the Editor. ViewTable is closed by clicking on the X in the upper

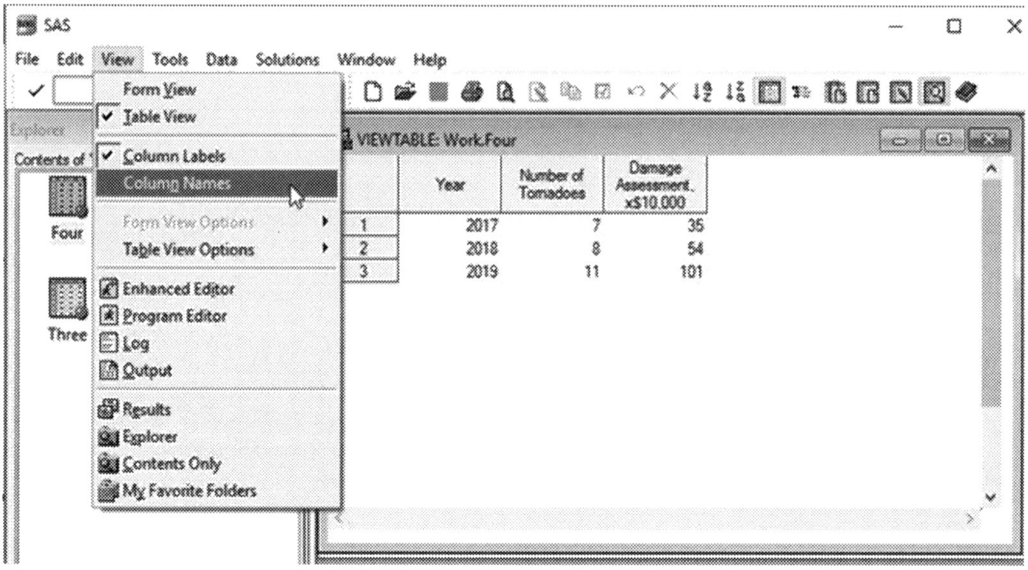

FIGURE 8.4
Changing the ViewTable display from column labels to column names.

right corner of ViewTable just below the X in the upper right corner that will close SAS, or right click the ViewTable window bar at the bottom of the SAS screen and select **Close** from the pop-up menu.

The Table Editor associated with ViewTable can be used to edit existing SAS data sets or create new ones. To create a new SAS data set using the Table Editor select **Tools – Table Editor** from the pull-down menus. A blank spreadsheet will open. The default column headers (A, B, C, …) are the default variable names. One can change these by right clicking on the column header to be changed and selecting **Column Attributes**. In the window that opens, one can enter a new variable name and even assign a label to the variable. See Figure 8.5. One must remember the guidelines for variable names (Section 2.2) and the 40-character limit for the label. There is also a box where the **Type** of variable can be selected, **Character** or **Numeric**. These changes take effect when **Close** or **Apply** is selected. Data values, character or numeric, can be entered into the cells of the table before or after the column attributes are assigned. When complete, the data can be saved as a SAS data set in the Work library by selecting **File – Save As – Work** and then enter a SAS data set name in the blank for **Member Name.** See Figure 8.6. As stated earlier in this section, Chapter 12 presents information on creating other SAS libraries in which SAS data sets can be saved.

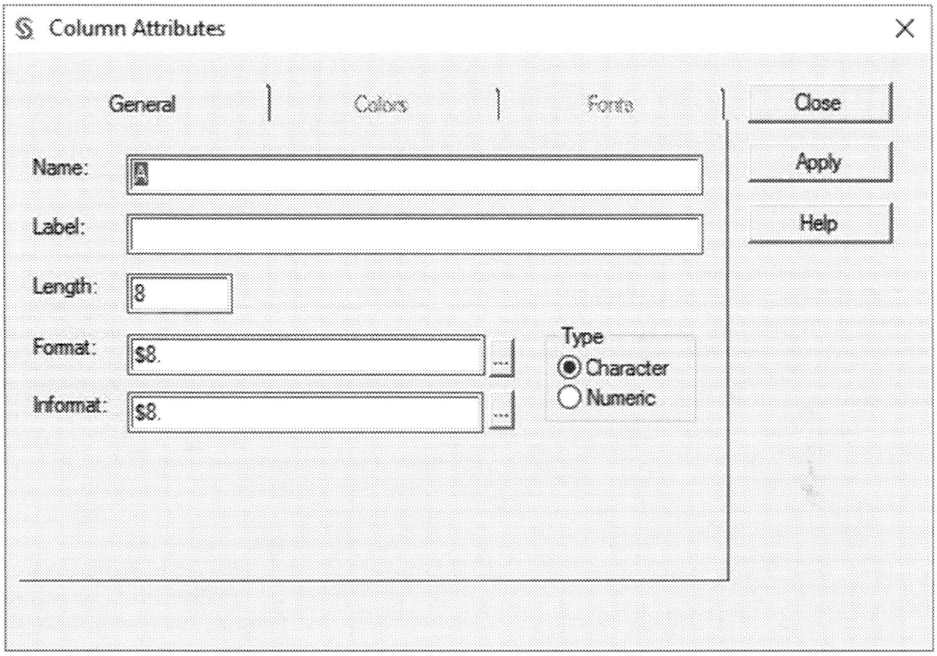

FIGURE 8.5
Column attributes options in the Table Editor.

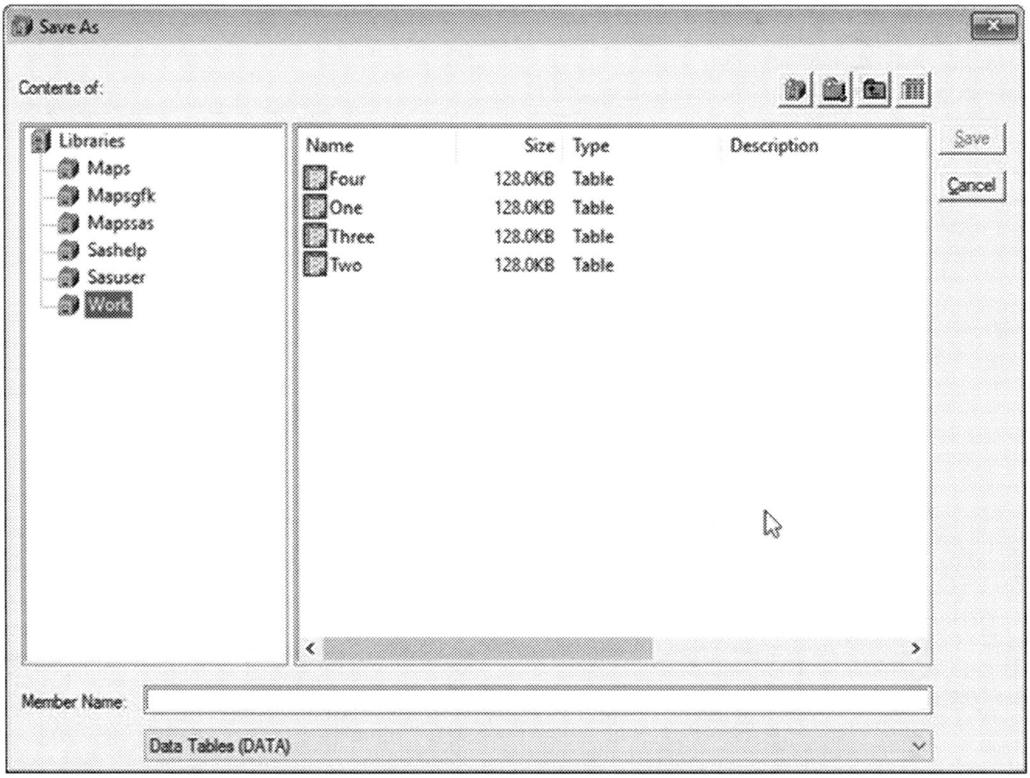

FIGURE 8.6
Saving a SAS data set to the work library from the Table Editor.

8.4 DROP, KEEP, and RENAME Statements

This is the third chapter on the DATA Step, and the options for creating and modifying SAS data sets is growing. In Chapter 4, creating new variables or modifying existing variables was demonstrated. Additionally, SET and MERGE statements were introduced as methods of combining SAS data sets. Variable management within these SAS data sets continues to evolve. Once a SAS data set is created, it may be desirable to remove a variable (and thus a column of data) or to change the name of the variable. Changing the name of the variable may be necessary if two or more SAS data sets are to be combined using SET or MERGE commands, and the SAS data sets have variable names that are not compatible with the DATA step command.

When managing variables, there are three useful DATA step commands: DROP, KEEP, and RENAME.

Example 1:

```
DATA one;
<list of DATA step commands>
DROP x y z;
```

The three variables (x, y, z) will be dropped from SAS data set *work.one*. All other variables will be retained.

Example 2:

```
DATA two;
<list of DATA step commands>
KEEP r s t;
```

Only three variables (r, s, t) will be kept in SAS data set *work.two*; all other variables will be deleted.

Example 3:

```
DATA three;
<list of DATA step commands>
RENAME oldname1 = newname1 oldname2=newname2;
```

Variables *oldname1* and *oldname2* will appear in *work.three* now as *newname1* and *newname2*.

The DROP and KEEP statements are straightforward and do not need any further explanation. An application of the RENAME statement appears in the following objective. Like the new variable assignment statements, LABEL, and IF-THEN statements, these three statements can be used in a DATA step after INPUT or SET or MERGE statements.

OBJECTIVE 8.4: Consider a plant greenhouse study that yielded two SAS data sets: *Work.Growth* and *Work.Germination*. The species, replication, temperature, and water information are the same in both SAS data sets. However, when the two SAS data sets were created, the variable names were not the same in both SAS data sets nor in the same order. Perhaps this was due to different people recording the data. Merge these two SAS data sets into a single SAS data set *Work.GHouse*.

Work.Growth

spec	rep	Temp	Water	Biomass	Leaf_N
A	1	30	N	34.2	18.2
A	2	30	S	38.9	22.1
B	1	20	N	25.6	16.0
C	1	30	N	28.4	19.2
B	2	30	S	22.4	14.6
⋮					

Work.Germination

Rep	Species	Temperature	Water	Germination
1	A	30	N	48
2	A	30	S	52
1	B	20	N	61
1	C	30	N	44
2	B	30	S	58
⋮				

```
DATA growth;
SET growth;
RENAME spec=Species   Temp=Temperature;

PROC SORT DATA=growth; BY Species Rep Temperature Water;
PROC SORT DATA=germination; BY Species Rep Temperature Water;
DATA GHouse;
MERGE growth germination;
BY Species Rep Temperature Water;
RUN;
```

Species	rep	Temperature	Water	Biomass	Leaf_N	Germination
A	1	30	N	34.2	18.2	48
A	2	30	S	38.9	22.1	52
B	1	20	N	25.6	16.0	61
B	2	30	S	22.4	14.6	58
C	1	30	N	28.4	19.2	44
⋮						

Note that the first SAS data set named in the MERGE statement determined the order of the variables held in common. "Rep" and "rep" are the same thing. One could have added rep=Rep to the RENAME statement to achieve proper case in the variable names. And, of course, a LABEL statement could have been added to the DATA step.

8.5 Chapter Summary

In this chapter, DATA step skills are further developed by instructing the reader how to access data in an external file and "read the data into SAS" for analysis. This book began with the DATA step where data lines were included in the program. Most often researchers record data in an external file of some type. The INFILE statement can access external text files that are space, tab, or comma delimited. Spreadsheet files created by popular software or data files created by other statistical programs can be accessed using the INFILE statement in a DATA step. The LABEL statement was introduced as another statement in the DATA step. The LABEL statement allows the programmer to include supporting text for the variable names used in the program. Three new and very useful statements, DROP, KEEP, and RENAME were added to the collection of DATA step syntax. More additions to the DATA step will be included in Chapter 12.

9

Frequency Analysis

When character or numeric data represent classifications of a response variable, the proportion or frequency of each class level of the response variable is of interest. Typically, the number of class levels is a small, finite value. The variables for these types of analyses are typically nominal or ordinal variables. Testing the distribution of the class levels of a single classification variable or a test of association between two classification variables are commonly of interest. The tests for these scenarios are Pearson χ^2 tests. Categorical data analysis includes these introductory methods. There are other methods that are more rigorous and require skills beyond a first statistical methods course, and hence, are beyond the scope of this book.

9.1 Statistical Tests

In Chapter 5, the GCHART procedure was used to produce frequency bar charts for categorical response variables. The χ^2 Goodness-of-Fit test (Freund, Wilson, and Mohr, 2010; Ott and Longnecker, 2016) can be used to assess whether or not the distribution of the categorical variable is modeled by a particular discrete distribution when the sample size is large enough. For two categorical variables, frequency counts are obtained for combinations of the class levels for two variables. The χ^2 analysis in this two-variable evaluation assesses whether or not the variables are independent of each other, or that there is an association between them.

9.1.1 One-Way Frequency Analysis – Goodness-of-Fit Test for Equal Proportion

A test of equal probability (or proportion) of occurrence of each class level can be done. This is a special case of the χ^2 Goodness-of-Fit test. The hypotheses and test procedure for this one-way frequency analysis are:

H_0: All class levels have the same proportion, $p_1 = \ldots = p_{k,}$ where $0 < p_i < 1$ for $i = 1, \ldots, k$.

H_1: At least one of the class level's p_i is different from the others.

Test statistic:

$$\chi^2 = \sum_{i=1}^{k} \frac{(n_i - E_i)^2}{E_i}$$

where:

k = number of classes (or levels)

n_i = observed frequency in class i

$$n = \sum_{i=1}^{k} n_i = \text{total sample size}$$

E_i = expected frequency for class i = n/k

One would reject H_0 if $\chi^2 \geq \chi^2_{\alpha,(k-1)}$. That is, when the observed frequency, n_i, differs sharply from the expected frequency, E_i, for one or more values of i, the value of χ^2 increases.

9.1.2 One-Way Frequency Analysis – Goodness-of-Fit Test for a Nominal Distribution

For a single classification variable, one can specify the probability of each class level, p_{i*}. The typical χ^2 Goodness-of-Fit Test evaluates whether or not this specified probability distribution is adequate. The hypotheses and test procedure for this one-way frequency analysis are:

H_0: $p_1 = p_{1*}, \ldots, p_k = p_{k*}$, where p_{i*} is specified, $0 < p_{i*} < 1$ for i = 1, …, k, and $\sum_{i=1}^{k} p_{i*} = 1$.

H_1: $p_i \neq p_{i*}$ for at least one class level i

Test statistic:

$$\chi^2 = \sum_{i=1}^{k} \frac{(n_i - E_i)^2}{E_i}$$

where:

k = number of classes (or levels)

n_i = observed frequency in class i

$$n = \sum_{i=1}^{k} n_i = \text{total sample size}$$

E_i = expected frequency for class i = $n \times p_{i*}$ (Because $\sum_{i=1}^{k} p_{i*} = 1$, then $\sum_{i=1}^{k} E_i = n$.)

One would reject H_0 if $\chi^2 \geq \chi^2_{\alpha,(k-1)}$, as before. Large values of χ^2 would indicate a departure from the specified probabilities, p_i.

9.1.3 Two-Way Frequency Analysis – χ^2 Contingency Analysis

When two classification variables are tested for whether they are independent variables or there is an association between them, this analysis is referred to as a contingency table test. Measures of association between the levels of two categorical variables can also be computed. A two-way table of observed frequencies is often produced in this analysis, as shown in Table 9.1. One variable determines the number of rows, r; one row for each class level of that variable. Similarly, the number of columns, c, are determined by the number of class levels for the second variable. Each n_{ij} entry in the table is referred to as a cell frequency. Like the previous methods, expected values, E_{ij}, are computed for each cell. These expected values are computed based on the assumption that the two variables are independent of each other. Pearson's χ^2 statistic is again used to measure the collective differences between the observed frequencies, n_{ij}, and the expected frequencies, E_{ij}.

TABLE 9.1

Form of a Two-way Frequency Table

		Column Variable			
		1	2	...	c
	1	n_{11}	n_{12}	...	n_{1c}
Row Variable	2	n_{21}	n_{22}	...	n_{2c}
	⋮	⋮	⋮		⋮
	r	n_{r1}	n_{r2}	...	n_{rc}

H_0: Two variables are independent.
H_1: Two variables are related or dependent.
Test statistic:

$$\chi^2 = \sum_{i=1}^{r}\sum_{j=1}^{c}\frac{(n_{ij} - E_{ij})^2}{E_{ij}}$$

where:

r = number of rows, c = number of columns
n_{ij} = observed frequency in row i and column j (cell ij)
$n = \sum_{i=1}^{r}\sum_{j=1}^{c} n_{ij}$ = total sample size
E_{ij} = expected frequency in row i, column j
= n × (observed proportion in row i) × (observed proportion in column j), or
= (row i frequency) × (column j frequency)/n

One would reject H_0 if $\chi^2 \geq \chi^2_{\alpha,(r-1)(c-1)}$.
All three of these χ^2 test procedures require:

1. Random samples
2. Large samples – indicated by the expected frequencies, E_{ij}, larger than 5

If the large sample size condition is not met, there are small sample test alternatives, such as Fisher's Exact test. Additionally, there are a number of analyses for a two-way frequency analysis and analyses among three or more classification variables using partial tables that are not covered here. These tests and calculations are beyond the scope of this book. See Agresti (2007, 2018) for more categorical data analysis topics.

9.2 The FREQ Procedure

The FREQ procedure can describe a SAS data set by producing frequency count tables for one variable and cross tabulation or contingency tables for two variables. The FREQ procedure can also produce tests of hypotheses for goodness of fit (equal proportion and a nominal distribution) for one variable and tests of independence between two variables.

The syntax of the FREQ procedure is as follows:

```
PROC FREQ DATA=SAS-data-set <options>;
TABLES requests </options>;
WEIGHT variable;
```

BY and WHERE statements can be added to this procedure as needed. In general, the data do NOT have to be sorted for the FREQ procedure to count the number of occurrences of each value of the variable. If the BY statement is used on the FREQ procedure, then the data would, of course, need to be compatibly sorted first, and the analysis conducted by the FREQ procedure would be performed for each level of the BY variable(s).

PROC FREQ statement

Some of the options for the PROC FREQ statement are as follows:

COMPRESS One-way tables typically begin on a new page. The COMPRESS option will place multiple one-way tables on a page if they will fit.

PAGE prints only one table per page. By default, PROC FREQ prints multiple tables per page, but begins a new page for a one-way table for the next variable in the analysis. PAGE and COMPRESS cannot be used together.

NLEVELS displays the number of levels of each variable specified in the TABLES statement. This information appears in a table separate from the frequency table produced by the TABLES statement.

ORDER = DATA | FREQ (pick one option only)

The ORDER = *option* specifies the order in which the variable levels are to be reported. ORDER = DATA places the levels of the table variable(s) in the order in which the levels were entered in the SAS data set.

ORDER = FREQ places the levels of the table variables(s) in descending order of frequency count; most frequent or popular levels are listed first.

Omitting the ORDER = *option* will result in the variable levels listed in alphabetic order or ascending numeric order.

TABLES statement

In a TABLES statement many requests can be made. More than one TABLES statement may be used in a single FREQ procedure, and each TABLES statement can have different requests and options. One-way and n-way tables can be requested using one or more TABLES statements.

To illustrate the allowable syntax for making table requests, suppose the data set contains the variables a, b, c, d, e, and f. These variables are read into the SAS data set in this order in the INPUT statement, that is: INPUT a b c d e f; assuming a small finite number of levels to each of the numeric variables. Of course, any or all of the variables could be character variables requiring "$" in the INPUT statement for each character variable.

For one-way frequency tables for variables a, b, and f, the syntax is:

```
TABLES a b f;
```

For one-way frequency tables for variables a, b, c, d, and f:

```
TABLES a b c d f;
or TABLES a - - d f;
```

The double dash (- -) can be used to specify a range of variables in the order that they are initialized in the DATA step. Section 20.1 contains more information about the double dash syntax.

For two-way frequency tables, an asterisk notation is used to indicate the two variables to be examined in the same table. The levels of the first variable determine the number of rows in the table, and the levels of the second variable determine the number of columns in the table. Generally the syntax is TABLES *rowvariable* * *columnvariable*;

The following list illustrates some equivalent syntax for two-way tables.

TABLES a*b a*c;	is equivalent to	TABLES a * (b c);
TABLES a*c a*d b*c b*d;	is equivalent to	TABLES (a b) * (c d);
TABLES a*d b*d c*d;	is equivalent to	TABLES (a - - c) * d;

For more than two variables, n-way tables can also be requested, where $n \geq 3$. This text will overview programming code and results for one-way and two-way tables.

The TABLES statement can end after the table requests such as those in the previous paragraph. One can include a number of options in the TABLES statement. A "/" (forward slash) after the last table request separates the requests from the selected options. There are a number of options available for the TABLES statement. Those options consistent with the χ^2 analyses in Section 9.1 are presented here. The options available for the TABLES statement control the printed results, produce statistical tests, or request addition calculations not included in the default set of statistics to be computed. A list of basic options is given in Table 9.2 and SAS Help and Documentation within the SAS software. SAS Institute (2013a) can also be consulted for the complete list of options.

The options in Table 9.2 are overviewed here.

TABLE 9.2

Summary of Basic Options for the TABLES Statement of the FREQ Procedure

		Applicable to	
Task	**Options**	**One-Way Tables**	**Two-Way Tables**
Control printed results	CROSSLIST		X
	LIST		X
	NOCOL		X
	NOCUM	X	X*
	NOFREQ		X
	NOPERCENT	X	X
	NOROW		X
	PLOTS=*requests*	X	X
Request further information	CELLCHI2		X
	EXPECTED		X
Specify statistical analysis	CHISQ	X	X
	TESTF = (*values*)	X	
	TESTP = (*values*)	X	

*works with the LIST option

CROSSLIST

prints two-way to n-way tables in a list format. CROSSLIST will include subtotals within the table. The LIST option does not include these subtotals. When the CHISQ option is included, the CROSSLIST output is not affected. NOPERCENT, NOROW, and NOCOL options can be used.

LIST

prints two-way to n-way tables in a list format (similar to the one-way tables) rather than as cross tabulation tables. When the CHISQ option is included, the LIST option has no effect on the output.

NOCOL

suppresses the column percentages in the cells of a two-way table.

NOCUM

suppresses printing of the cumulative frequencies and cumulative percentages of one-way frequencies and when the LIST option is used on a two-way table.

NOFREQ

suppresses the display of observed frequencies in the cells of two-way tables. Row total frequencies are also suppressed by this option. When using the LIST option, the NOFREQ option has no effect.

NOPERCENT

suppresses printing of percentages and cumulative percentages for a one-way frequency analysis and for two-way frequency tables with the LIST option. It also suppresses the printing of all cell percentages (and the marginal row and column percentages) for two-way tables.

NOROW

suppresses the row percentages in the cells of a two-way table.

PLOTS = ALL | NONE | FREQPLOT | DEVIATIONPLOT (pick one only)

There are several ODS Graphics available in the FREQ procedure depending upon the calculations requested in a TABLES statement. One can suppress all of the plots (PLOTS=NONE) or request all available plots (PLOTS=ALL) for the type of analysis requested. Two of the basic plots available are the deviation plot and the frequency plot.

PLOTS=DEVIATIONPLOT is the default for a one-way table when CHISQ is also specified on the TABLES statement.

Frequency plots may be requested for one-way and two-way tables. PLOTS=FREQPLOT produces frequency plots for each level of the variable in a one-way table, and produces frequency plots for each level of variable2 (column variable) within each level of variable1 (row variable) when TABLES variable1*variable2 is written. Switch "variable1" with "variable2" to reverse the roles of the variables in the graph. The frequency plots are bar charts by default.

```
PLOTS = FREQPLOT(TYPE=BARCHART)
```

Dot plots can also be requested with the option:

```
PLOTS = FREQPLOT(TYPE=DOTPLOT).
```

PLOTS(ONLY) = (*list of plot requests*) the ONLY modifier will suppress the default graphics and only those plots requested in parenthesis using ODS Graphics names (Section 19.2) will be produced.

CELLCHI2

prints each cell's contribution to the total χ^2 statistic, $\frac{(n_{ij} - E_{ij})^2}{E_{ij}}$, for the table cell in row i and column j. This option has no effect on one-way tables or when the LIST option is used for a two-way analysis.

EXPECTED

prints the expected cell frequencies, E_{ij}, under the hypothesis of independence in a two-way table (Section 9.1.3). Expected frequencies are not printed for one-way frequency tables nor when the LIST option is used for two-way tables.

CHISQ

For one-way tables, a χ^2 test of equal proportions (Section 9.1.1) across the class levels of the specified variables. Using the TESTF or TESTP option, one can conduct a goodness-of-fit test for a nominal distribution (Section 9.1.2) identified using either of these options. For two-way contingency tables, χ^2 tests of independence (Section 9.1.3) and some measures of association based on χ^2 are computed.

TESTF = (*values*)

The χ^2 Goodness-of-Fit test for the specified frequencies of each class level of the variable in a one-way table. The order of the frequencies must be the same as the order of the levels in the table. See the ORDER option on the PROC FREQ statement to control this.

TESTP = (*values*)

The χ^2 Goodness-of-Fit test for the specified percentages of each class level of the variable in a one-way table. These percentage values must sum to 100. Probabilities between zero and one can also be specified and must sum to 1.00. In either case the order of the values must be the same as the order of the levels in the table. See the ORDER option on the PROC FREQ statement to control this. The TESTP option cannot be used in the same TABLES statement as the TESTF option. This option is most generally used in conjunction with the CHISQ option, but this is not required.

WEIGHT statement

Normally, each observation in the SAS data set contributes a value of 1 to the frequency counts. When a WEIGHT statement is used, each observation contributes the weighting variable's value for the observation. The WEIGHT statement is typically used when the SAS data set already contains frequency counts. Only one WEIGHT statement can be used in a single FREQ procedure. The effect of this statement is illustrated in Objective 9.1. See Section 9.3 for more information about the WEIGHT statement.

Frequency Analysis Data – Right to Work laws differ across states in the United States. Right to Work implies that an employee does not have to join a labor union in a workplace that is unionized. A random sample of 202 residents is taken, and each resident is surveyed regarding his/her opinion on Right to Work issues, and the employment status of the resident is also recorded. The results of the survey appear in Table 9.3.

OBJECTIVE 9.1:

- Create the *rtw* data set containing the variables: EmpClass, Opinion, and Y, where Y is the frequency given in Table 9.3.
- LABEL the variables in the *rtw* data set.

- Conduct a chi-square Goodness-of-Fit test for equal probability of each level of Opinion on Right to Work. Use $\alpha = 0.05$ in the conclusion.
- Investigate the effects of the ORDER=DATA option on the PROC FREQ statement and the inclusion of the WEIGHT statement.

TABLE 9.3

Data for the Frequency Analysis

	Opinion on Right to Work			
Employment Classification	**Favor**	**Do Not Favor**	**Undecided**	**Totals**
Industry	20	24	16	60
Business	40	51	9	100
Unemployed	20	15	7	42
Totals	80	90	32	202

```
DM 'LOG; CLEAR; ODSRESULTS; CLEAR; ';

DATA rtw;
INPUT EmpClass $ Opinion $ Y @@;
LABEL EmpClass = "Employment Classification"
      Opinion = "Opinion on Right to Work"
      Y = "Observed Frequency";
DATALINES;
I  F 20     I  DNF   24   I   U  16
B  F 40     B  DNF   51   B   U   9
U  F 20     U  DNF   15   U   U   7
;
* The following FREQ procedure will count the number of occurrences *;
* of the levels of the variables CLASS and OPINION                   *;
PROC FREQ DATA=rtw;
TABLES Opinion;
TITLE "Objective 9.1 - No WEIGHT Statement";

* The WEIGHT statement is necessary in order to get correct frequencies;
* when the counts for each level of a variable are included in the data;
PROC FREQ DATA=rtw ORDER=DATA;
TABLES Opinion / CHISQ;
WEIGHT Y;
TITLE "Objective 9.1 - with a WEIGHT Statement";
RUN;
QUIT;
```

Objective 9.1 - No WEIGHT Statement

The FREQ Procedure

Opinion on Right to Work

Opinion	Frequency	Percent	Cumulative Frequency	Cumulative Percent
DNF	3	33.33	3	33.33
F	3	33.33	6	66.67
U	3	33.33	9	100.00

In the first FREQ procedure, there is no WEIGHT statement. In the one-way frequency table produced, each level of Opinion has a frequency of 3 rather than the values given in Table 9.3. Without the WEIGHT statement, each occurrence of each Opinion level (DNF, F, U) is counted. It is also observed that the values of Opinion appear in alphabetical order in this table which may not be the optimal arrangement in a presentation.

Objective 9.1 - with a WEIGHT Statement

The FREQ Procedure

Opinion on Right to Work

Opinion	Frequency	Percent	Cumulative Frequency	Cumulative Percent
F	80	39.60	80	39.60
DNF	90	44.55	170	84.16
U	32	15.84	202	100.00

Chi-Square Test for Equal Proportions

Chi-Square	28.5545
DF	2
Pr > ChiSq	<.0001

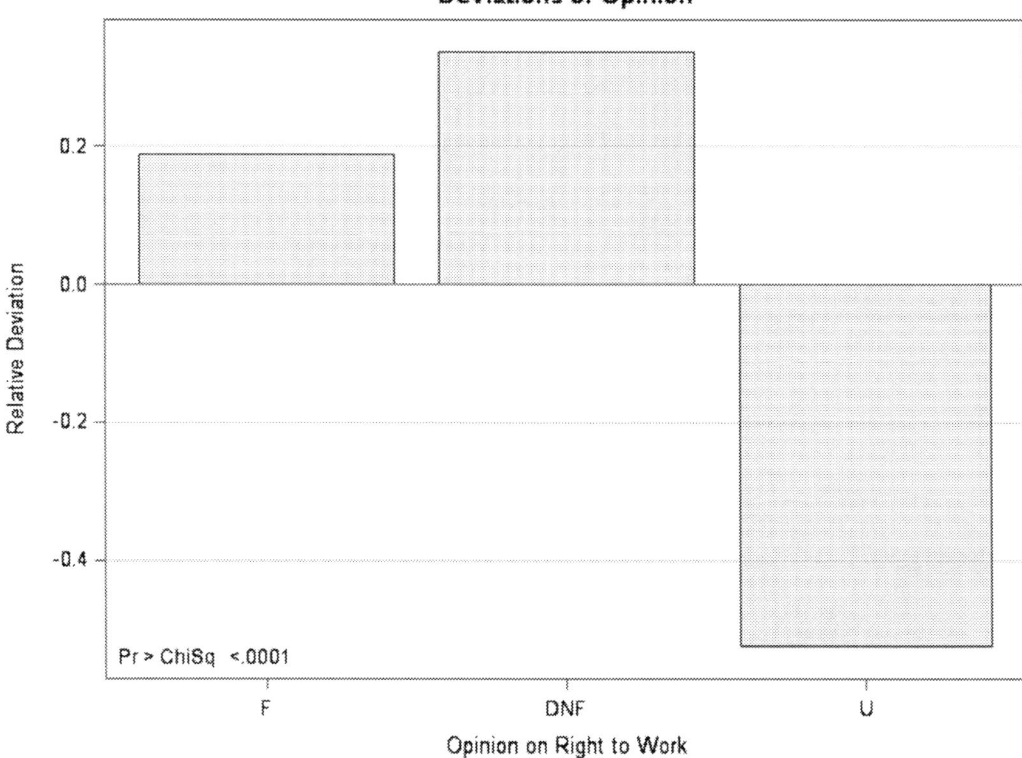

Sample Size = 202

The second FREQ procedure produced a one-way frequency table with the correct frequencies as a result of including the WEIGHT statement. One should note that the ORDER=DATA option used in the second FREQ procedure lists the values of the variables in the frequency table in the order the variables were entered in the data lines. With the correct frequencies, a test of equal proportions was included (CHISQ) option.

Following the one-way frequency table for the variable Opinion is a **Chi-Square Test for Equal Proportion** table. The χ^2 test statistic in Section 9.1.1, degrees of freedom, and observed significance level have been computed. One concludes that the three opinions on Right to Work do not occur with equal probability ($\alpha = 0.05$, $\chi^2_2 = 28.5545$, $p < 0.0001$).

The default ODS Graphic from this χ^2 analysis is a deviation plot, titled as **Deviations of Opinion** illustrates this conclusion. The sign of the relative deviation indicates whether the observed frequency was larger than expected (+) or smaller than expected (–) under the assumption of equal probability. Shorter bar lengths for all levels of opinion would have been the result if the opinions occurred with equal probability. Finally, the sample size of 202 is printed at the end of the analysis.

In both of the procedures, the variable label was used in the tables of results. No additional options in the FREQ procedure were needed for the label to appear in the output.

OBJECTIVE 9.2: Test whether or not the employee classification in this population is 45% Industry, 50% Business, and 5% Unemployed. Use $\alpha = 0.05$ in the Goodness-of-Fit test.

```
PROC FREQ DATA=rtw ORDER=DATA;
TABLES EmpClass / TESTP = (45 50 5);
WEIGHT y;
TITLE 'Objective 9.2 - Goodness of Fit Test for Employee Classification';
RUN;
```

Objective 9.2 - Goodness of Fit Test for Employee Classification

The FREQ Procedure

Employment Classification

EmpClass	Frequency	Percent	Test Percent	Cumulative Frequency	Cumulative Percent
I	60	29.70	45.00	60	29.70
B	100	49.50	50.00	160	79.21
U	42	20.79	5.00	202	100.00

Chi-Square Test for Specified Proportions

Chi-Square	111.2673
DF	2
Pr > ChiSq	<.0001

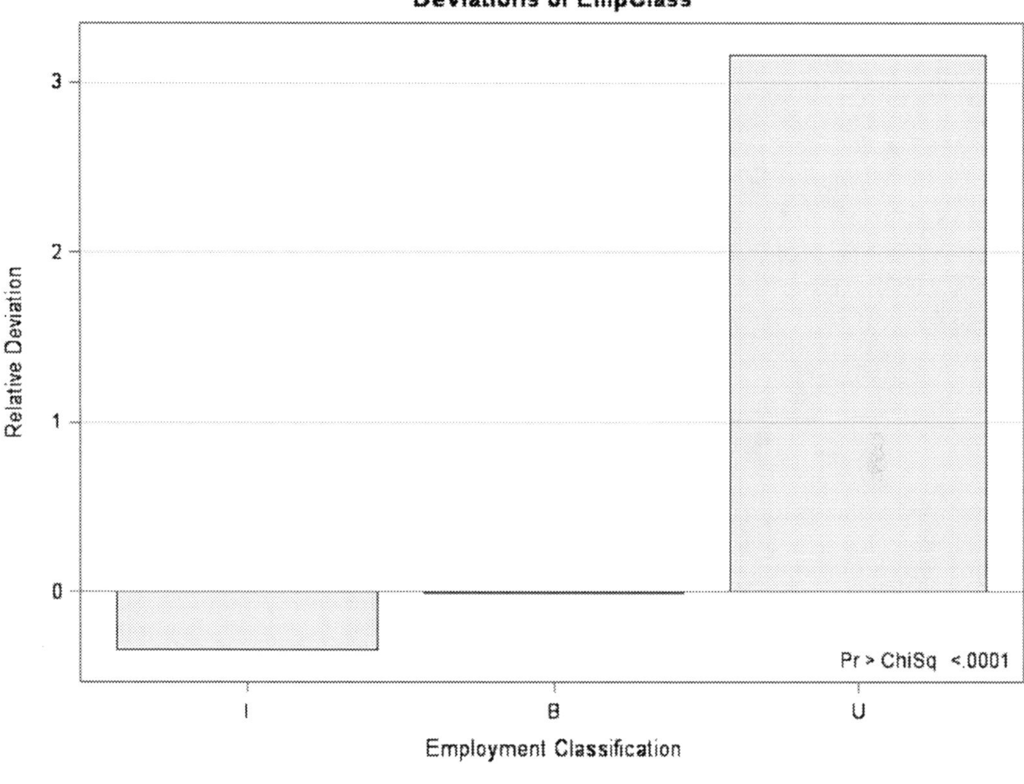

Sample Size = 202

In the **Objective 9.2 – Goodness-of-Fit Test for Employee Classification** results, the TESTP option was used to specify the probabilities. The order of the values in TESTP = (45 50 5) was determined by the order of the data, hence ORDER=DATA was used in the PROC FREQ statement. If the ORDER=DATA option had not been used, one would have to rewrite the TESTP option as TESTP = (50 45 5) to correspond to the alphabetical order of the levels B, I, and U. In the **Employment Classification** table, there is an additional column in the objective in comparison to Objective 9.1. The Test Percent column is the result of the TESTP option. One should confirm the values in this column to make certain that the correct probabilities for each class level are used. TESTP = (45 50 5) could have been written as TESTP = (0.45 0.50 0.05) where the values sum to 1. It makes no difference for this one-way analysis whether or not the CHISQ option is used.

Based on this data, the Employee Classification does not follow the specified probability distribution ($\alpha = 0.05$, $\chi^2_2 = 111.2673$, $p < 0.0001$). The deviation plot illustrates which of the classifications was extreme. In the deviation plot for Employment Classification (**Deviation of EmpClass**) the length of the bars and the direction of the bars assist the researcher in determining which of the classes deviate from the null hypothesis probabilities. Each bar height is the value of relative deviation computed by $(n_i - E_i)/E_i$ where $E_i = np_i$. For the Unemployed, $E_3 = 202(0.05) = 10.1$, and $(n_3 - E_3)/E_3 = (42 - 10.1)/10.1 = 3.158$. In this deviation plot U (Unemployed) has a positive relative deviation implying a larger than expected percentage of the sample data was unemployed. For I (Industry) the relative deviation is negative but much smaller in magnitude while B (Business) is near zero. This large, positive relative deviation for the Unemployed category appears to account for the significance of this χ^2 test. Since neither of the other two categories has a relative deviation that is large in magnitude, one possible explanation is that both Industry and Business in this study have experienced job losses resulting in the larger than expected Unemployment frequency.

OBJECTIVE 9.3: Is there evidence to indicate that a person's opinion concerning Right to Work depends on his or her employment status? That is, can it be concluded that the two variables are dependent? Use $\alpha = 0.05$.

```
PROC FREQ DATA=rtw;
TABLES EmpClass * Opinion / LIST;
WEIGHT y;
TITLE 'Objective 9.3 - LIST Option';
RUN;

PROC FREQ DATA=rtw ORDER=DATA;
TABLES EmpClass*Opinion / CHISQ;
WEIGHT y;
TITLE 'Objective 9.3 - Ordered Data in a Two-way Table';
RUN;
```

Objective 9.3 - LIST Option

The FREQ Procedure

EmpClass	Opinion	Frequency	Percent	Cumulative Frequency	Cumulative Percent
B	DNF	51	25.25	51	25.25
B	F	40	19.80	91	45.05
B	U	9	4.46	100	49.50
I	DNF	24	11.88	124	61.39
I	F	20	9.90	144	71.29
I	U	16	7.92	160	79.21
U	DNF	15	7.43	175	86.63
U	F	20	9.90	195	96.53
U	U	7	3.47	202	100.00

Objective 9.3 - Ordered Data in a Two-way Table

The FREQ Procedure

Table of EmpClass by Opinion

Frequency Percent Row Pct Col Pct				
EmpClass (Employment Classification)	**Opinion (Opinion on Right to Work)**			
	F	DNF	U	Total
I	20	24	16	60
	9.90	11.88	7.92	29.70
	33.33	40.00	26.67	
	25.00	26.67	50.00	
B	40	51	9	100
	19.80	25.25	4.46	49.50
	40.00	51.00	9.00	
	50.00	56.67	28.13	
U	20	15	7	42
	9.90	7.43	3.47	20.79
	47.62	35.71	16.67	
	25.00	16.67	21.88	
Total	80	90	32	202
	39.60	44.55	15.84	100.00

Statistics for Table of EmpClass by Opinion

Statistic	DF	Value	Prob
Chi-Square	4	10.6405	0.0309
Likelihood Ratio Chi-Square	4	10.4423	0.0336
Mantel-Haenszel Chi-Square	1	3.5085	0.0611
Phi Coefficient		0.2295	
Contingency Coefficient		0.2237	
Cramer's V		0.1623	

Sample Size = 202

In **Objective 9.3 – LIST Option,** the effect of the LIST option is evident in the two-way table for EmpClass by Opinion. The first variable listed in the two-way request in the TABLES statement is EmpClass, and that variable determined the first column of the frequency list. The LIST option suppressed all default ODS Graphics. If the CHISQ option and the LIST option were both included on the TABLES statement, the CHISQ option would supersede the LIST option, and the output would look like the second part of the output for this objective.

In **Objective 9.3 – Ordered Data in a Two-way Table,** a 3 × 3 table is produced. (Each variable has three levels, hence 3 × 3.) In the upper left corner of the **Table of EmpClass by Opinion** is a legend for each cell of the table. The first number in each cell is the observed frequency. Since the ORDER=DATA option was used in the PROC FREQ statement and EmpClass appeared first in the TABLES statement identifying EmpClass as the row variable, these observed frequencies should be in the same order as that given in Table 9.3. The second value in each cell of the table is Percent determined by (observed frequency)/n × 100; which is 20/202 × 100 = 9.90 in the first cell. The third number is the row percent determined by (observed frequency)/(row total) × 100; which is 20/60 × 100 = 33.33 in the first cell. And finally, the fourth value in the cell is the column percent determined by (observed frequency)/(column total) × 100, which is 20/80 × 100 = 25.00 in the first cell. One can suppress one or more of these cell values using the NOPERCENT, NOROW, or NOCOL options on the TABLES statement. CELLCHI2 and EXPECTED will result in additional values computed for each cell of the table.

The **Statistics for Table EmpClass by Opinion** are the result of the CHISQ option. The χ^2 test for independence is the first test statistic (Chi-Square) printed in the table. Thus, one could conclude that a person's employment status is related to his/her opinion on the Right to Work issue ($\alpha = 0.05$, $\chi^2_4 = 10.6405$, $p = 0.0309$). The remaining test statistics produced by the FREQ procedure (Likelihood Ratio Chi-Square, and Mantel-Haenszel Chi-Square) and the measures of association (Phi Coefficient, Contingency Coefficient, and Cramer's V) were not overviewed here. A categorical data analysis reference, such as Agresti (2007), is recommended for these topics. There are no default ODS Graphics for a two-way classification.

OBJECTIVE 9.4: The frequency plot is a graphic illustrating the observed frequencies for the cells in the two-way table. Produce a frequency plot by adding the option PLOTS=FREQPLOT to the TABLES statement of the second FREQ procedure in Objective 9.3.

```
PROC FREQ DATA=rtw ORDER=DATA;
TABLES EmpClass*Opinion/ CHISQ PLOTS=FREQPLOT;
WEIGHT Y;
TITLE "Objective 9.4";
RUN;
```

Objective 9.4

The FREQ Procedure

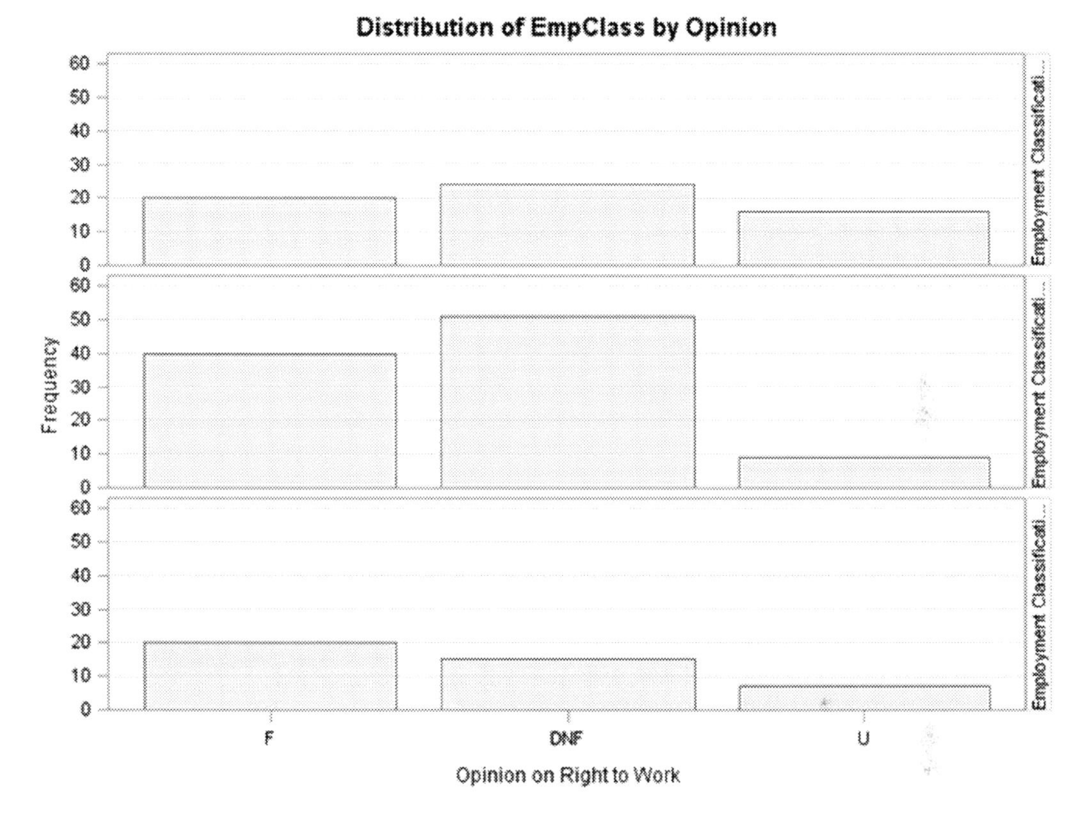

In the Objective 9.4 analysis repetitive results from Objective 9.3 have been suppressed, and only the frequency plot **Distribution of EmpClass by Opinion** requested is shown. Each row of the frequency plot is a level of the row variable (first variable in the TABLES statement request) EmpClass. Generally, a LABEL for the variables is helpful. In this particular graphic, the label "Employment Classification" is too long for the space SAS has allotted for the row variable at the right edge of the graph, and the label has been truncated. Thus, using the assigned labels the graph is not well-labeled since the values of the EmpClass variable are not shown on the graph. If the variables in the TABLES request were switched (Opinion*EmpClass), the same issue occurs for the label "Opinion on Right to Work". In this instance, no labels in the DATA step might be a better alternative for the frequency plot, as shown in the **Objective 9.4 Alternative – No LABEL** results. These results are produced by removing or disabling the LABEL statement in the DATA step creating data set *rtw*; The SAS code not shown.

Objective 9.4

Alternative – No LABEL

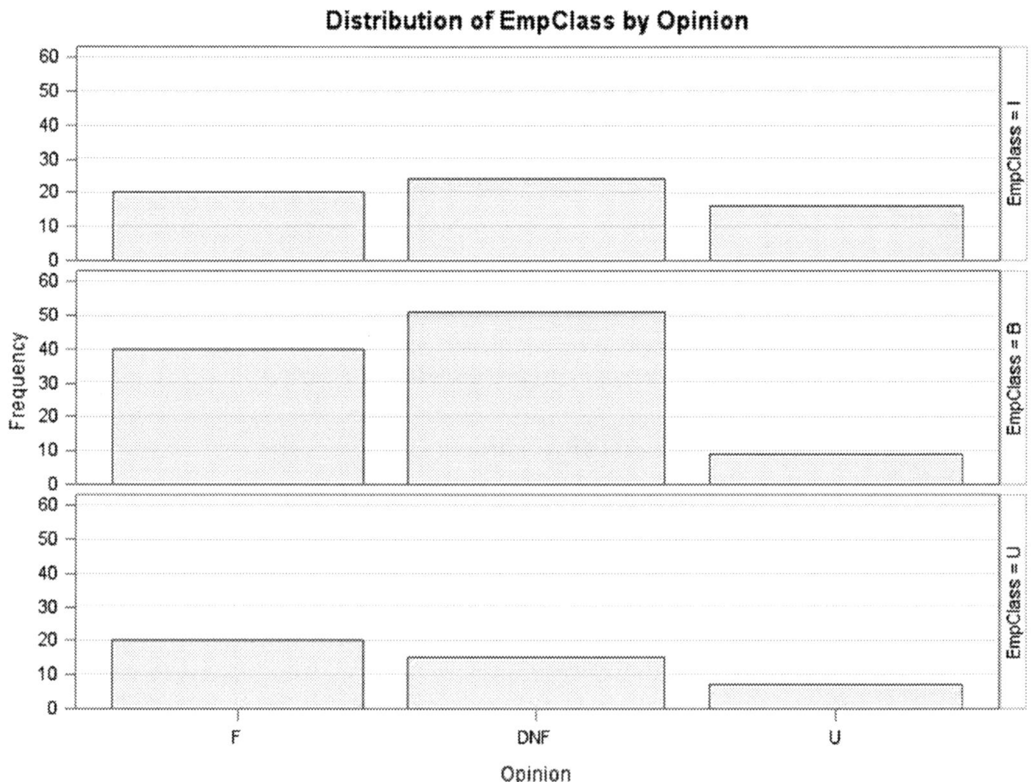

OBJECTIVE 9.5: Repeat Objective 9.4 changing the frequency plot to a dot plot. Are there similar issues with the length of variable labels?

```
PROC FREQ DATA=rtw ORDER=DATA;
TABLES EmpClass*Opinion/ CHISQ PLOTS=FREQPLOT(TYPE=DOTPLOT);
WEIGHT Y;
TITLE "Objective 9.5";
RUN;
```

Objective 9.5

The FREQ Procedure

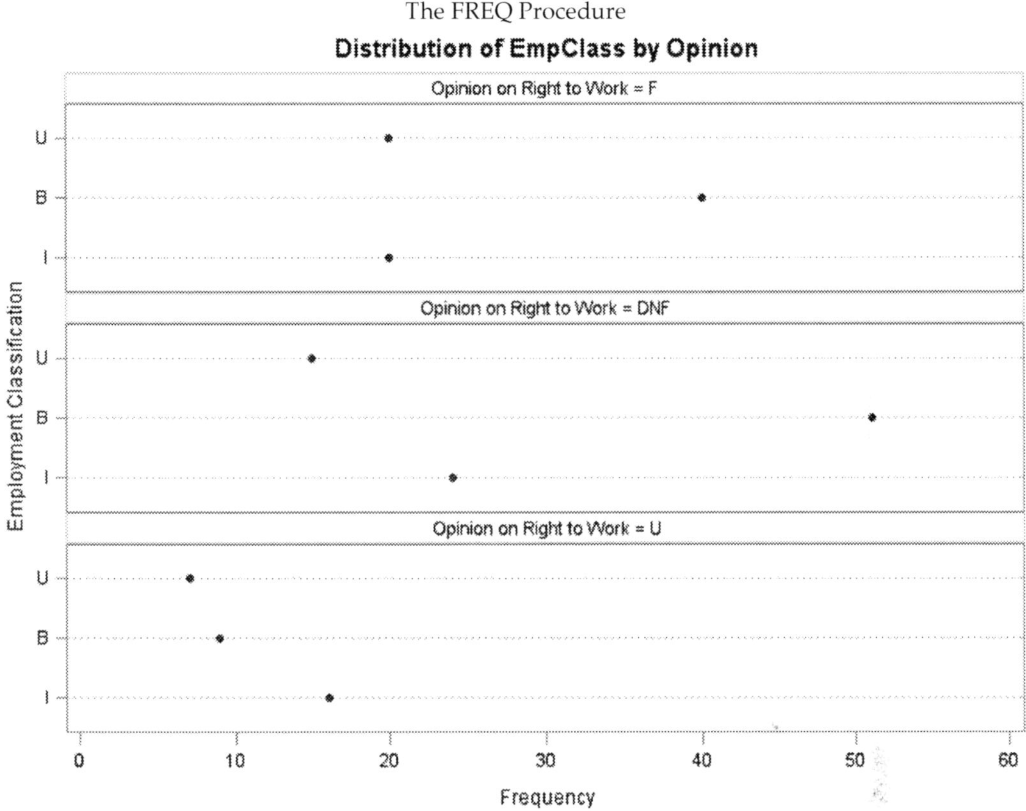

In the **Objective 9.5** output repetitive results from Objective 9.3 have again been suppressed, and only the dot plot requested is shown. In the dot plot **Distribution of EmpClass by Opinion** the frequency is located on the horizontal axis. On the vertical axis is the row variable from the TABLES statement. The complete label for the row variable appears in the left margin. Each row panel of the dot plot is determined by the levels of second variable, Opinion, in the TABLES statement. At the top of each panel for the class levels of Opinion the complete label and value of Opinion correctly appear.

OBJECTIVE 9.6: Repeat Objective 9.3 for the second FREQ procedure suppressing all of the default cell contents except the observed frequency. Include the expected frequency in the χ^2 for independence calculation. Produce only two-way analyses and no graphics.

```
PROC FREQ DATA=rtw ORDER=DATA;
TABLES EmpClass*Opinion/ CHISQ NOROW NOCOL NOPERCENT EXPECTED;
           *The options on the TABLES statement can be in any order;
WEIGHT y;
TITLE "Objective 9.6";
RUN;
```

Objective 9.6

The FREQ Procedure

Table of EmpClass by Opinion

Frequency Expected	Opinion (Opinion on Right to Work)			
EmpClass (Employment Classification)	**F**	**DNF**	**U**	**Total**
I	20 23.762	24 26.733	16 9.505	60
B	40 39.604	51 44.554	9 15.842	100
U	20 16.634	15 18.713	7 6.6535	42
Total	80	90	32	202

In the **Objective 9.6** results the two-way table, **Table of EmpClass by Opinion**, has only two statistics per cell of the table: the observed frequency and the expected value for that cell under the assumption of independence. The χ^2 test for independence assesses whether the collective differences between those two values are large enough to conclude that the two variables are dependent or related. The inclusion of the CELLCHI2 option on the TABLES statement would have computed a third value in each cell of the table. This value is $\frac{(n_{ij} - E_{ij})^2}{E_{ij}}$ for each row i and column j cell, and the sum of these values is the reported χ^2 test statistic.

Multiple analyses or TABLES statements can be included in single FREQ procedure. If a WEIGHT statement is included, it is applicable to every TABLES statement in the procedure. Here is an example of how three of the previous objectives could be written in a single FREQ procedure:

```
PROC FREQ DATA=rtw ORDER=DATA;
TABLES Opinion / CHISQ;                  *Objective 9.1;
TABLES EmpClass / TESTP = (45 50 5);     *Objective 9.2;
TABLES EmpClass * Opinion / LIST;        *Objective 9.6;
WEIGHT y;
RUN;
```

Any TITLEn statements in the procedure shall apply to all of the tables produced. Single line comments (Section 20.2) after each TABLES statement are used to identify the associated objectives. If any one of the TABLES statements requires a different WEIGHT or no WEIGHT statement, then more than one block of PROC FREQ code is needed.

Consider the following FREQ procedure code:

```
PROC FREQ DATA=rtw ORDER=DATA;
TABLES Opinion EmpClass EmpClass * Opinion / CHISQ PLOTS=NONE;
WEIGHT y;
RUN;
```

Test statistics for equal proportions of Opinion and EmpClass levels and the test statistic for independence between the two variables would be computed. The default one-way deviation plots are suppressed by the PLOTS=NONE option. If the TESTP = (45 50 5) option was included, both Opinion and EmpClass would be tested for those proportions, and the TESTP option would have no effect on the two-way analysis.

It is left as an exercise for the reader to investigate options that control output (Table 9.1), examining additional computations and changing the order of the variables in the TABLES statement for both the one-way and two-way analyses.

The chi-square tests presented in this chapter require large samples. When this requirement is not met, SAS still computes the test statistic but also prints a warning message beneath the table containing the test statistics indicating what percent of the cells do not meet the large sample condition, such as:

WARNING: 33% of the cells have expected counts less than 5. Chi-Square may not be a valid test.

When the sample size condition is met, no warning statement is generated in the results. These χ^2 tests are reliable when the sample size is large. If the test is interpreted when a sample size warning has been given, the reliability of that conclusion is in question. Other categorical data analysis methods are recommended in cases when the sample size is too small for the χ^2 test. See Agresti (2007, 2018) for other test options. Syntax options for alternate methods are also available in the FREQ procedure. Many of these are TABLES statement options.

9.3 A Note about the WEIGHT Statement

For the correct analysis and description of a data set that does not require a WEIGHT statement, a large data set is necessary. In these larger data sets, each observation is separately entered. In the Right to Work example introduced in Table 9.3, there would be 202 observations (or lines) in the *rtw* data set if there were no Y variable in the data set. That is, one observation or line for each of the survey respondents would have to be in the data set. This SAS data set would contain 20 observations (lines) with the EmpClass I and Opinion F combination, 24 with EmpClass I and Opinion DNF, and so on. The FREQ procedure would count the number of occurrences of each level of each of the variables or variable combinations requested in the TABLES statement. If some preliminary counting of the classification variables has been done, such as in the case of the data set *rtw*, then a WEIGHT statement is necessary to name the variable that represents that count in a correct frequency analysis.

When there are more classification variables to be analyzed than shown in the Right to Work survey example in this chapter, the data will need to be entered without preliminary counts. These data sets tend to be large. The analysis using the FREQ procedure would not need a WEIGHT statement, but the TABLES statement and all of the options overviewed for that statement are still applicable.

9.4 Chapter Summary

This chapter reviewed χ^2 testing procedures for one-way and two-way frequency tables. The FREQ procedure can be used for counting the frequency of levels of categorical response variables. χ^2 test options for one-way and two-way frequency tables can also be requested along with the frequency tables. The FREQ procedure also can produce graphics that support these analyses. Advanced categorical analyses methods such as Fisher's Exact Test, measures of association, agreement analysis, odds ratios, and more can also be obtained using the FREQ procedure.

10

Summarizing a Data Table in a Formal Report

Previously the MEANS, UNIVARIATE, and FREQ procedures were used to compute some summary statistics for a SAS data set. All or part of the tabular output from these procedures may be selected for a formal report. Reports are very often a simple reorganization of the data that even a PRINT procedure could achieve. PRINT, FREQ, and MEANS procedures, and the DATA step will be revisited in this chapter. Additionally, the REPORT procedure will be introduced. The REPORT procedure can also perform some basic statistics calculations and create ordered tables for a formal report.

10.1 Revisiting the PRINT, FREQ, and MEANS Procedures

Consider the data in Table 10.1 containing student names, ID numbers, project group #, homework (HW) scores, and exam scores (EX) for a college course. The "Total" row, identifies the total points possible for the grade item in that column. (These data differ from the student grade data in Chapters 2 and 4.)

OBJECTIVE 10.1:

- Create a SAS data set *gradebook* for Table 10.1. Keep the first line as the total possible point values for the graded items.
- Using the DATA step compute the total scores for the homework, exams, and for the course for each student. Missing scores should be counted as a zero for that item.
- Label each of the new variables.
- Print the SAS data set *gradebook* with labels and examine the results for accuracy. Suppress the observation numbers.

```
DATA gradebook;
INPUT Student $9.    ID    Group HW1    HW2    EX1    HW3    HW4    EX2;
HWTotal = SUM(HW1, HW2, HW3, HW4);
EXTotal = SUM(EX1, EX2);
CourseTL = SUM(HWTotal, EXTotal);
LABEL HWTotal = "Homework Point Total"
      EXTotal = "Exam Score Total"
            CourseTL= "Point Total for the Course";
DATALINES;
Total            .     .     75     110    100    50    25    100
Dave        101  1     71    88     93     46     23    88
Lynn        381  2     64    96     95     48     25    .
Michael     987  2     68    75     97     35     12    60
```

```
Leslie          579  3    55    75    81     .    17    82
Andrew          239  1    70    79    77    38    23    77
Elizabeth       128  3    67    103   94    42    20    92
;
PROC PRINT DATA=gradebook LABEL NOOBS;
TITLE 'Objective 10.1';
TITLE2 'PRINT Procedure with LABEL Option';
RUN;
QUIT;
```

TABLE 10.1

Gradebook Data for the REPORT Procedure

Student	ID#	Group	HW1	HW2	EX1	HW3	HW4	EX2
Total			75	110	100	50	25	100
Dave	101	1	71	88	93	46	23	88
Lynn	381	2	64	96	95	48	25	
Michael	987	2	68	75	67	35	12	60
Leslie	579	3	55	75	81		17	82
Andrew	239	1	70	79	77	38	23	77
Elizabeth	128	3	67	103	94	42	20	92

The name Elizabeth has more than eight characters. SAS will truncate it to eight characters (Elizabet) unless more space for the character string is specified in the INPUT statement. Hence, $9. was included in the INPUT statement after the variable Student. Alternatively, $1–9 column pointer controls could have been included. More space could have been specified. Using column pointer controls, one could modify the "Total" entry to "Total Possible". However, one must be careful to position the ID value for all students in a column that has not been designated for the Student character string; otherwise one or more digits of the ID would be included with that student's name. See Chapter 2, Objective 2.2 for more information about column pointer controls.

Objective 10.1

PRINT Procedure with LABEL Option

Student	ID	Group	HW1	HW2	EX1	HW3	HW4	EX2	Homework Point Total	Exam Score Total	Point Total for the Course
Total	.	.	75	110	100	50	25	100	260	200	460
Dave	101	1	71	88	93	46	23	88	228	181	409
Lynn	381	2	64	96	95	48	25	.	233	95	328
Michael	987	2	68	75	97	35	12	60	190	157	347
Leslie	579	3	55	75	81	.	17	82	147	163	310
Andrew	239	1	70	79	77	38	23	77	210	154	364
Elizabeth	128	3	67	103	94	42	20	92	232	186	418

The HWTotal and EXTotal calculations could have been calculated using arithmetic operations rather than using the SUM function. Using simple addition these statements are:

```
HWTotal = HW1 + HW2 + HW3 + HW4;
EXTotal = EX1 + EX2;
```

However, missing values for any of the variables on the right-hand side of the equations will result in a missing value for the new variable in the SAS data set. The SUM function will sum all of the non-missing values specified in the parenthesis. Missing values will be regarded as zero which is needed in this context. See Objective 4.2 for a more detailed examination of this topic for a different set of student scores, SAS data set *grades*. Additionally, using the arithmetic operation rather than the SUM function, the CourseTL variable would be missing for each student who has a missing value for HWTotal or EXTotal.

OBJECTIVE 10.2:

- Modify *gradebook* in a new DATA step.
- Assign letter grades for each student using IF – THEN – ELSE syntax based on the grading scale:
  Lowest A 90%
  Lowest B 80%
  Lowest C 70%
  Lowest D 60%
  Recover the CourseTL from the line for the Total possible in the output from the previous objective to compute this.
- Keep the Total record in this SAS data set.
- Count the frequency of each letter grade and include the number of grade levels observed. CAUTION: Do not include the TOTAL entry as a student in the course.

```
DATA gradebook;
SET gradebook;                                          *max points = 460;
IF CourseTL GE 414 THEN Grade = "A";           *460 x 0.90 = 414 Lowest A;
ELSE IF 368 LE CourseTL LE 413 THEN Grade="B"; *460 x 0.80 = 368 Lowest B;
ELSE IF 322 LE CourseTL LE 367 THEN Grade="C"; *460 x 0.70 = 322 Lowest C;
ELSE IF 276 LE CourseTL LE 321 THEN Grade="D"; *460 x 0.60 = 276 Lowest D;
ELSE Grade="F";
IF Student="Total" THEN Grade=" ";

PROC FREQ DATA=gradebook NLEVELS;
WHERE student NE "Total";
TABLES Grade;
TITLE 'Objective 10.2';
TITLE2 'FREQ Procedure with NLEVELS Option';
TITLE3 'One-way Table';
RUN;
```

There are different ways one might program this second DATA step to assign letter grades to the students. This is just one. The single-line comments to the right of the IF – THEN and ELSE IF statements are included to support how the intervals in the programming statements were determined. The last statement, `IF Student="Total" THEN Grade=" ";` is included so that no letter grade is assigned to the total possible record. One would want to position this statement after the creation of the other letter grades. If this statement were positioned as the first IF – THEN statement, a letter grade of A would still have been assigned to Total. The reason is that the statement `IF CourseTL GE 414 THEN Grade = "A";` would have executed second, superseding the previous statement where no letter grade was assigned.

In the FREQ procedure, the WHERE statement excluding the Total record in *gradebook* from the frequency analysis is used. A WHERE statement in a procedure can be used to exclude one or more observations from the procedure. One could have deleted the Total student record in *gradebook*, but it is not necessary to do so. Additionally, this information is an important part of the data. Deleting one or more observations is generally not recommended for a procedure unless the observations in question are in error.

Objective 10.2

FREQ Procedure with NLEVELS Option

One-way Table

The FREQ Procedure

Number of Variable Levels

Variable	Levels
Grade	4

Grade	Frequency	Percent	Cumulative Frequency	Cumulative Percent
A	1	16.67	1	16.67
B	1	16.67	2	33.33
C	3	50.00	5	83.33
D	1	16.67	6	100.00

The NLEVELS option on the PROC FREQ statement produced the **Number of Variable Levels** table. This is useful when there are many observed levels of the variable in the TABLES statement. The last entry in the table of one-way frequencies indicates a cumulative frequency of six which accounts for all students in the class. In this objective, the letter grades were created and assigned, then counted in the FREQ procedure. These results do not identify which letter grade was earned by each student in the class. One would need to PRINT *gradebook* or examine this SAS data set in ViewTable (Section 8.3).

OBJECTIVE 10.3: Examine the cross-tabular results of letter grades earned by each project group. Include the number of levels of each variable. In the two-way table, print only the observed frequencies.

```
PROC FREQ DATA=gradebook NLEVELS;
WHERE student NE "Total";
TABLES Group*Grades / NOPERCENT NOROW NOCOL;
TITLE 'Objective 10.3';
TITLE2 'Two-way Table with NLEVELS Option';
RUN;
```

Some of the options for the TABLES statement that restrict or modify the information in each cell of the table are found in Section 9.2.

Objective 10.3

Two-way Table with NLEVELS Option

The FREQ Procedure

Number of Variable Levels

Variable	Levels
Group	3
Grade	4

Frequency **Table of Group by Grade**

	Grade				
Group	A	B	C	D	Total
1	0	1	1	0	2
2	0	0	2	0	2
3	1	0	0	1	2
Total	1	1	3	1	6

The NLEVELS option produces the number of observed levels for each variable in the two-way table. For a very large data set, the number of grade levels should not increase unless someone has an F letter grade. (Hopefully no one has an F!) However, as the number of students increases, the number of groups may also increase. So, it is good to list the Group as the row variable since the two-way table may grow in size.

OBJECTIVE 10.4: Compare the CROSSLIST and LIST options for the output of a two-way frequency table for Group by Grade. The CROSSLIST option appeared in Chapter 9 (Table 9.2) but was not covered in that chapter's objectives.

```
PROC FREQ DATA=gradebook;
WHERE student NE "Total";
TABLES Group*Grade / CROSSLIST;
TITLE 'Objective 10.4';
TITLE2 'FREQ Procedure - Two-way Table with CROSSLIST Option';

PROC FREQ DATA=gradebook;
WHERE student NE "Total";
TABLES Group*Grade / LIST;
TITLE2 'FREQ Procedure - Two-way Table with LIST Option';
RUN;
```

These two options could have been programmed using two TABLES statements in the same FREQ procedure. Because of the change in the TITLEn statements, two FREQ procedures were used.

Objective 10.4

FREQ Procedure - Two-way Table with CROSSLIST Option

The FREQ Procedure

Table of Group by Grade

Group	Grade	Frequency	Percent	Row Percent	Column Percent
1	A	0	0.00	0.00	0.00
	B	1	16.67	50.00	100.00
	C	1	16.67	50.00	33.33
	D	0	0.00	0.00	0.00
	Total	2	33.33	100.00	
2	A	0	0.00	0.00	0.00
	B	0	0.00	0.00	0.00
	C	2	33.33	100.00	66.67
	D	0	0.00	0.00	0.00
	Total	2	33.33	100.00	
3	A	1	16.67	50.00	100.00
	B	0	0.00	0.00	0.00
	C	0	0.00	0.00	0.00
	D	1	16.67	50.00	100.00
	Total	2	33.33	100.00	
Total	A	1	16.67		100.00
	B	1	16.67		100.00
	C	3	50.00		100.00
	D	1	16.67		100.00
	Total	6	100.00		

Objective 10.4

FREQ Procedure - Two-way Table with LIST Option

The FREQ Procedure

Group	Grade	Frequency	Percent	Cumulative Frequency	Cumulative Percent
1	B	1	16.67	1	16.67
1	C	1	16.67	2	33.33
2	C	2	33.33	4	66.67
3	A	1	16.67	5	83.33
3	D	1	16.67	6	100.00

The CROSSLIST option produces the **Table of Group by Grade**. All of the default information from a two-way frequency table is produced in this table. For each level of the row variable (first variable in the TABLES statement request) there are subtotals for frequency and percent produced. The LIST option produces a table for the observed combinations of the levels of the two variables, but does not have subtotals for each level of the row variable, Group. The reader should also investigate the CHISQ, NOPERCENT, and other options that control printed results with the CROSSLIST option.

As the first four objectives demonstrate, one can produce tabular arrangements and summaries of a data set for a formal report using the PRINT and FREQ procedures. Here, these tables are shown as they appear in the Results Viewer (HTML format). HTML, RTF, and PDF file creation is presented in Section 19.5.

In addition to the PRINT and FREQ procedures, the MEANS procedure also allows summary data to be displayed. The sample statistics that are presented can be controlled for the entire sample or for subgroups within the sample. The output from all of these procedures is enhanced by the usage of variable labels (Section 8.2) and effective TITLE statements (Section 2.4).

OBJECTIVE 10.5: Using the MEANS procedure, compute the mean, minimum, maximum, and sample size for homework, exam, and course totals for all students.

```
PROC MEANS DATA=gradebook MEAN MIN MAX N;
WHERE student NE "Total";
VAR HWTotal EXTotal CourseTL;
TITLE 'Objective 10.5';
TITLE2 'MEANS Procedure with no CLASS Statement';
RUN;
```

Since first introducing the MEANS Procedure in Chapter 3, the LABEL statement was introduced in Chapter 8. The effect of the variable labels can be observed in the table of results produced by the MEANS Procedure. One does not have to request the MEANS procedure to use the created labels.

Objective 10.5

MEANS Procedure with no CLASS Statement

The MEANS Procedure

Variable	Label	Mean	Minimum	Maximum	N
HWTotal	**Homework Point Total**	206.6666667	147.0000000	233.0000000	6
EXTotal	**Exam Score Total**	156.0000000	95.0000000	186.0000000	6
CourseTL	**Point Total for the Course**	362.6666667	310.0000000	418.0000000	6

One can control the number of decimal places using a FORMAT statement in the MEANS procedure. See Chapter 18 for the FORMAT statement.

OBJECTIVE 10.6: Redo Objective 10.5 computing the summary statistics for each group in a single table.

```
PROC MEANS DATA=gradebook MEAN MIN MAX N;
WHERE student NE "Total";
CLASS group;
VAR HWTotal EXTotal CourseTL;
TITLE 'Objective 10.6';
TITLE2 'Group is identified as the CLASS variable';
RUN;
```

The CLASS statement in the MEANS Procedure is used to create subdivisions in a single table of results. Using a BY statement (BY group;) instead of the CLASS statement would result in multiple tables of results, one for each level of the BY variable. Additionally, the SAS data set would have to be sorted by group prior to running the MEANS Procedure by group.

Objective 10.6

Group is identified as the CLASS variable

The MEANS Procedure

Group	N Obs	Variable	Label	Mean	Minimum	Maximum	N
1	2	HWTotal	Homework Point Total	219.0000000	210.0000000	228.0000000	2
		EXTotal	Exam Score Total	167.5000000	154.0000000	181.0000000	2
		CourseTL	Point Total for the Course	386.5000000	364.0000000	409.0000000	2
2	2	HWTotal	Homework Point Total	211.5000000	190.0000000	233.0000000	2
		EXTotal	Exam Score Total	126.0000000	95.0000000	157.0000000	2
		CourseTL	Point Total for the Course	337.5000000	328.0000000	347.0000000	2
3	2	HWTotal	Homework Point Total	189.5000000	147.0000000	232.0000000	2
		EXTotal	Exam Score Total	174.5000000	163.0000000	186.0000000	2
		CourseTL	Point Total for the Course	364.0000000	310.0000000	418.0000000	2

10.2 The REPORT Procedure

The REPORT procedure is a procedure that can create an organized table of results for each of the lines in the data set or can summarize information according to subgroups in the data set. Computationally, the procedure is limited, but it organizes the results much more easily. Several objectives will illustrate the capability of the procedure.

A simple syntax of the REPORT procedure appears below. This is merely an introduction to the procedure. There are some statements and options available that are not shown here. For more information about the REPORT procedure one can consult *Base SAS 9.4 Procedure Guide* (SAS Institute, 2013a).

```
PROC REPORT DATA=SAS-data-set <options>;
COLUMN specify column variables – order is important;
DEFINE report-item/<options>;
RBREAK BEFORE | AFTER/<options>;
```

BY and WHERE statements can be added to this procedure as needed.

The options on the PROC REPORT statement include:

WINDOWS|NOWINDOWS

WD or NOWD are the aliases for these options. NOWINDOWS is the default. There is a windowing environment one can use to obtain REPORT results by selecting WINDOWS or WD, but it not recommended by the author at this time. Either use NOWD or omit this option to produce report results. (Information about "REPORT procedure windows" can also be found in SAS Help and Documentation or SAS Institute (2013a).

COLUMN statement

This statement selects the variables to be included in the report. The order of the variables specified in this statement determines the order (left to right) the column variables appear in the report.

DEFINE statement

This statement modifies the basic column reporting by allowing summary statistics to be computed for numeric variables, specify column headers and the width of columns in the report, and change the order of the rows in the report. Options for the DEFINE statement include:

ORDER orders the rows for the defined variable in ascending order.

DESCENDING ORDER orders the rows for the defined variable in descending order.

WIDTH = # specifies the width of the column for the report-item. This has no effect in HTML output but is useful when using LISTING output.

GROUP defines the report-item as a grouping variable. GROUP variable(s) are typically listed as one of the first variables in the COLUMN statement.

"column-header" labels the column with the specified text. This replaces the variable name or its LABEL from the DATA step.

SUM, MEAN, N, MAX, MIN are selected statistics. Choose only one statistic per variable to summarize across the group. SUM is the default.

RBREAK statement

This statement will include a grand total in the report. Options include:

BEFORE | AFTER pick one of these options. This option places the grand total at the top (BEFORE) or at the bottom (AFTER) of columns for each of the variables in the COLUMN.

SUMMARIZE requests the total to be printed

OL prints a single line *over* the total

DOL prints a double line *over* the total

UL prints a single line *under* the total

DUL prints a double line *under* the total

The options OL, DOL, UL, and DUL are not effective in HTML Output, but they are effective in LISTING Output (Section 19.1).

OBJECTIVE 10.7: Produce a report containing the totals of the homework and exams, the course total, and the letter grade for each student in the course. Include a grand total for each of the numeric variables at the bottom of the list. Compare the output with a PRINT procedure (Section 2.3) used to accomplish the same task.

```
PROC REPORT DATA=gradebook;
WHERE student NE "Total";
COLUMN student HWTotal EXTotal CourseTL grade;
RBREAK AFTER / SUMMARIZE;
TITLE 'Objective 10.7';
TITLE2 'REPORT Procedure - Sum the columns with RBREAK Statement';

PROC PRINT DATA=gradebook NOOBS LABELS;
WHERE student NE "Total";
VAR student HWTotal EXTotal CourseTL grade;
SUM HWTotal EXTotal CourseTL;
TITLE2 'PRINT Procedure with a SUM Statement';
RUN;
```

The COLUMN statement in the REPORT procedure requests all of the variables to be included in the report and the order in which these variables are to appear. This is analogous to the VAR statement of the PRINT procedure. No DEFINE statements were included in the REPORT procedure in this objective. With the exception of the RBREAK statement producing column totals, this is the default output for the REPORT procedure.

Objective 10.7

REPORT Procedure – Sum the columns with RBREAK Statement

Student	Homework Point Total	Exam Score Total	Point Total for the Course	Grade
Dave	228	181	409	B
Lynn	233	95	328	C
Michael	190	157	347	C
Leslie	147	163	310	D
Andrew	210	154	364	C
Elizabeth	232	186	418	A
	1240	936	2176	

Objective 10.7

PRINT Procedure with a SUM Statement

Student	Homework Point Total	Exam Score Total	Point Total for the Course	Grade
Dave	228	181	409	B
Lynn	233	95	328	C
Michael	190	157	347	C
Leslie	147	163	310	D
Andrew	210	154	364	C
Elizabeth	232	186	418	A
	1240	**936**	**2176**	

Both procedures produce tables or reports with the same information. The REPORT procedure columns appear to adjust for wider column headers in HTML Output. The PRINT procedure will not use the variable labels without the LABELS option on the PROC PRINT statement. The sum of the values in the columns is produced by both procedures. The REPORT procedure computes these sums for all numeric variables in the COLUMN statement while the PRINT procedure requires the SUM statement to identify the numeric variables in the VAR statement to be summed. Both procedures can also produce ordered lists of data as seen in the next objective.

OBJECTIVE 10.8: Modify the report in Objective 10.7 so that the (letter) grades are listed from highest to lowest; that is, A's first, B's second, and so on, and the course total is highest to lowest within each letter grade. Subtotal the numeric variables for each letter grade. Again, compare REPORT versus PRINT. Use a BY statement in each of the procedures.

```
PROC SORT DATA=gradebook; BY grade DESCENDING coursetl;
PROC REPORT DATA=gradebook;
WHERE student NE "Total";
BY grade;
```

```
COLUMN student hwtotal extotal coursetl grade;
RBREAK AFTER / SUMMARIZE;
TITLE 'Objective 10.8';
TITLE2 'REPORT Procedure - Ordered Grade List';
RUN;

PROC PRINT DATA=gradebook NOOBS LABEL;
WHERE student NE "Total";
BY grade;
VAR student hwtotal extotal coursetl grade;
SUM  hwtotal extotal coursetl ;
TITLE2 'PRINT Procedure - Ordered Grade Lists';
RUN;
```

Since the BY statement is used in each of the procedures, the data must be compatibly sorted prior to the first procedure requiring the BY statement. One may recall the effect of a BY statement on the PRINT procedure which is a separate list for each value of the BY variable. A BY statement on the REPORT procedure generates a report for each value of the BY variable also.

Objective 10.8

REPORT Procedure - Ordered Grade List

Grade=A

Student	Homework Point Total	Exam Score Total	Point Total for the Course	Grade
Elizabeth	232	186	418	A
	232	186	418	

Objective 10.8

REPORT Procedure - Ordered Grade List

Grade=B

Student	Homework Point Total	Exam Score Total	Point Total for the Course	Grade
Dave	228	181	409	B
	228	181	409	

Objective 10.8

REPORT Procedure - Ordered Grade List

Grade=C

Student	Homework Point Total	Exam Score Total	Point Total for the Course	Grade
Andrew	210	154	364	C
Michael	190	157	347	C
Lynn	233	95	328	C
	633	406	1039	

Objective 10.8

REPORT Procedure - Ordered Grade List

Grade=D

Student	Homework Point Total	Exam Score Total	Point Total for the Course	Grade
Leslie	147	163	310	D
	147	163	310	

Objective 10.8

PRINT Procedure - Ordered Grade Lists

Grade=A

Student	Homework Point Total	Exam Score Total	Point Total for the Course	Grade
Elizabeth	232	186	418	A

Grade=B

Student	Homework Point Total	Exam Score Total	Point Total for the Course	Grade
Dave	228	181	409	B

Grade=C

Student	Homework Point Total	Exam Score Total	Point Total for the Course	Grade
Andrew	210	154	364	C
Michael	190	157	347	C
Lynn	233	95	328	C
Grade	**633**	**406**	**1039**	

Grade=D

Student	Homework Point Total	Exam Score Total	Point Total for the Course	Grade
Leslie	147	163	310	D
	1240	**936**	**2176**	

A quick look at the effect of the BY statement on the REPORT procedure is likely not what was intended. First, one notes the repetition of the titles for each report. These reports appear on different pages of the output also. Perhaps if the data set had more observations, the appearance of these reports with the repeating titles would improve. The BY statement results in the table delimiters of "Grade = A" and so on. For those letter grades with only one observation, there is a total computed, which is, of course, the same value of the single observation.

The effect of the BY statement on the PRINT procedure is a bit more desirable in that the titles are not repeated for each level of grade (the BY variable). The SUM statement operates on each list for each grade level unless there is only one observation. This is also different from the REPORT procedure with a BY statement. For the final table, "Grade = D", the sum of all of the observations of that variable is computed. Since there is only one observation in the last grade level, no subtotal is computed. If the last level of the BY variable has more than one observation, then a subtotal for the level and a grand total would be computed. To illustrate this, suppose the "Grade=D" were not included in the PRINT procedure. The last table would look like this:

Grade=C

Student	Homework Point Total	Exam Score Total	Point Total for the Course	Grade
Andrew	210	154	364	C
Michael	190	157	347	C
Lynn	233	95	328	C
Grade	**633**	**406**	**1039**	
	1093	**773**	**1866**	

A subtotal for "Grade = C" is produced in the second to last row of the table, and a grand total for each variable in the SUM statement is produced in the last row of the table. Another option on the PROC PRINT statement one may want to investigate is the N option (Section 2.3) which produces the number of observations in the printed list and for each level of the BY variable.

So far, not much difference between the REPORT and the PRINT procedures has been demonstrated. REPORT does, indeed, have more capability than the PRINT procedure. One may still need grouped results, and the REPORT procedure allows for the definition of a GROUP variable in a DEFINE statement.

OBJECTIVE 10.9: Modify Objective 10.8 for the REPORT procedure only. Group the results in a single table or report by letter grade (A, B, …) and descending course totals within each letter grade. Use a DEFINE statement with a GROUP option. Do not use a BY statement, and do not sum any of the numeric entries in this report.

```
PROC REPORT DATA=gradebook;
WHERE student NE "Total";
COLUMN student hwtotal extotal coursetl grade;  *put columns in this order;
DEFINE Grade / GROUP ;                        *declare GRADE as the group variable;
DEFINE coursetl / ORDER DESCENDING;           *puts coursetl in descending order;
TITLE 'Objective 10.9';
TITLE2 'REPORT Procedure - One Ordered Grade List';
TITLE3 'Option 1';
RUN;

PROC REPORT DATA=gradebook;
WHERE student NE "Total";
COLUMN grade student hwtotal extotal coursetl;        *grade is listed first;
DEFINE Grade / GROUP;
DEFINE coursetl / ORDER DESCENDING;
TITLE 'Objective 10.9';
TITLE2 'REPORT Procedure - One Ordered Grade List';
TITLE3 'Option 2';
RUN;
```

The single line comments (Section 20.2) in the code emphasize the items for the reader. In the Option 1 code, there are two DEFINE statements. There can only be one variable in each DEFINE statement. There can be multiple DEFINE statements in a single REPORT procedure. It is not necessary for all variables in the COLUMN statement to have a corresponding DEFINE statement. The second DEFINE statement indicates that the course total should be arranged in descending order.

Option 2 moves the grouping variable "Grade" to the first column. The group variable can be in any position in the COLUMN statement, and the column position can affect the GROUP option. The RBREAK statement has been removed from both options since the sum of the columns is no longer requested.

The GROUP and ORDER DESCENDING options on the DEFINE statements will order the rows in the report. It is not necessary to SORT the data set when using the GROUP or ORDER options.

Objective 10.9

REPORT Procedure – One Ordered Grade List

Option 1

Student	Homework Point Total	Exam Score Total	Point Total for the Course	Grade
Elizabeth	232	186	418	A
Dave	228	181	409	B
Andrew	210	154	364	C
Michael	190	157	347	C
Lynn	233	95	328	C
Leslie	147	163	310	D

Objective 10.9

REPORT Procedure – One Ordered Grade List

Option 2

Grade	Student	Homework Point Total	Exam Score Total	Point Total for the Course
A	Elizabeth	232	186	418
B	Dave	228	181	409
C	Andrew	210	154	364
	Michael	190	157	347
	Lynn	233	95	328
D	Leslie	147	163	310

These two options for this report are quite similar. The difference is the Grade column. In both options it is observed that the letter grades are in ascending order, and within each letter grade the Point Total for the Course is in descending order. No SORT procedure is necessary to do this. By moving the Grade variable from the last column to the first column one notices that Michael and Lynn do not have a grade listed to the left of their names. The GROUP option on the DEFINE statement produces this effect. With a longer list of observations in a report, the blank entries provide a visual break between the levels of the GROUP variable. A third option not included here would be to change the GROUP option to ORDER in the DEFINE statement for the grade variable in Option 2 and keep the new order of the variables in the COLUMN statement.

OBJECTIVE 10.10: Produce a report of the homework scores and homework total. HW1 and HW2 were not labeled in the initial DATA step. Label these columns in the REPORT procedure.

```
PROC REPORT DATA=gradebook;
WHERE student NE "Total";
COLUMN student hw1 hw2 hwtotal;
DEFINE hw1 / 'Homework 1';
DEFINE hw2 / 'Homework 2';
TITLE 'Objective 10.10';
TITLE2 'Report of Homework Scores';
RUN;
```

The *gradebook* data in its original order was used to generate this output. HWTotal was labeled earlier in this chapter so no label was necessary in this objective unless one wanted to change that label in this procedure. The labels created in the DEFINE statements are not available outside of this procedure.

Objective 10.10

Report of Homework Scores

Student	Homework 1	Homework 2	Homework Point Total
Dave	71	88	228
Lynn	64	96	233
Michael	68	75	190
Leslie	55	75	147
Andrew	70	79	210
Elizabeth	67	103	232

OBJECTIVE 10.11: Redo the task in Objective 10.10. Insert a third DEFINE statement to:

```
DEFINE hwtotal / "Ordered HW Total" ORDER;
```

This will put the observations in order from low to high according to the values in the HWTotal column.

Objective 10.11

Report of Homework Scores

Student	Homework 1	Homework 2	Ordered HW Total
Leslie	55	75	147
Michael	68	75	190
Andrew	70	79	210
Dave	71	88	228
Elizabeth	67	103	232
Lynn	64	96	233

OBJECTIVE 10.12: Create a list report of students where the student names, ID's, course totals, and grades are listed according to the project group variable. Label the id variable, "ID #".

```
PROC REPORT DATA=gradebook;
WHERE student NE "Total";
COLUMN group student id coursetl grade;
DEFINE group / GROUP 'Project Group';
DEFINE id / "ID #";
TITLE 'Objective 10.12';
RUN;
```

The uppercase GROUP in the DEFINE statement is an option for that statement, and it can appear either before or after the quoted text for the column header. In that same DEFINE statement "group" (lowercase) is a variable in the *gradebook* SAS data set.

Objective 10.12

Project Group	Student	ID #	Point Total for the Course	Grade
1	Dave	101	409	B
	Andrew	239	364	C
2	Michael	987	347	C
	Lynn	381	328	C
3	Elizabeth	128	418	A
	Leslie	579	310	D

If a variable is mistakenly left out of the COLUMN statement but a DEFINE statement for the variable has been written, the procedure will generate a report using only the variables in the COLUMN statement. For example, if the Group variable is omitted from the COLUMN statement, a warning is issued in the Log, and the report will not contain the Group variable.

```
615   PROC REPORT DATA=gradebook;
616   WHERE student NE "Total";
617   COLUMN student id coursetl grade;
618   DEFINE group / GROUP 'Project Group';
619   DEFINE id / "ID #";
620   TITLE 'Objective 10.12';
621   RUN;
WARNING: group is not in the report definition.
```

It is suggested that the reader experiment with a different ordering of variables in the COLUMN statement, such as: `COLUMN student id grade coursetl group;` while keeping the same DEFINE statements and observe the effect of the reordered COLUMN statement.

OBJECTIVE 10.13: Create a report of the average course total score for each project group. Relabel the CourseTL variable for this report to indicate it is an average score for the project groups.

```
PROC REPORT DATA=gradebook;
WHERE student NE "Total";
```

```
COLUMN group coursetl;
DEFINE group / GROUP 'Project Group';
DEFINE coursetl / "Project Group Course Average" MEAN;
TITLE 'Objective 10.13';
RUN;
```

When there is a GROUP variable, by default numeric variables are summed for the group levels, thus the MEAN option is required for this report. One can verify the values in the second column are means of for each project group in the Objective 10.12 report. In addition to the MEAN option on the DEFINE statement, one should also investigate the effects of the options: SUM, N, MAX, and MIN.

Objective 10.13

Project Group	Project Group Course Average
1	386.5
2	337.5
3	364

Note: If ID or any other numeric variable were included in the COLUMN statement and no DEFINE statement for the ID variable is included, then the sum of the numeric variables would be reported. Such as, COLUMN group id coursetl;

Project Group	ID	Project Group Course Average
1	340	386.5
2	1368	337.5
3	707	364

The ID variable is summed for each group (no, this does not make sense), and the course total for each student is averaged for the project groups.

If a character variable were included in the COLUMN statement, then no means are computed for the groups. Such as, COLUMN group coursetl grade;

Project Group	Point Total for the Course	Grade
1	409	B
	364	C
2	347	C
	328	C
3	418	A
	310	D

Caution: If DEFINE coursetl/"Project Group Course Average"; is part of the REPORT code submitted, the header for the second column would mistakenly label the individual student course totals using the column header in the DEFINE statement.

10.3 Chapter Summary

While the computational difficulty involved in the REPORT procedure is not rigorous, the procedure has the capability of organizing both small and large data sets into logical summary report displays. Statements and options within the procedure affect the visual display of the output. There are many more options for the REPORT procedure and DEFINE statements that generate more sophisticated color reports. The PRINT, FREQ, and MEANS procedures were also revisited. With the accumulated information about the DATA step and a bit more SAS programming experience, one can more fully investigate these introductory procedures.

11

Regression and Correlation Analysis

Often researchers collect several columns of data and are interested in whether one or more of the column variables is a good predictor of a response measured in another column. When a single variable is used for prediction purposes, a simple linear regression model is fit to the data. There are several ways to assess how well the model fits the data both graphically and statistically. When more than one variable is used to predict a response variable, this is multiple linear regression.

Multiple random response variables can be measured, and the linear association between pairs of the variables can be investigated. When both of the variables are considered random variable responses, then a correlation analysis is often used to investigate the linear associations between pairs of the random response variables.

This chapter will overview simple linear regression analysis and correlation analysis and the SAS programming needed to perform those analyses.

11.1 Simple Linear Regression

When two variables are observed for each observation in a data set, whether or not one variable can be used to predict the other can be investigated. A simple linear regression line can be fit to the data. The values of the independent and dependent variables are given by (x_i, y_i). The statistical model for simple linear regression is given by:

$$y_i = \beta_0 + \beta_1 x_i + \varepsilon_i,$$

where i = 1, … n and y_i is the value of the dependent variable, β_0 is the y-intercept, β_1 is the slope of the line, x_i is the value of the independent variable, and ε_i is the unobservable random error about the regression line. Simple linear regression assumes that the ε_i are normally distributed with a mean of zero and with a common variance, σ^2. That is, $\varepsilon_i \sim N(0, \sigma^2)$. The values of the y-intercept and slope are population parameters that can be estimated using ordinary least square regression techniques (Draper and Smith, 1998). The estimates of β_0 and β_1 are given by $\hat{\beta}_0$ and $\hat{\beta}_1$, respectively. The estimated regression line is given by $\hat{y}_i = \hat{\beta}_0 + \hat{\beta}_1 x_i$ where $\hat{y}_i$ is the predicted value of the dependent variable at a given x_i; that is, $(x_i, \hat{y}_i)$ is a point on the regression line.

Of interest is whether or not the regression line adequately models the data. That is, is x a good predictor of y? If so, then it is expected that the value of the slope would be different from zero. A line with a zero slope is a horizontal line and is equivalent to using $\bar{y}$ to predict the outcome for y. Therefore, it is of interest to test whether or not the slope is zero. Another phrasing for this is, "Is the slope significant?" Additionally, a confidence interval for the slope can be computed. Similarly, the y-intercept can be tested or estimated. Depending upon the application, one may need to test whether or not either of the simple linear regression parameters differs from a nominal value, b_0. The t-tests and confidence intervals methods are summarized in Table 11.1.

TABLE 11.1

Testing Regression Model Parameters

Hypotheses	Test Statistic	Reject H_0 if	(1–α)100% CI for β_i
H_0: $\beta_i = 0$ H_1: $\beta_i \neq 0$	$t = \frac{\hat{\beta}_i}{se(\hat{\beta}_i)}$	$\lvert t\rvert \geq t_{\alpha/2,df}$ or $t^2 > F_{\alpha,1,df}$	$\hat{\beta}_i \pm t_{\alpha/2,df}\, se(\hat{\beta}_i)$
H_0: $\beta_i = b_0$ H_1: $\beta_i \neq b_0$, where $b_0 \neq 0$	$t = \frac{\hat{\beta}_i - b_0}{se(\hat{\beta}_i)}$	$\lvert t\rvert \geq t_{\alpha/2,df}$ or $t^2 > F_{\alpha,1,df}$	

where i = 0 or 1, $se(\hat{\beta}_i)$ is the standard error of the $\hat{\beta}_i$ parameter estimate, and df is the error degrees of freedom. In simple linear regression df = n – 2.

As in previous chapters, a rejection region is specified, but in practice, one reads the observed significance level or p-value for these tests as determined by the t_{df} distribution or $F_{1,df}$ distribution. And, as before, a significant difference is concluded when this observed significance level is less than or equal to the nominal value of α for the hypothesis test.

When the estimated regression line, $\hat{y}_i = \hat{\beta}_0 + \hat{\beta}_1 x_i$, is computed, $\hat{y}_i$ is an estimate of $\mu_{Y|X} = \beta_0 + \beta_1 x$ where $\mu_{Y|X}$ = "expected value or mean of Y given a particular value of X". Confidence intervals for the mean response at a given value of $X = x_0$, $\mu_{Y|X}$, and prediction intervals for an individual Y response at a given value of $X = x_0$, $Y|X$, can also be computed. These intervals are centered at the point prediction, $\hat{y}_i$. In a simple linear regression case, these interval calculations are:

$$(1-\alpha)100\%\ \text{CI for}\ \mu_{Y|X} \qquad (\hat{\beta}_0 + \hat{\beta}_1 x_0) \pm t_{\alpha/2,n-2}\sqrt{MSE\left(\frac{1}{n} + \frac{(\bar{x}-x_0)^2}{\sum (x_i - \bar{x})^2}\right)}$$

and

$$(1-\alpha)100\%\ \text{PI for}\ Y|X \qquad (\hat{\beta}_0 + \hat{\beta}_1 x_0) \pm t_{\alpha/2,n-2}\sqrt{MSE\left(1 + \frac{1}{n} + \frac{(\bar{x}-x_0)^2}{\sum (x_i - \bar{x})^2}\right)}$$

where MSE is the error variance from the ANOVA and is the estimate of σ^2.

When there is only one independent variable, the regression model is a simple linear regression model. More independent variables and hence, more regression parameters could be included in a regression analysis. The resulting model is a multiple regression model. If there are k different independent variables used to predict Y, then there are k slope parameters and a y-intercept (k + 1 parameters total). The Table 11.1 formulas for the hypotheses tests and the confidence intervals for β_i apply for each of the possible k + 1 parameters in the model, and df = n – k – 1 in the multiple regression application. However, a knowledge of Type I (sequential) testing, Type III testing, and model selection are some of the necessary concepts needed when fitting multiple regression models. See Draper and Smith, 1998; Montgomery, Peck, and Vining, 2012; Myers, 2000; or Kutner, Nachtsheim, and Neter, 2004 for these applied regression topics.

11.2 The REG Procedure

When either simple linear or multiple regression model equations are to be estimated for a set of data, the REG procedure can provide estimates of the parameters, tests of their significance, residual analysis, plots of data, and much more.

The syntax of the REG procedure:

PROC REG DATA = *SAS-data-set* <options>;
<*label:*> MODEL *dependents* = *regressors* </*options* >;
ID *variable(s)*;
<*label:*> TEST *equation;*
RUN;

BY and WHERE statements can be included, of course, when subsetting the data for a regression analysis is needed. This is not a complete list of statements for the REG procedure but a few of the essential statements to get started in a simple linear regression analysis.

PROC REG Statement options include:

PLOTS= NONE | DIAGNOSTICS | FIT (pick one plot option)
ODS Graphics must be enabled (Section 19.1) to produce these graphics. By default, the REG procedure produces a panel of regression DIAGNOSTICS, plot(s) of the residuals versus each independent variable, and a graph of the simple linear regression equation FIT to the identified dependent and independent variables as part of the default output. If there is more than one independent variable, there is no FIT graph produced, only the DIAGNOSTICS panel and residual plot(s).

PLOTS=NONE will suppress all graphs.

PLOTS(ONLY) = FIT or PLOTS(ONLY) = DIAGNOSTICS will select only the indicated graph for the output.

PLOTS(UNPACK) = DIAGNOSTICS or PLOTS = DIAGNOSTICS (UNPACK) will produce larger versions of the eight plot diagnostics that, by default, appear in a 3 × 3 panel. The ninth item in the panel is a small table of regression statistics which is not produced with the UNPACK option.

SIMPLE
Simple statistics for each variable in the MODEL statement are printed. (Mean, summation, uncorrected sum of squares, ...)

MODEL statement
The MODEL statement is required in the REG procedure. In the MODEL statement the dependent and independent variables are identified. The dependent variables always occur to the left of the equality symbol, and the independent variable(s) are always listed to the right of the equality symbol. If only one independent variable is identified, the result is a simple linear regression analysis as is taught in introductory statistics courses. For the same regressor or independent variable, one can specify multiple dependent variables in a single MODEL statement, such as MODEL a b c = x; .

More than one MODEL statement can be used in a single regression procedure. Each MODEL statement can have different options. Each MODEL statement can have a different *label* which can assist in identifying the output associated with it, but the usage of the *label* is optional.

MODEL statement options include:

ALPHA = p specifies the Type I error rate for the confidence and prediction intervals that may be requested in the MODEL statement options. $0 < p < 1$. If the ALPHA option is not specified, all intervals are computed with the default setting p = 0.05.

CLI requests the (1-p)100% upper- and lower-confidence limits for an individual predicted value. (Note: CL**I** for **individual**.)

CLM requests the (1-p)100% upper- and lower-confidence limits for the expected value or mean response. (Note: CL**M** for **mean** response.)

CLB requests the (1-p)100% confidence limits for the regression parameters. Typically, regression parameters are identified as β's in statistical literature, hence the CL**B**.

P calculates the predicted values from the input data and the estimated model.

For the CLI, CLM, and P options on the MODEL statement, it is recommended that one use the ID statement to facilitate reading the output results. Objective 11.3 will demonstrate this.

The following options are listed here for those readers who have some familiarity with multiple regression topics.

COLLIN computes variance inflation factors (VIF's) and other collinearity diagnostics.

INFLUENCE computes influence diagnostics for each observation in the analysis. A plot of leverage statistics is one of the graphs in the DIAGNOSTICS panel.

NOINT all regression models have a y-intercept by default. This option forces the regression line through the origin; $\beta_0 = 0$.

R requests a listing of the residuals, as well as studentized residuals and Cook's D for each observation. Cook's D is another one of the graphs in the DIAGNOSTICS panel.

SELECTION = *name* identifies the model selection method to be used when determining a multiple regression model, where *name* can be FORWARD (or F), BACKWARD (or B), STEPWISE, MAXR, MINR, RSQUARE, ADJRSQ, CP. Only one method can be specified in a MODEL statement.

SS1 prints the Type I or sequential sums of squares for each term in the model.

SS2 prints the Type II or partial sums of squares for each term in the model. Type III SS are the default output in the REG procedure.

ID statement

When one requests predicted values, residuals, confidence limits or any other of the options that produce output for each of the observations in the data set, an ID statement identifies one or more variables to serve as reference column(s) on the left side of the produced table. This statement is optional, but it can facilitate an easier identification of results for each observation in the data.

TEST statement

The TEST statement is an optional statement in a regression analysis. If it is of interest to test whether a regression parameter differs from a specified non-zero value, the TEST statement can accomplish this. The independent variable name used in the *equation* must also appear in the MODEL statement, or one can test the *intercept*. As an example,

```
MODEL y = x;
Slope_eq_5: TEST x = 5;
```

The default output from the MODEL statement contains a test of whether the regression parameters are zero or not. This TEST statement requests a test of whether the slope coefficient of x is 5 or not; that is, H_0: $\beta_1 = 5$ versus H_1: $\beta_1 \neq 5$. The actual output for this test is an F-test. Objective 11.2 will demonstrate how to interpret this. "Slope_eq_5" is the programmer's choice of text to label or identify the requested test in the results.

To demonstrate the REG procedure, an example from the beef cattle industry will be used. These data are found in Table 11.2. For the study period, the daily average dry matter intake (DMI) and average daily gain (ADG) are measured on 56 steers. The amount of feed intake and rate of weight gain impact the size and end weight of the animal. Prior to harvest cattle producers control the feed intake of the animals (DMI) and also can monitor

TABLE 11.2

Beef Cattle Data (56 observations)

DMI	ADG	CWT	BackFat	REA	DMI	ADG	CWT	Backfat	REA
17.0090	3.05	819.000	0.54571	13.3821	21.1642	3.59	836.000	0.51452	14.7292
17.0183	2.87	857.000	0.53974	13.7779	21.2228	3.49	866.878	0.52544	14.8798
17.4826	3.12	826.000	0.46000	13.5262	22.2457	3.61	888.000	0.54220	14.6725
18.0043	2.73	831.256	.	.	23.3023	3.56	892.000	0.48324	13.6784
16.2731	2.83	799.000	0.54622	12.8589	24.5569	3.68	885.000	0.47611	13.5639
17.5451	2.98	850.000	0.50550	13.8188	19.7097	3.47	854.000	0.58245	13.9235
16.6000	2.78	790.000	0.50280	13.4800	20.1899	3.37	842.000	0.58304	13.1376
17.1309	2.80	805.000	0.51200	12.6460	17.9807	2.92	798.000	0.47350	13.4375
17.8689	2.85	814.000	0.49434	13.4662	17.8008	2.84	817.000	0.52578	12.8422
17.8994	3.03	838.000	0.53535	13.2404	17.3405	2.88	817.000	0.48162	13.6371
18.4669	3.04	806.563	.	.	18.2908	2.76	796.000	0.47315	13.4815
16.7834	2.86	835.000	0.50667	13.8126	18.7467	2.88	842.000	0.50133	14.5440
17.7115	2.94	844.000	0.45829	13.6686	19.0026	3.84	939.000	0.58390	14.4117
16.5675	2.94	810.000	0.57653	12.6427	18.4399	2.80	831.000	0.46667	14.1912
17.2382	2.90	822.000	0.54649	12.9730	17.0606	1.53	701.000	0.39456	12.3437
18.9442	3.12	826.000	0.53853	13.4189	18.2209	2.98	848.000	0.52294	14.4610
17.6088	2.96	850.000	0.53351	13.0730	16.5719	2.92	807.000	0.49563	14.1172
19.3759	3.19	842.000	0.46176	13.8103	17.8575	2.80	792.000	0.46196	13.6490
16.5377	2.82	829.000	0.54061	13.6045	18.2689	3.10	847.000	0.49333	14.8190
17.6711	2.98	807.000	0.48278	12.9250	18.8171	3.04	846.000	0.46959	13.4986
18.5135	3.09	824.000	0.50319	13.1232	19.8960	2.90	800.000	0.49308	13.7439
16.3334	2.86	823.000	0.50861	13.7305	20.8472	3.36	858.000	0.51261	14.2328
16.9373	2.91	818.000	0.52319	13.1000	22.6299	3.38	896.000	0.53629	14.6718
17.5444	2.91	816.000	0.45947	13.4447	21.9390	3.44	876.000	0.51627	14.1254
18.2698	2.98	831.172	0.51020	13.2291	23.7774	3.62	922.000	0.53076	15.4403
18.6745	3.16	844.000	0.45442	14.3259	23.6753	3.77	965.000	0.56481	15.9731
17.7172	3.01	848.000	0.54147	13.7000	21.0995	3.25	848.000	0.50533	13.4050
19.8536	3.21	820.397	0.46944	13.9285	21.4372	3.25	859.000	0.52818	13.9649

ADG. At harvest, the carcass weight (CWT) and back fat thickness (Backfat) are measured. Another characteristic of the harvested meat is the rib eye area (REA).

For the objectives in this chapter, the SAS data set *beef* has been created using all five columns of data in Table 11.2.

OBJECTIVE 11.1: Regress carcass weight on feed intake (DMI) and observe the default output. Use $\alpha = 0.05$ to interpret the significance of the slope parameter.

```
PROC REG DATA=beef;
MODEL cwt = dmi;
TITLE 'Objective 11.1';
RUN;
```

The MODEL statement is order important in that it specifies the dependent variable = independent variable, in that order. If these were reversed, there are output results, but they are not meaningful. In this case the reversal would mean predicting the DMI based on the end measure of CWT.

Objective 11.1
The REG Procedure
Model: MODEL1
Dependent Variable: CWT

Number of Observations Read	56
Number of Observations Used	56

Analysis of Variance

Source	DF	Sum of Squares	Mean Square	F Value	Pr > F
Model	1	41996	41996	49.88	<.0001
Error	54	45462	841.88001		
Corrected Total	55	87458			

Root MSE	29.01517	**R-Square**	0.4802
Dependent Mean	837.39762	**Adj R-Sq**	0.4706
Coeff Var	3.46492		

Parameter Estimates

Variable	DF	Parameter Estimate	Standard Error	t Value	Pr > |t|
Intercept	1	590.35542	35.19188	16.78	<.0001
DMI	1	13.10479	1.85545	7.06	<.0001

The REG Procedure
Model: MODEL1
Dependent Variable: CWT

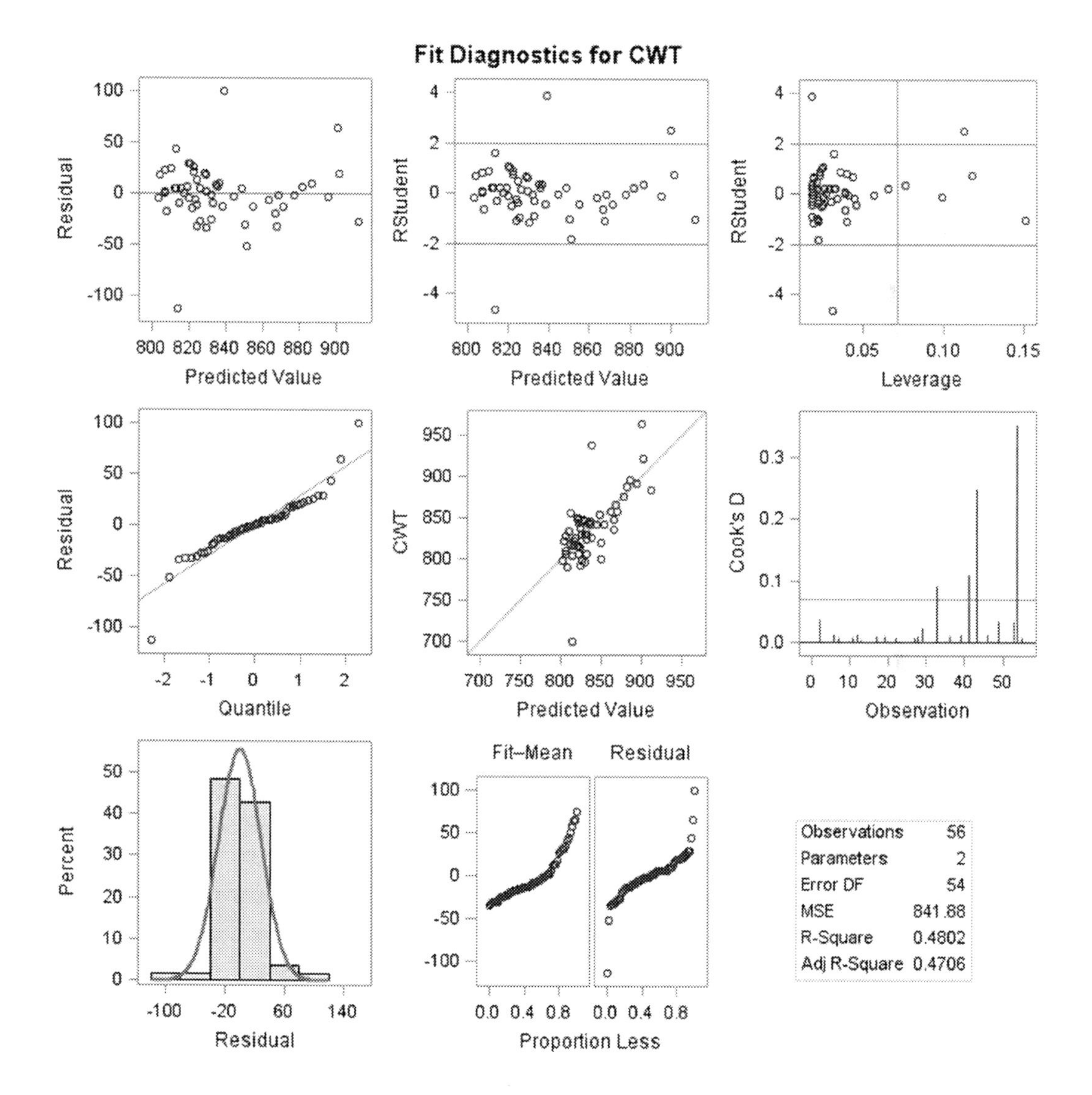

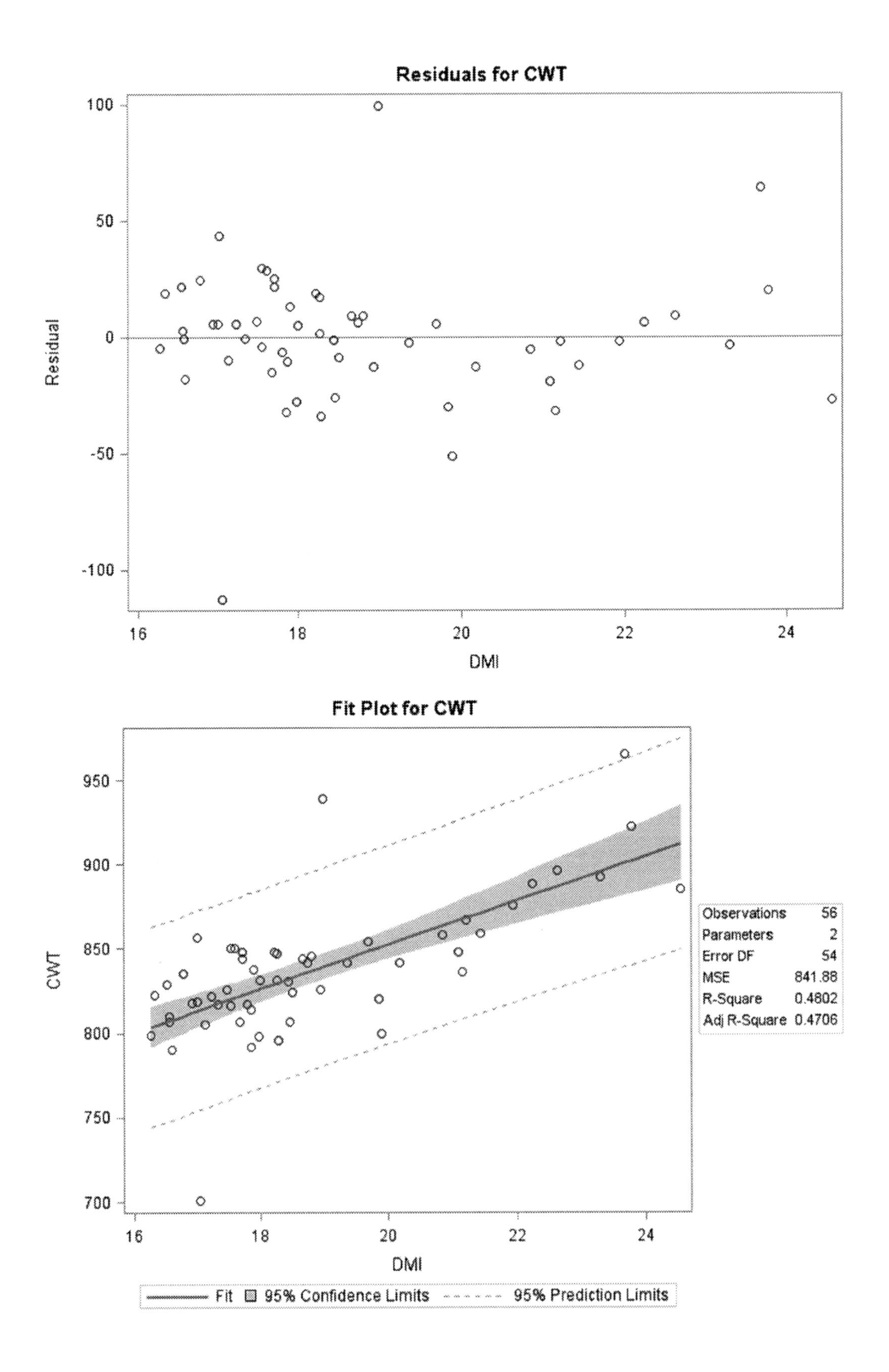
Residuals for CWT
Residual
DMI
Fit Plot for CWT
CWT
DMI
Observations 56
Parameters 2
Error DF 54
MSE 841.88
R-Square 0.4802
Adj R-Square 0.4706
Fit
95% Confidence Limits
95% Prediction Limits

In the output for this simple linear regression analysis, the dependent variable CWT is identified in the first lines of the results followed by the Number of Observations Read (the number of observations in the SAS data set *beef*) and the Number of Observations Used, how many of the observations in *beef* that are complete records for the variables involved in this analyses.

In the **Analysis of Variance** table, the F-value is the test statistic for the significance of the slopes in the model. Below the **Analysis of Variance** table are fit statistics for the regression analysis, the coefficient of determination (**R-square**) and the coefficient of variation or CV (**Coeff Var**) among them.

In the table of **Parameter Estimates** one obtains the estimates of the y-intercept ($\hat{\beta}_0$) and slope ($\hat{\beta}_1$) coefficient of the estimated regression line in the Parameter Estimate column of the table. In this objective, the regression line is $\widehat{CWT} = 590.3554 + 13.10479DMI$, and the reader should identify where these coefficients are found in the **Parameter Estimates** table. The values in the Standard Error column are the standard errors of the estimates, denoted as $se(\hat{\beta}_i)$, and the t-Value is the test statistic for testing whether or not the parameter in question differs from zero, both as shown in Table 11.1. The test for the significance of the slope parameter in the case of simple linear regression is also the test being done in the **Analysis of Variance** table. One notes that $(7.06)^2 = 49.88$, or more generally, $t_{df}^2 = F_{1,df}$ for the simple linear regression slope. If the SAS data set *beef* had labels for the DMI and CWT variables, there would be a Label column between the Variable and DF columns in the table of **Parameter Estimates**.

Following the table of parameter estimates and t-tests, is the **Fit Diagnostics for CWT** the panel graph of DIAGNOSTICS for this analysis. The second graph, **Residuals for CWT**, is a residual plot of the residuals by each independent variable in the model. Lastly, using the legend at the bottom of the **Fit Plot for CWT**, the fitted regression line among the observed data points and 95% confidence and prediction limits about the regression line can be identified. One or more of these plots can be suppressed in the analysis using the PLOTS = option in the PROC REG statement.

The SIMPLE option on the PROC REG statement is a quick way to obtain summary statistics for all variables in the MODEL statement. Including the SIMPLE option in Objective 11.1 would produce the **Descriptive Statistics** table.

Descriptive Statistics

Variable	Sum	Mean	Uncorrected SS	Variance	Standard Deviation
Intercept	56.00000	1.00000	56.00000	0	0
DMI	1055.67257	18.85130	20145	4.44620	2.10860
CWT	46894	837.39762	39356605	1590.14312	39.87660

The columns for the Sum, Mean, Variance, and Standard Deviation for the variables DMI and CWT should be familiar to the reader. These statistics are also available using the MEANS Procedure. The Uncorrected SS column contains the sum of the squared values of the variables. That is, $\Sigma(DMI)^2 = 20145$ and $\Sigma(CWT)^2 = 39356605$. The values in the Intercept row indicate the sample size (n = 56).

OBJECTIVE 11.2: Rerun the simple linear regression analysis suppressing all of the graphs, producing 99% confidence intervals for each of the parameters in the model. Test whether the slope in this analysis differs from 9.2. Use $\alpha = 0.01$ in your conclusions.

```
PROC REG DATA=beef PLOTS=NONE;
MODEL cwt = dmi / CLB ALPHA=0.01;
Slope9_2: TEST dmi = 9.2;
TITLE 'Objective 11.2';
RUN;
```

The CLB option will compute the confidence intervals for the parameters. Ninety five percent confidence is the default. Other confidence levels can be computed by using the ALPHA=p option. The *label* on the TEST statement is optional and can be omitted. However, if multiple TEST statements are used in an analysis, labels are strongly recommended since the output results use these labels. The first part of these results is identical to the results in Objective 11.1 and is not repeated here.

Objective 11.2

<Number of observations, ANOVA, and Fit statistics tables suppressed by the author>

Parameter Estimates

Variable	DF	Parameter Estimate	Standard Error	t Value	Pr > |t|	99% Confidence Limits	
Intercept	1	590.35542	35.19188	16.78	<.0001	496.39363	684.31722
DMI	1	13.10479	1.85545	7.06	<.0001	8.15076	18.05881

Model: MODEL1

Test Slope9_2 Results for Dependent Variable CWT

Source	DF	Mean Square	F Value	Pr > F
Numerator	1	3728.60286	4.43	0.0400
Denominator	54	841.88001		

The CLB option produced the additional two columns in the **Parameter Estimates** table under the "99% Confidence Limits" column heading. With 99% confidence, one concludes that the true population slope is between 8.15 and 18.06, and similarly interpret the interval estimate of the y-intercept.

When testing whether the slope is some non-zero value, the TEST statement can be used. The *label* "Slope9_2" was used to identify the test of H_0: $\beta_1 = 9.2$ versus H_0: $\beta_1 \neq 9.2$, and this label can be found in the heading of the last table of results **Test Slope9_2 Results for Dependent Variable CWT**. When creating a label for a TEST statement (and for a MODEL statement) use the usual naming conventions for a SAS variable name (Section 2.2). That is, the label must begin with a character value and not a number and cannot contain special characters except the underscore. If the label is absent, the table of test results is still printed, but is labeled **Test 1 Results for Dependent Variable CWT** if there is only one TEST statement. The TEST statement conducts an F-test for this test of hypotheses. Recall the association between t and F mentioned in the remarks of Objective 11.1, $t_{df}^2 = F_{1,df}$. Here, there is no evidence that the slope is different from 9.2 ($\alpha = 0.01$, $F_{1,54} = 4.43$, $p = 0.0400$). This is also supported by the 99% confidence interval for the slope (8.15, 18.06) which captures 9.2.

The TEST statement can be used to test the y-intercept parameter also. For example, past models have used 500 as the y-intercept. Has the y-intercept parameter changed? Including the following TEST statement in the PROC REG block of statements would produce a table containing an F-test for this test also:

```
Intercept500: TEST intercept = 500 ;
```

Test Intercept500 Results for Dependent Variable CWT

Source	DF	Mean Square	F Value	Pr > F
Numerator	1	5549.75282	6.59	0.0130
Denominator	54	841.88001		

Conclusion: No, the y-intercept has not changed from 500 ($\alpha = 0.01$, $F_{1,54} = 6.59$, $p = 0.0130$).

In these TEST statements, the numerator df is one in each case. For those readers with more experience in regression topics, the TEST statement can also be used to produce simultaneous tests of more than one parameter, and these tests have more than one numerator df. For these multiple equation simultaneous tests, a comma separates the equations in a single TEST statement.

OBJECTIVE 11.3: Predict the carcass weight when the dry matter intake is 18, 20, or 22. Compute 99% confidence intervals for the mean response and 99% prediction intervals also. Use an ID statement. Suppress all plots.

First one must review the data in Table 11.2. There are no DMI observations of 20 or 22. Thus, these must be incorporated in analysis. Create a small SAS data set containing all three of these DMI values. Then concatenate this small SAS data set to the SAS data set *beef*. Run a regression analysis on this modified *beef* data set requested predicted value and the 99% intervals.

```
DATA obj11_3;
INPUT DMI @@;
DATALINES;
18 20 22
;

DATA beef2;
SET obj11_3 beef;
RUN;
PROC REG DATA=beef2 PLOTS(ONLY)=FIT;
MODEL cwt = dmi / P CLI CLM ALPHA=0.01;
ID dmi ;
TITLE 'Objective 11.3';
RUN;
```

The SAS data set *obj11_3* is listed first in the SET statement. This will put the three DMI values 18, 20, and 22 "at the top" of the *beef2* SAS data set which will assist in reading the predicted values in the results.

Objective 11.3
The REG Procedure
Model: MODEL1
Dependent Variable: CWT

Number of Observations Read	59
Number of Observations Used	56
Number of Observations with Missing Values	3

< ANOVA, Fit statistics tables suppressed by the author>

Parameter Estimates

Variable	Label	DF	Parameter Estimate	Standard Error	t Value	Pr > |t|
Intercept	Intercept	1	590.35542	35.19188	16.78	<.0001
DMI	DMI	1	13.10479	1.85545	7.06	<.0001

Output Statistics

Obs	DMI	Dependent Variable	Predicted Value	Std Error Mean Predict	99% CL Mean		99% CL Predict		Residual
1	18.0	.	826.2416	4.1867	815.0631	837.4200	747.9692	904.5140	.
2	20.0	.	852.4511	4.4245	840.6378	864.2645	774.0855	930.8167	.
3	22.0	.	878.6607	7.0118	859.9393	897.3822	798.9606	958.3608	.
4	17.0	819.0000	813.2543	5.1690	799.4531	827.0555	734.5645	891.9441	5.7457
5	17.0	857.0000	813.3771	5.1575	799.6066	827.1476	734.6927	892.0615	43.6229
6	17.5	826.0000	819.4606	4.6350	807.0852	831.8360	741.0083	897.9129	6.5394
7	18.0	831.2558	826.2982	4.1837	815.1278	837.4686	748.0270	904.5695	4.9576
8	16.3	799.0000	803.6103	6.1578	787.1691	820.0515	724.4148	882.8057	−4.6103
⋮									
57	23.7	965.0000	900.6150	9.7544	874.5709	926.6591	818.8843	982.3457	64.3850
58	21.1	848.0000	866.8600	5.6951	851.6541	882.0659	787.9117	945.8083	−18.8600
59	21.4	859.0000	871.2858	6.1689	854.8150	887.7567	792.0842	950.4875	−12.2858

Sum of Residuals	0
Sum of Squared Residuals	45462
Predicted Residual SS (PRESS)	49157

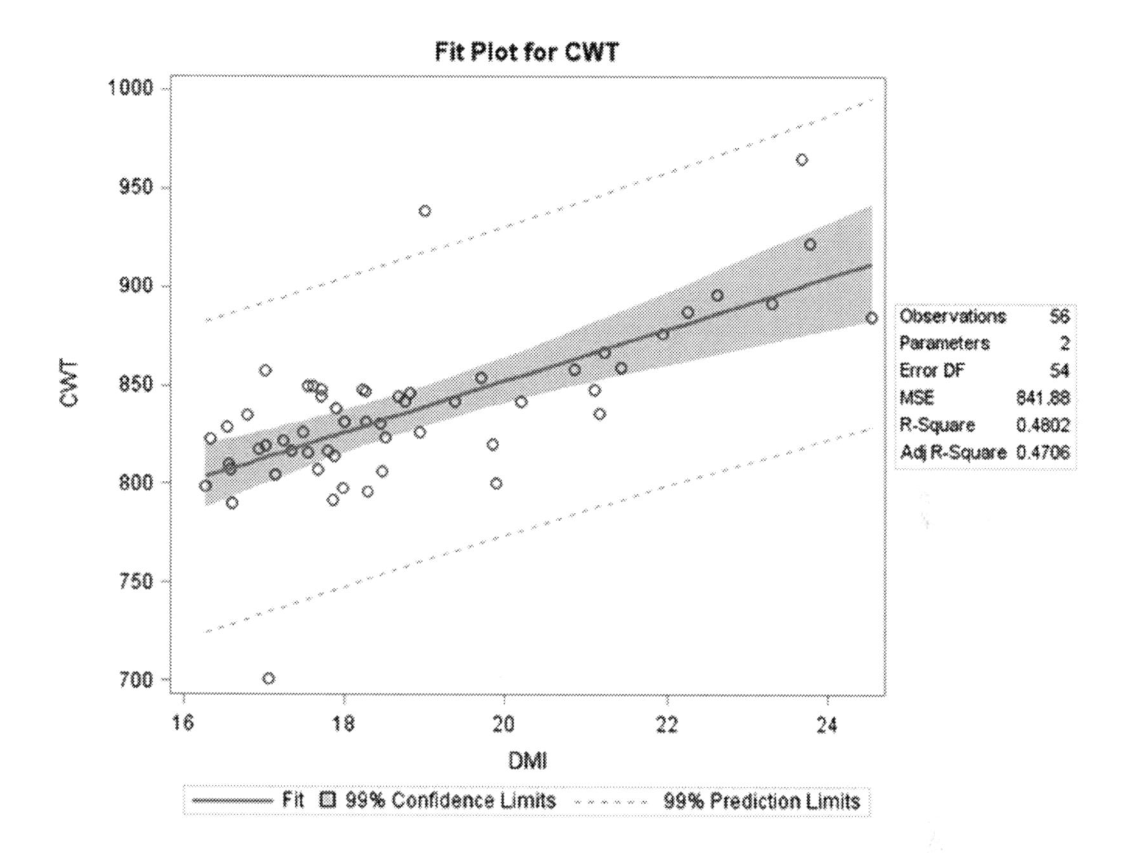

The REG procedure will only use observations in the data set that have both CWT and DMI to determine the equation of the regression line. In the **Number of Observations Used** that number did not change from 56 as seen in Objective 11.1. In the table of **Parameter Estimates**, these are the same as in Objective 11.1 also. Thus, the addition of the three new lines of data which only have DMI values, do not corrupt the analysis. The P option on the MODEL produces the predicted values on the estimated regression line for all of the values of the independent variable, even those without an observed dependent variable. The CLI and CLM options create the requested intervals about those predicted values also.

In the table of **Output Statistics,** one notices that the values of 18, 20, and 22 that were appended to the *beef* data set appear in the first three rows, and that the last row of the table indicates 59 observations now. The DMI column in this table is the result of the ID statement. This table of predicted values is difficult to read without the column(s) of independent variables included in the ID statement. When DMI = 18, the predicted CWT is 826.2416. Observation 7 also has DMI = 18, and the predicted value of CWT is same, of course. It should be. Observation 7 in the table does have an observed CWT (Dependent Variable column). The CLM option on the MODEL produces the two columns headed by 99% CL Mean, and the CLI option produces the 99% CL Predict information for all values of DMI. The three appended DMI values do not have a value in the Residual column since Residual = Dependent Variable – Predicted Value.

When predicting for selected values, a short data set, such as *obj11_3*, and the ID statement are quick solutions for enhancing the output. One could also take advantage of sorting the original SAS data set, *beef*, by the independent variable, DMI, before concatenation.

In the **Fit Plot for CWT**, the observed data (open circles) and the regression line are the same as in Objective 11.1. The confidence limits and prediction limits are wider since the confidence level is now 99%.

11.3 Correlation Coefficient

A measure of linear association between two random variables is a correlation coefficient. Correlation differs from regression analysis in that there are no independent and dependent variable associations. The two variables that are analyzed are both random variable responses. When the data are normally distributed, a Pearson correlation coefficient can be computed. When the data are not normally distributed, a Spearman correlation coefficient (rank-based calculation) can be computed. In either case, the sample correlation coefficient, r, estimates the population correlation coefficient, ρ. Both $-1 \le r \le 1$ and $-1 \le \rho \le 1$. When the correlation coefficient is near zero, the two variables are not linearly related. If one variable increases as the other increases, the correlation is positive. And, if one variable decreases as the other increases, the correlation is negative. For either the Pearson or Spearman measurements, it may be of interest to test whether or not there is a linear association between the two variables. That is, is the correlation non-zero? This test is overviewed in Table 11.3.

A rejection region is specified, but in practice, one reads the observed significance level or p-value for these tests as determined by the t_{n-2} distribution where n = sample size. A significant correlation is concluded when this observed significance level is less than or equal to the nominal value of α for the hypothesis test. See Ott and Longnecker, 2016 or Freund, Wilson, and Mohr, 2010 for information about both Pearson and Spearman correlation coefficients.

There are other non-parametric correlations, partial correlations, and other measures of association one can compute given non-normal data or conditional data. (SAS Institute, 2013a). These are topics beyond the scope of this book.

TABLE 11.3

Test for the Pearson or Spearman Correlation Coefficient

Hypotheses	Test Statistic	Reject H_0 if
H_0: $\rho = 0$ (No linear association.) H_1: $\rho \neq 0$ (Linear association)	$t = \frac{r\sqrt{n-2}}{\sqrt{1-r^2}}$	$\lvert t \rvert \ge t_{\alpha/2,n-2}$

11.4 The CORR Procedure

The CORR Procedure can be used to compute correlations for pairs of variables specified. The syntax of the CORR procedure is as follows:

```
PROC CORR DATA=setname <options>;
VAR variablelist1;
WITH variablelist2;
RUN;
```

Options on the PROC CORR statement include:

PEARSON requests that the Pearson correlation coefficient and its test of significance be computed. If no options are specified, the Pearson coefficient prints by default.

SPEARMAN requests that the Spearman correlation coefficient and its test of significance be computed. The Spearman correlation coefficient is appropriate when no distribution assumptions are placed on the two random variables.

PLOTS = NONE suppresses all plots.

ODS Graphics (Section 19.1) must be enabled to produce the following plots. Many types of graphics are available. Only the MATRIX and SCATTER plots are overviewed here.

PLOTS = MATRIX<(*matrix options*) >
requests scatter plots for all pairs of variables arranged in a matrix form. If only a VAR statement is used in the procedure, a square matrix of scatter plots is produced. If a WITH statement is also used with the VAR statement, a rectangular array of plots is produced where the rows are determined by the variables in the WITH statement and the columns are determined by the variables in the VAR statement.

Matrix options include:
HISTOGRAM produces a histogram for each variable on the diagonal of the scatter plot matrix (VAR statement only). HIST can be used instead of HISTOGRAM.
NVAR=ALL or NVAR= *n* (*n* is a specified value) specifies the maximum number of variables in the VAR statement to be in the matrix plot. For a large number of variables in the VAR statement, one will want to list the variables to be included in the plot matrix first in the VAR statement. NVAR=5 is the default setting. For *n* values greater than 5 the resulting scatter plot matrix may have plots that are too small to be useful.

PLOTS=SCATTER <(*scatter-options*)>
requests scatter plots for pairs of variables that are not arranged in a matrix. That is, the procedure displays a scatter plot for each applicable pair of distinct variables from the VAR list if a WITH statement is not specified. Otherwise, the procedure displays a scatter plot for each applicable pair of variables, one from the WITH list and the other from the VAR list. A confidence ellipse (not covered here) is part of the default output. ELLIPSE=NONE will suppress this ellipse and is the recommendation at this time.

If the resulting maximum number of variables in the VAR or WITH list is greater than 10, only the first 10 variables in the list are displayed in the scatter plots.

The following two blocks of code will produce different correlation results.

1

```
PROC CORR DATA=one;
VAR a b;
WITH x y z;
RUN;
```

2

```
PROC CORR DATA=one;
VAR a b x y z;
RUN;
```

In **1**, six correlation coefficients will be computed. A table of correlations and their significances will be computed. Significance levels for the test of no linear association versus some linear association (a two-sided test) will be computed below each of the correlations. The value of the test statistic, t, is not part of the printed output. The general form of the output will look like this:

	A	B
X	corr(X,A)	corr(X,B)
Y	corr(Y,A)	corr(Y,B)
Z	corr(Z,A)	corr(Z,B)

where corr(X,Y) represents the correlation coefficient for the variables X and Y.

In **2**, 15 correlation coefficients will be computed. Significance levels will again be computed below each of the correlations. A correlation matrix of the following form will be computed.

	A	B	X	Y	Z
A	corr(A,A)=1	corr(A,B)	corr(A,X)	corr(A,Y)	corr(A,Z)
B	corr(B,A)	corr(B,B)=1	corr(B,X)	corr(B,Y)	corr(B,Z)
X	corr(X,A)	corr(X,B)	corr(X,X)=1	corr(X,Y)	corr(X,Z)
Y	corr(Y,A)	corr(Y,B)	corr(Y,X)	corr(Y,Y)=1	corr(Y,Z)
Z	corr(Z,A)	corr(Z,B)	corr(Z,X)	corr(Z,Y)	corr(Z,Z)=1

OBJECTIVE 11.4: Test whether or not all pairs of the variables in *beef* are correlated. Produce the default scatterplot matrix. Use $\alpha = 0.05$ in the conclusions.

```
PROC CORR DATA=beef PLOTS=MATRIX ;
VAR dmi adg cwt backfat rea;
TITLE 'Objective 11.4';
RUN;
```

Objective 11.4

The CORR Procedure

5 Variables: DMI ADG CWT BackFat REA

Simple Statistics

Variable	N	Mean	Std Dev	Sum	Minimum	Maximum
DMI	56	18.85130	2.10860	1056	16.27305	24.55687
ADG	56	3.07054	0.35583	171.95000	1.53000	3.84000
CWT	56	837.39762	39.87660	46894	701.00000	965.00000
BackFat	54	0.50917	0.03808	27.49513	0.39456	0.58390
REA	54	13.74598	0.70175	742.28295	12.34369	15.97315

Pearson Correlation Coefficients
Prob > |r| under H0: Rho=0
Number of Observations

	DMI	ADG	CWT	BackFat	REA
DMI	1.00000	0.74309	0.69296	0.12068	0.59204
		<.0001	<.0001	0.3847	<.0001
	56	56	56	54	54
ADG	0.74309	1.00000	0.86215	0.48513	0.62398
	<.0001		<.0001	0.0002	<.0001
	56	56	56	54	54
CWT	0.69296	0.86215	1.00000	0.49071	0.71914
	<.0001	<.0001		0.0002	<.0001
	56	56	56	54	54
BackFat	0.12068	0.48513	0.49071	1.00000	0.16516
	0.3847	0.0002	0.0002		0.2327
	54	54	54	54	54
REA	0.59204	0.62398	0.71914	0.16516	1.00000
	<.0001	<.0001	<.0001	0.2327	
	54	54	54	54	54

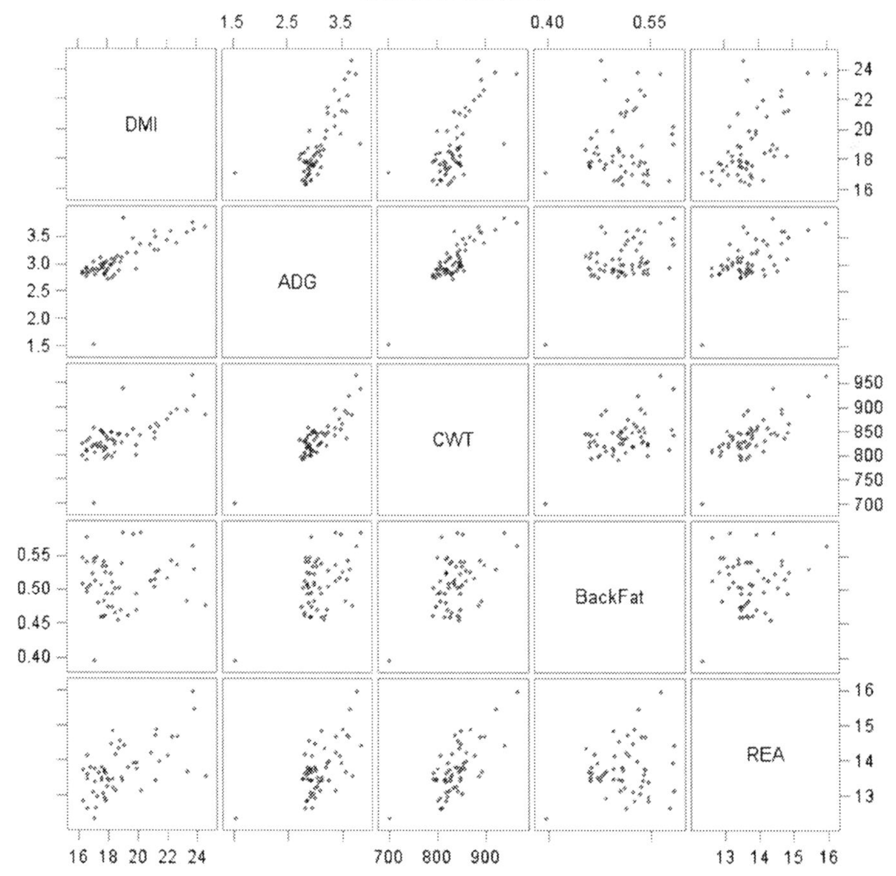

In the table of **Simple Statistics**, each of the variables in the VAR statement are briefly summarized. One notes that there are two missing observations for Backfat and REA resulting in sample sizes of 54 instead of 56. Variable labels (Section 8.2) would appear in this table if they were used in the creation of SAS data set *beef*.

In the second table, **Pearson Correlation Coefficients**, the complete heading text is a legend for the contents of each cell in the table of results. In the DMI row, the first cell has a correlation of 1.0000 because every variable is perfectly correlated with itself, and one observes that each diagonal entry in the table is a 1.00. In the DMI row, ADG column, one concludes that $r = 0.74309$ and $p < 0.0001$ for the hypothesis test in Table 11.3 is based on $n = 56$ pairs of data. Similarly, one reads the Pearson correlation coefficient in the top spot in each cell of the table. The value of the t-statistic in Table 11.3 is not printed in this table; only the observed significance level or p-value. The third entry in each cell is the sample size used for computing this correlation. If all of the sample sizes for the entire table were equal, say 56, then "N = 56" would be in the heading for this table. (See Objective 11.6.) This matrix is symmetric. That is, the correlation between DMI and ADG is the same as the correlation between ADG and DMI, and so on. The values above the diagonal of the table also appear below the diagonal. Given all this information, Backfat is not linearly associated with DMI nor REA ($p \geq 0.2327$), but all other pairs of variables are significantly linearly related ($p \leq 0.0002$).

The **Scatter Plot Matrix** illustrates the linear associations computed in the previous table, and the variable names appear on the diagonal for labeling purposes. In the fourth row (and also fourth column) of the Scatter Plot Matrix, one sees that the observations for Backfat with the variables DMI, CWT, and REA is more of a random scatter, whereas DMI versus ADG and DMI versus CWT illustrate a linear association.

OBJECTIVE 11.5: Test whether or not CWT is correlated with Backfat and REA. Produce individual scatterplots suppressing the confidence ellipse in each. Use $\alpha = 0.05$ in the conclusions.

```
PROC CORR DATA=beef PLOTS=SCATTER(ELLIPSE=NONE);
VAR cwt;
WITH backfat rea;
TITLE 'Objective 11.5';
RUN;
```

It does not matter whether the VAR statement is before the WITH statement or not. One should note that the WITH statement will determine the number of rows in the table of results. Also, this objective could have been programmed as:

```
PROC CORR DATA=beef PLOTS=SCATTER(ELLIPSE=NONE);
VAR backfat rea;
WITH cwt;
```

Objective 11.5
The CORR Procedure

2 With Variables:	BackFat REA
1 Variables:	CWT

Simple Statistics

Variable	N	Mean	Std Dev	Sum	Minimum	Maximum
BackFat	54	0.50917	0.03808	27.49513	0.39456	0.58390
REA	54	13.74598	0.70175	742.28295	12.34369	15.97315
CWT	56	837.39762	39.87660	46894	701.00000	965.00000

Pearson Correlation Coefficients
Prob > |r| under H0: Rho=0
Number of Observations

	CWT
BackFat	0.49071 0.0002 54
REA	0.71914 <.0001 54

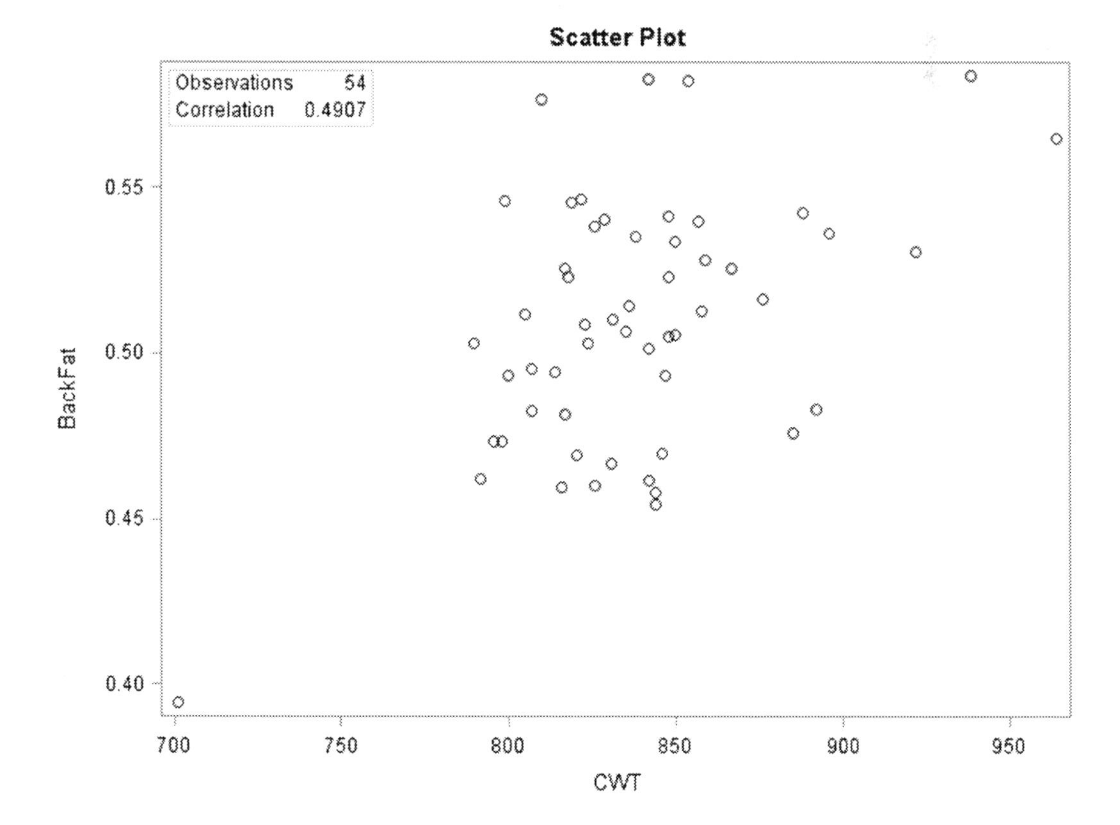

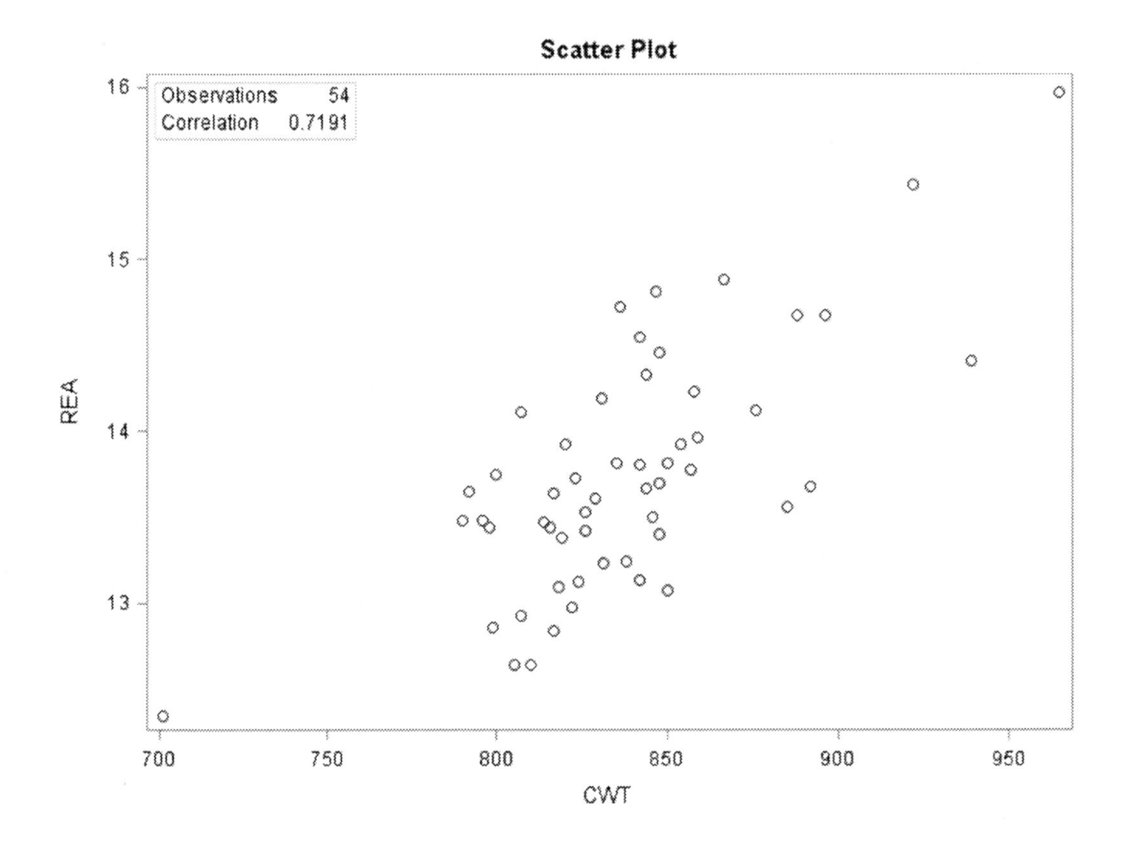

The larger scatter plots for the pairs of variables have the computed correlation coefficient and number of observations recorded in a corner of the graph. There are fewer graphs when one uses the WITH statement. If only a VAR statement had been used, three scatterplots would have been produced. The additional plot would be for Backfat versus REA. With the PLOTS=SCATTER(ELLIPSE=NONE) option in Objective 11.1, nine larger scatterplots with the annotations would have been produced.

OBJECTIVE 11.6: Compute the Spearman correlation coefficients for all pairs of variables in the set: DMI, ADG, and CWT. Use a scatter plot matrix and include histograms for each response.

```
PROC CORR DATA=beef SPEARMAN PLOTS=MATRIX(HISTOGRAM) ;
VAR dmi adg cwt ;
TITLE 'Objective 11.6';
RUN;
```

The SPEARMAN calculation is not the default and must be requested in the PROC CORR statement. The HISTOGRAM matrix option does not work when a WITH statement is used in the procedure.

Objective 11.6
The CORR Procedure

3 Variables: DMI ADG CWT

Variable	N	Mean	Std Dev	Sum	Minimum	Maximum
DMI	56	18.85130	2.10860	1056	16.27305	24.55687
ADG	56	3.07054	0.35583	171.95000	1.53000	3.84000
CWT	56	837.39762	39.87660	46894	701.00000	965.00000

Spearman Correlation Coefficients, N = 56
Prob > |r| under H0: Rho=0

	DMI	ADG	CWT
DMI	1.00000	0.75915 <.0001	0.63654 <.0001
ADG	0.75915 <.0001	1.00000	0.74060 <.0001
CWT	0.63654 <.0001	0.74060 <.0001	1.00000

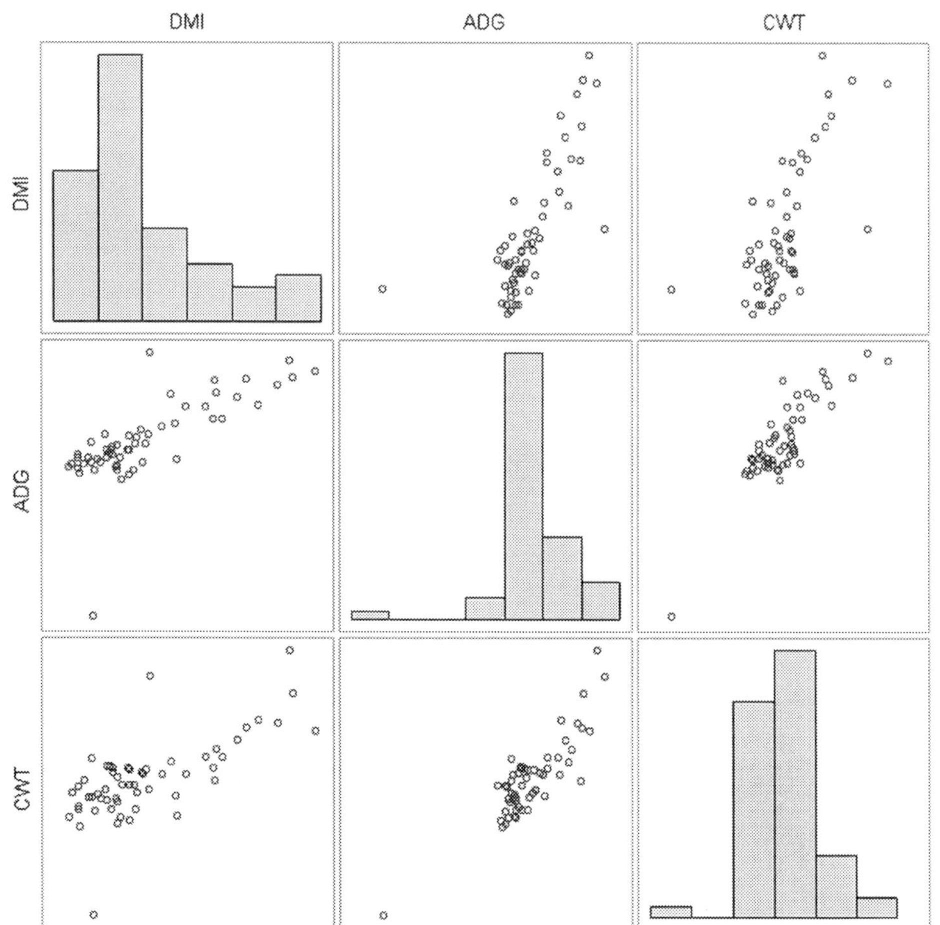

For these three variables in these pairwise calculations, the sample sizes are all the same. In the table for the **Spearman Correlation Coefficient**, this common sample size "N = 56" appears in the header for the table rather than the third entry in each cell. The HISTOGRAM selected for the diagonal of the scatterplot matrix allows one to view the distribution of the values for each of the variables in the analysis in addition to examining the linear associations.

11.5 Chapter Summary

In this chapter, two introductory methods of investigating linear associations between numeric variables were overviewed. In simple linear regression analysis, there is an independent or explanatory variable that is investigated as a good predictor of a dependent variable. Correlation analysis does not have these independent and dependent variable roles in a linear association analysis. Both variables in a correlation analyses are regarded as random responses. Both analyses are aided by plots of the data on a two-dimensional set of axes. The REG procedure can be used to estimate predictive equations when there is more than one independent variable, and the CORR procedure can also be used for more advanced topics in assessing linear association.

12

SAS Libraries and Permanent SAS Data Sets

When programming in the SAS Editor environment, SAS data sets can be created in a DATA step using either an INPUT or INFILE statement (Sections 2.2 and 8.1). Thus far in this book, SAS data sets are temporary and exist only during the SAS session in which they are created. When the SAS session is closed, these SAS data sets are gone. The data may be stored in a SAS program or an external file, but a SAS data file does not exist once SAS is closed. A permanent SAS data set is a file that SAS correctly recognizes, and the contents can be opened and used by SAS without the use of a DATA step or other method (Chapter 14 – The IMPORT Procedure) to create a SAS data set.

12.1 The Work Library

So far in this book, the programming done in SAS Editor required SAS data sets be created during the current SAS session. If a program creating a SAS data set is closed, and one or more programs are created or opened in the SAS Editor, the SAS data sets created in the first program are still "available" for the programmer to use provided that the SAS data set name has not been reused. These SAS data sets created by the methods demonstrated thus far in this book are temporarily stored in a library called WORK.

OBJECTIVE 12.1: Run the following short program in SAS Editor creating the temporary SAS data set *one*. Observe the contents of the LOG window.

```
DATA one;
INPUT X Y Z;
DATALINES;
25  27  34
28  31  29
41  58  29
37  28  83
RUN;
```

All that has been requested in this program is that a SAS data set called *one* be created. After executing this program, the SAS Log window gives feedback regarding the SAS data set *one*.

```
NOTE: The data set WORK.ONE has 4 observations and 3 variables.
```

In the above program, the SAS data set was named *one*. This SAS data set is temporarily stored in a library called WORK. The complete specification of the SAS data set is WORK. ONE. WORK.ONE is referred to as the two-level SAS data set name. In general, SAS data sets are specified by libname.setname where libname is the library name, and setname is the name of the SAS data set.

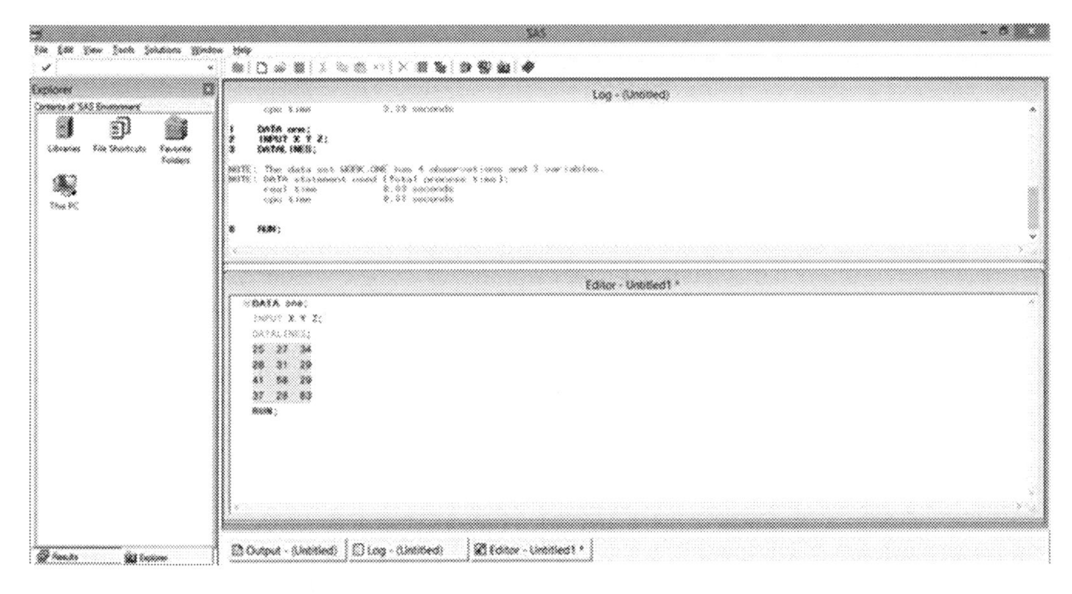

FIGURE 12.1
The Explorer window.

To view the WORK library and its contents, the Explorer tab on the left side of the SAS screen should be used. There are tabs at the lower left of the SAS screen for **Results** and **Explorer**. After selecting the **Explorer** tab, **Libraries**, marked by a file cabinet icon, should be selected. See Figure 12.1. The contents of the **Explorer** were briefly introduced in Section 8.3 in the ViewTable topic; however, the subject of libraries was not yet introduced.

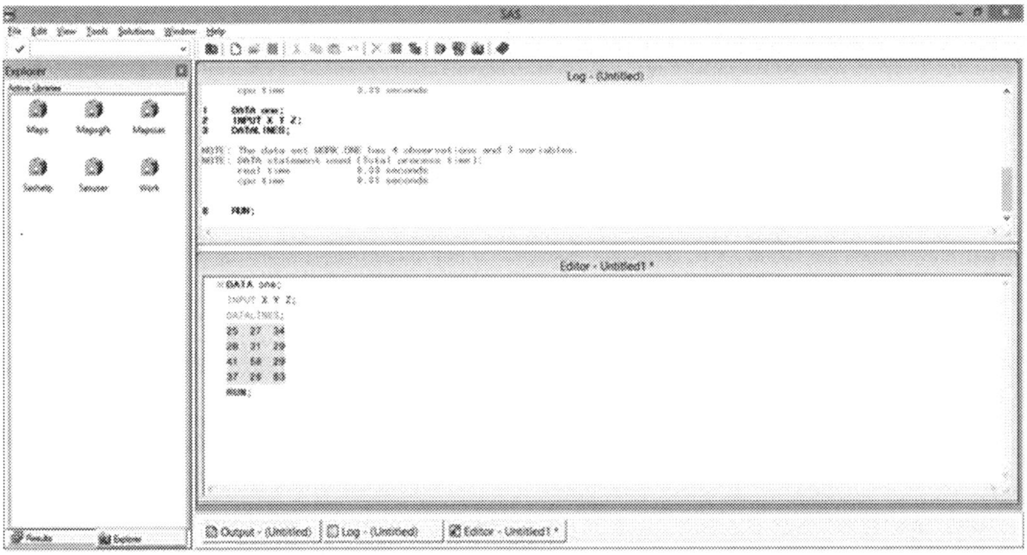

FIGURE 12.2
Active libraries in Explorer.

Active Libraries in SAS are visible in this **Explorer** window. SAS sets up a few default libraries: Sashelp, Sasuser, and Work. See Figure 12.2. There may be other libraries in the **Explorer**, such as those shown in Figure 12.2 that are associated with SAS/GRAPH procedures that utilize geographic map information. For the SAS products installed, this list of default libraries and their contents may change, but the Work library will always be available. Moving from the **Editor** to **Explorer** one should note that the buttons on the toolbar at the top of the SAS screen change. For example, the Submit ("running man") button does not appear on the toolbar when **Explorer** is the active window, but it is active when the **Editor** is the active window. The first button on the **Explorer** toolbar is the "Up One Level" button which is likely the most used button while in **Explorer.** This allows one to go back or exit the view of a SAS library.

Clicking on the **Work** icon (single file drawer) will reveal the SAS data sets in the **Work** library. Thus far, in this chapter, only the SAS data set WORK.ONE has been created in Figure 12.3. In Section 8.3 Figure 8.2 illustrated more than one SAS data set in the Work library.

When data sets are large and reading in the data takes a long time, or when data sets are frequently used, the creation of a permanent SAS data set may be recommended. A permanent SAS dataset is a file that SAS recognizes. SAS recognizes the variables, the values they take, and the labels assigned to those variables. To create a permanent SAS data set, a new library (other than WORK) must be created. A permanent SAS data set is a file that is saved to the computer. The library names one creates are not permanently recognized by SAS by default. Similar to SAS data set names, once a library is created during a SAS session, it is active until the SAS session closes. When SAS closes, library names created during the current SAS session will be erased. This keeps the SAS program from becoming too cluttered with libraries. There is a method of keeping the library permanently recognized discussed later in this chapter.

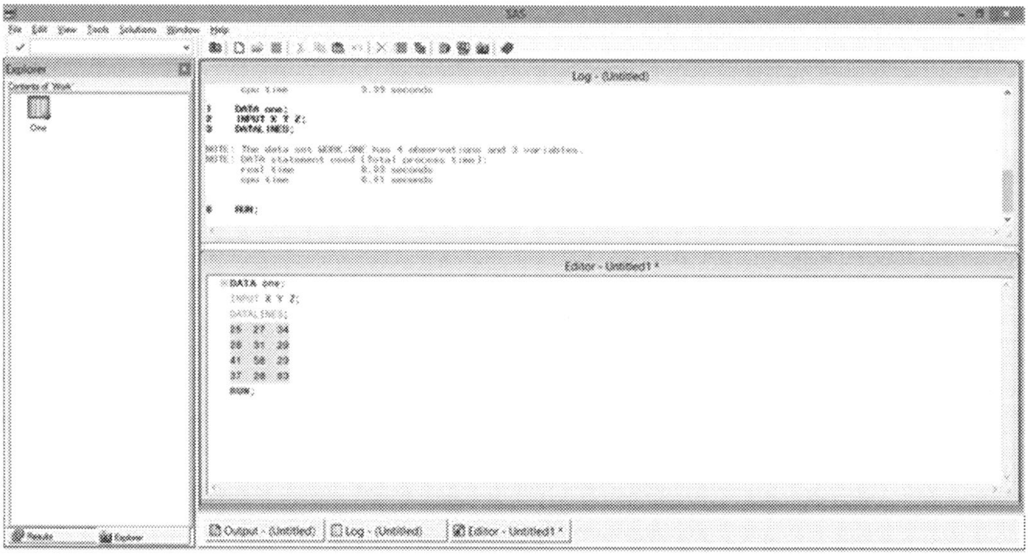

FIGURE 12.3
Contents of the Work library.

12.2 Creating a New SAS Library

There are multiple ways to create a new SAS library. One method is to use SAS code in the Editor, two are point-and-click methods, and a fourth uses the command line in the top left corner of the SAS screen.

12.2.1 The LIBNAME Statement

The LIBNAME statement is used to assign a library name to a folder on the computer hard drive or external file storage device. The simple syntax of the LIBNAME statement is:

```
LIBNAME libref 'drive:\folder';
```

where *libref* is the name to be assigned to the library using SAS naming conventions and only eight characters, and *drive:\folder* is the directory and folder(s) in which the permanent SAS data set will be saved as a file. The quotation marks can be either single or double quotation marks. The *libref* is a short name or nickname for the drive:\folder specification. The new library will be referred to by this *libref* during the current SAS session. There are other syntax options for the LIBNAME statement, but this brief introduction is sufficient for now.

A single LIBNAME statement can be submitted from the SAS Editor to create a new library, or the LIBNAME statement can be submitted as one of the lines in a longer SAS program. This new library will be active after the LIBNAME statement successfully executes. A note in the SAS log confirms the new library assignment. This will be demonstrated in Objective 12.2. Multiple LIBNAME statements can be submitted to identify multiple folders on the computer in which permanent SAS data sets are stored.

The default libraries in **Explorer** should be left alone except for the Work library, of course. SAS utilizes information in the other default libraries. The content of those libraries should not be altered.

12.2.2 Toolbar Function: Add New Library

The **Add New Library** button on the toolbar at the top of the SAS screen (file drawer icon with a starburst on it) is an alternative to the LIBNAME statement. This button is available with **Explorer** and the **Editor**. When using the **Add New Library** button, a prompt to enter a library name and the path (drive:\folder) appears in a new dialog window. **Name** is the library name to be assigned. This is the same as the *libref* in the LIBNAME statement. Most users will leave the **Engine** set at **Default**. The default engine is the version of SAS that is running. (Figure 12.4) The box for **Enable at startup** can be checked if SAS is to automatically assign this library when a SAS session begins although this is not recommended when beginning to use SAS libraries. This action can later be reversed by clicking on the box again to deselect the option. **Browse** allows the perusal of the available folders or directories on the computer which can be used as an assigned library. In this introduction to assigning libraries, the **Options** will be left blank. Once the **Name** and **Path** are specified, then **OK** is selected to create the library.

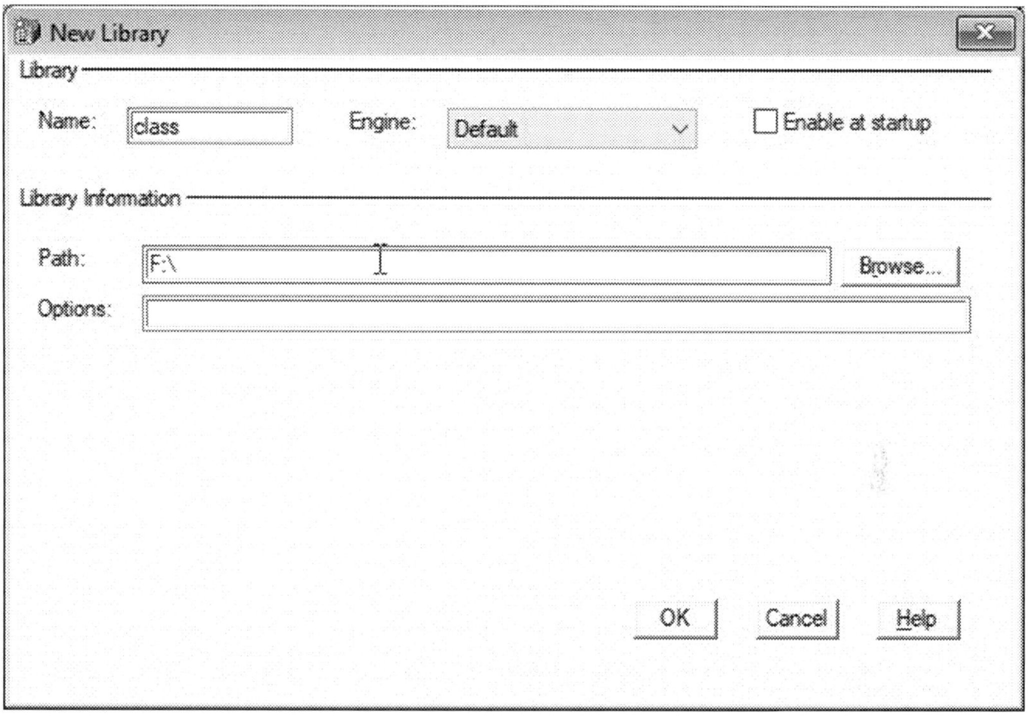

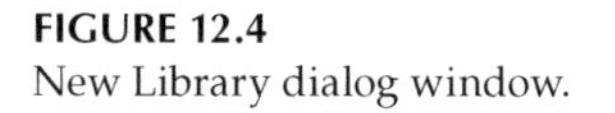

FIGURE 12.4
New Library dialog window.

12.2.3 Explorer View

In the **Explorer** window that appears at the left of the SAS screen a third method of creating a new library can be done. If the **Explorer** has been closed earlier in the SAS session, it can be reopened. To reopen, **View – Explorer** can be selected from the pull-down menu. In the **Explorer**, **Libraries** can be selected and then view **Active Libraries**. No active libraries should be selected. **File – New** are selected from the pull-down menu, and the **New Library** dialog box shown in Figure 12.4 will open. This dialog box should be completed as in Section 12.2.2.

12.2.4 The LIBASSIGN Command

The command line is the blank in the upper left corner of the SAS screen. The command line has only been referenced once in this book so far, and the command was the SUBMIT command in Section 1.2. The origin of the command line pre-dates the Windows platform (that is, it pre-dates point-and-click capability). The command line was exclusively used to submit programs, file (or save) SAS programs, and many other tasks in older versions of SAS.

For the task of creating a new library, LIBASSIGN is entered on the command line. Once entered, the New Library dialog box in Figure 12.4 will open and should be completed as in Sections 12.2.2.

For each of the point-and-click approaches and the command line approach, a note will also appear in the SAS log indicating whether the library was successfully defined or not, but there is no program code produced or recorded by these approaches.

12.3 Creating a Permanent SAS Data Set

When creating a permanent SAS data set, one should avoid all of the default library names SAS has created including WORK, and select a user-defined library. After the LIBNAME statement defining a new library, the DATA Step can be used to create the permanent SAS data set. The DATA statement must use the two-level SAS name as demonstrated in the following objective.

OBJECTIVE 12.2: Create a SAS library named CLASS on the f:\drive (or specify a drive:\ folder for your use) and store the small 4 observation 3 variable data set in Objective 12.1 in that location. Print the contents of the permanent SAS data set.

```
LIBNAME class 'f:\';

DATA class.one;
INPUT X Y Z;
DATALINES;
25 27 34
28 31 29
41 58 29
37 28 83
;
PROC PRINT DATA=class.one;
TITLE 'Objective 12.2';
RUN;
QUIT;
```

Note the messages in the Log Window.

```
NOTE: Libref CLASS was successfully assigned as follows:
      Engine:        V9
      Physical Name: f:\

NOTE: The data set CLASS.ONE has 4 observations and 3 variables.
```

Engine refers to the version of SAS that creates the SAS data set. V9 indicates that this objective was performed using version 9 of SAS. More notes on this are in the next section.

After running this program, the contents of the f:\drive can be checked. A new file named *one.sas7bdat* will now be in that directory or folder. The name chosen for the SAS data set has now become the file name. The "sas7bdat" extension is assigned by SAS versions 7 and later to identify permanent SAS data sets. In general, the file name will have the form *setname.sas7bdat.* Files with the sas7bdat extension cannot be opened in the Editor or Program Editor windows of SAS, but these data sets can be called into SAS procedures. The file *one.sas7bdat* will remain in f:\folder until it is over written or deleted.

If the first two lines of the SAS code were:

```
LIBNAME class 'f:\ ;

DATA one;
```

Then the CLASS library would have been created, but the failure to include the two-level SAS name in the DATA statement would produce a temporary SAS data set WORK.

ONE. That is, though the CLASS library was created, the DATA step did not put a data set in it.

This objective used the simple DATA Step with the INPUT and DATALINES statements to create the permanent SAS data set. The DATA Step with an INFILE statement could also be used to create permanent SAS data sets. The two-level SAS name in the DATA statement must be used to do this.

12.4 How to Identify the Location of Libraries

Since several libraries can be defined in a single SAS session, and since SAS has a few libraries that are named and active in every SAS session, the names of the active libraries and their locations can be identified in the following ways.

LIB can be entered on the command line. A new window that contains the active libraries, the Engine (the version of SAS in which the files were created), and the Host Pathname (the directory:\folder) are identified. See Figure 12.5.

Clicking on the library icon in the LIBNAME window (Figure 12.5) will open the library and list the SAS data sets in that library. In Figure 12.6 the contents of the Class library are listed in two ways. (It should be observed that there are two other SAS data sets already in the Class library prior to creating CLASS.ONE.) On the left, the **Class** library was selected in the **Explorer**. Only the SAS data set name is given in the **Explorer**. On the right is the information obtained by clicking on the Class library icon in the LIBNAME window in Figure 12.5. More details are given in the LIBNAME window on the right.

These engine and path name details can be obtained in the **Explorer** in an additional step after opening the library. By right clicking on the icon for a SAS data set and selecting **Properties** from the pop-up menu, a **Properties** window will open. See Figure 12.7 for the CLASS.ONE properties. There are tabs across the top of the **Properties** window. Selecting the **Details** tab will yield further information about the permanent SAS data set, specifically the Filename as shown in Figure 12.8.

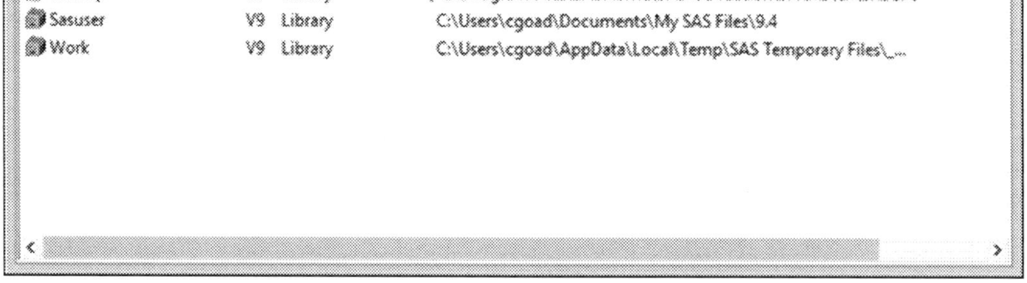

LIBNAME

Active Libraries

Name	Engine	Type	Host Pathname	Modified
Class	V9	Library	f:\	
Maps	V9	Library	C:\Program Files\SASHome\SASFoundation\9.4\maps	
Mapsgfk	V9	Library	C:\Program Files\SASHome\SASFoundation\9.4\mapsgfk	
Mapssas	V9	Library	C:\Program Files\SASHome\SASFoundation\9.4\maps	
Sashelp	V9	Library	('C:\Program Files\SASHome\SASFoundation\9.4\nls\en\SASCFG	
Sasuser	V9	Library	C:\Users\cgoad\Documents\My SAS Files\9.4	
Work	V9	Library	C:\Users\cgoad\AppData\Local\Temp\SAS Temporary Files_...	

FIGURE 12.5
Active libraries and their descriptions.

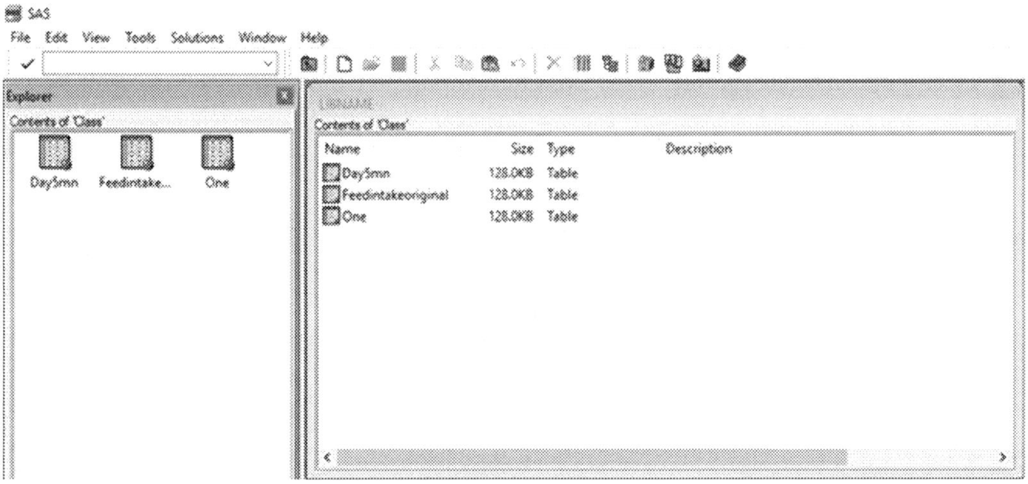

FIGURE 12.6
Comparison of Explorer and LIBNAME windows.

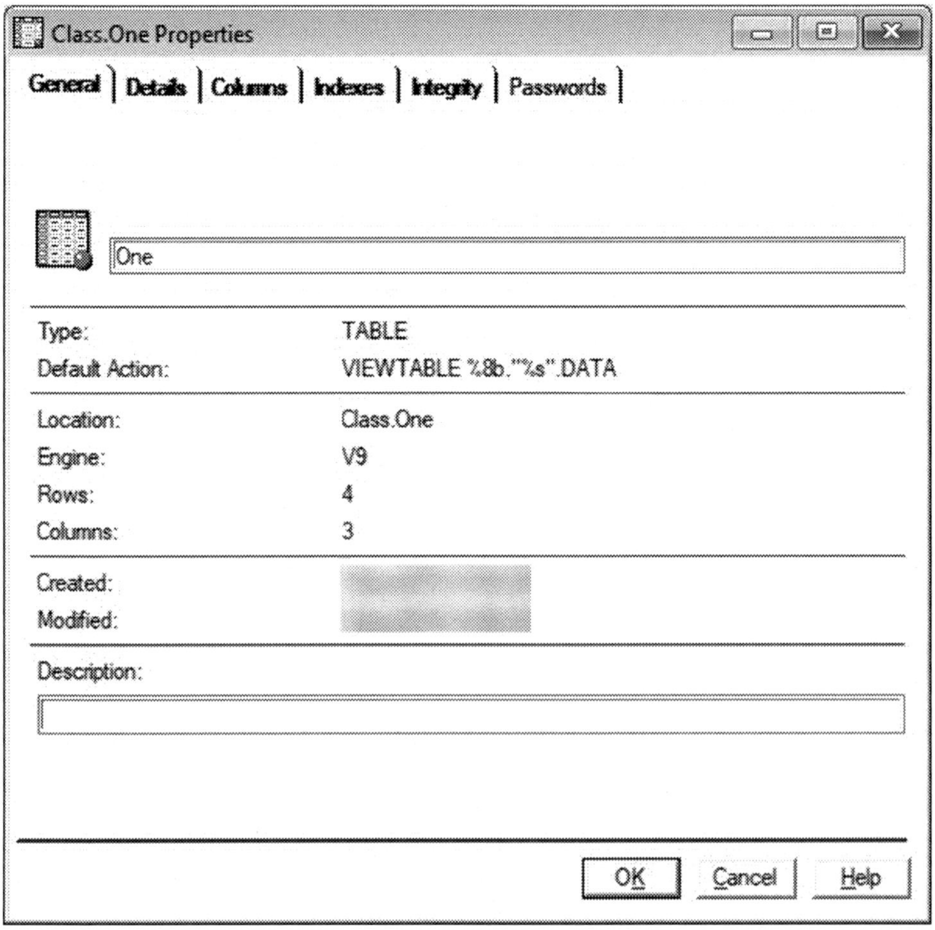

FIGURE 12.7
General properties for the SAS Data Set CLASS.ONE.

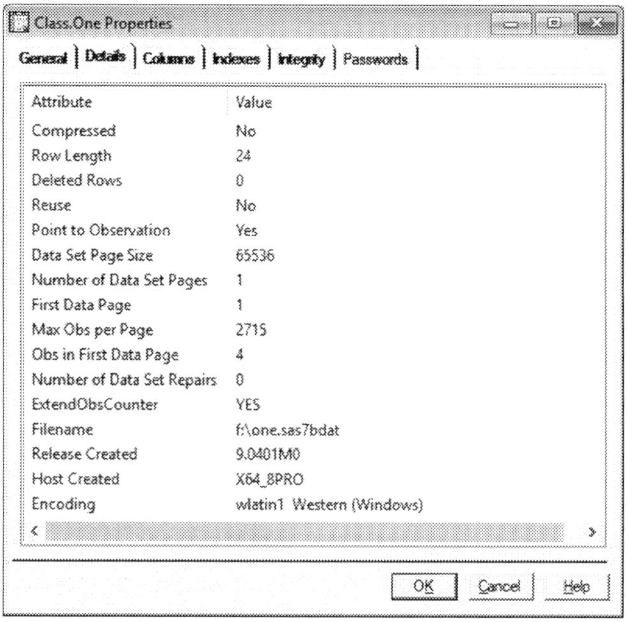

Class.One Properties

General | Details | Columns | Indexes | Integrity | Passwords

Attribute	Value
Compressed	No
Row Length	24
Deleted Rows	0
Reuse	No
Point to Observation	Yes
Data Set Page Size	65536
Number of Data Set Pages	1
First Data Page	1
Max Obs per Page	2715
Obs in First Data Page	4
Number of Data Set Repairs	0
ExtendObsCounter	YES
Filename	f:\one.sas7bdat
Release Created	9.0401M0
Host Created	X64_8PRO
Encoding	wlatin1 Western (Windows)

OK Cancel Help

FIGURE 12.8
Property **Details** for the SAS Data Set CLASS.ONE.

As in Section 8.3, the contents of any of the temporary or permanent SAS data sets can be opened in ViewTable by double-clicking on the icon for the SAS data set. This will open the ViewTable product within SAS and display the SAS data set in a spreadsheet type of format. Before running a SAS program requiring this SAS data set, it is recommended that ViewTable be closed.

12.5 Using a Permanent SAS Data Set in a SAS Program

Within a SAS program, SAS data sets can be called into a procedure by their two-level names. For each procedure where the DATA = *SAS-data-set* option appears in the procedure (PROC) statement, the value of *SAS-data-set* is given by libname.setname. If the library is active, then any permanent SAS data file can be called into a program in the Editor. If the library is not active, then a LIBNAME statement needs to be submitted or one of the other library creation methods in Section 12.2 needs to be performed. The assigned library name is only valid during the current SAS session even though the permanent SAS data sets exist after the SAS session closes.

For example, in Figure 12.6 the files *one.sas7bdat*, *Day5mn.sas7bdat*, and *Feedintakeoriginal.sas7bdat* are three permanent SAS data sets located on the F:\drive. Any one of these SAS data sets can be called into a SAS program for an analysis or modification of the SAS data set. By using the two-level SAS name for a SAS data set, one can create permanent SAS data sets using a DATA step or create OUTPUT data sets as seen for procedures in earlier chapters of this book. In a new SAS session, a library would first have to be defined. Consider the following example.

```
LIBNAME class 'F:\';

PROC MEANS DATA=class.one;
VAR x y z;
OUTPUT OUT=two MEAN=xmean ymean zmean RANGE=xrange yrange zrange;

DATA class.Feedintake30;
SET class.feedintakeoriginal;
IF _N_ > 30 THEN DELETE;                 *keep the first 30 observations.;
RUN;
QUIT;
```

In the PROC MEANS code, the permanent SAS data set *CLASS.ONE* is called into the MEANS procedure. One may have had to PRINT the data set or open it in ViewTable to identify the variable names before writing the procedure code. In the OUTPUT statement the means and ranges for each of the three variables are saved in a new ***temporary*** data set *WORK.TWO*. This is a temporary SAS data set since no library name was specified in the OUTPUT statement. If a two-level SAS name, such as *CLASS.TWO*, were specified, these summary statistics would be saved to a permanent SAS data set in the CLASS library.

In the DATA step, the permanent SAS data set *Feedtakeoriginal* is called into the DATA step using the SET command. A new permanent SAS data set *CLASS.FEEDINTAKE30* is created and the file *feedintake30.sas7bdat* will be written to F:\after this program successfully runs. Other variable assignment statements, transformations, labels, or variable manipulations (Chapters 4 and 8) can be added after the SET statement, and these changes will be recorded in *CLASS.FEEDINTAKE30*.

12.6 Engine – A Quick Note

In Objective 12.2 the CLASS library was created for this first time, and the contents of the SAS log window were observed.

```
NOTE: Libref CLASS was successfully assigned as follows:
      Engine:        V9
      Physical Name: f:\

NOTE: The data set CLASS.ONE has 4 observations and 3 variables.
```

For this book, SAS (version 9.4) has been used in all of the illustrations. The engine is "V9" as seen in the Log window message. SAS is sensitive to the engine that created the permanent SAS data sets that have been saved or accumulated. Using the LIBNAME syntax given at the beginning of this chapter, the default engine is V9. Though somewhat unlikely to be encountered at this point in time, permanent SAS data sets exist in some data archives that may have a different file extension. Files with an sd2 extension, for example, were created using version 6 of SAS in a Windows environment, and thus the engine is V6. Older permanent SAS data sets have an ssd extension.

The engine type can be declared by including it in the LIBNAME statement between the *libref* and the "drive:\folder".

LIBNAME class 'f:\';	will recognize permanent SAS data files created by the current engine, V9. (That is, files with sas7bdat extension are recognized.)
LIBNAME class1 V6 'f:\';	will recognize permanent SAS data files in f:\that were created using version 6 of SAS. (These files will have an sd2 extension.)

Both of these libraries refer to the same drive:\folder, but each library contains different items. That is, SAS data sets in CLASS1 are not in the CLASS library since they are recognized by different engines. SAS versions 7, 8, and 9 have the same engine, and V9 encompasses all three. All permanent SAS data files created in these three versions of SAS have sas7bdat extensions.

There are several other engine options. More generally, the SAS engine is a component of SAS software that allows the software to access files (typically data files) of varying formats. The specified engine controls the read/write access to a file. Specifying an engine other than the default engine for SAS in a Windows platform either in a LIBNAME statement or in the New Library dialog (Figure 12.4), is really beyond the scope of this book. One should consult SAS Institute (2013b) for more information about SAS engines.

12.7 Chapter Summary

In this chapter, SAS libraries for the Windows platform were introduced. SAS libraries are necessary when creating permanent SAS data sets. There is a syntax option and a point-and-click option for creating a SAS library. Once a library has been assigned, SAS can recognize permanent SAS data files contained in that drive:\folder. Permanent SAS data sets are advantageous when the data sets are large and take considerable time to create, or when the data sets are used with a high frequency, and it is helpful to be able to start a SAS session immediately working with a successfully created SAS data set. When multiple SAS users are working with a data set, sharing a common data file can save time and reduce the number of errors that can occur in the creation of the data set. Some point-and-click SAS products, such as SAS Enterprise Guide and SAS Enterprise Miner, begin by identifying one or more libraries containing the SAS data sets to be analyzed.

13

DATA Step Information 4 – SAS Probability Functions

There are many DATA step functions in SAS that compute probabilities from known distributions. There are also inverse functions that can compute the value of the random variable associated with a cumulative probability. As in Chapter 4, these SAS probability functions can be included in DATA step operations. This chapter will present some discrete and continuous probability functions from among the many, many DATA step functions available. SAS Institute (2016a) lists the SAS DS2 functions, or SAS Help and Documentation can also be searched for DATA step functions among Base SAS product information.

13.1 Discrete Probability Distributions

Common families of discrete probability distributions are binomial, Poisson, hypergeometric, and negative binomial. SAS functions computing probabilities from the binomial and Poisson distributions are demonstrated here.

13.1.1 The Binomial Distribution

A binomial random variable, Y, can have values of 0, 1, 2, …, n where n is the number of independent trials. Each trial has a probability of success defined by p, and $0 \leq p \leq 1$. The formula for the probability of m success in n trials is given by:

$$P(Y = m \mid n) = \binom{n}{m} p^{m} (1-p)^{n-m}$$

where $\binom{n}{m}$ is the computed number of combinations of n items chosen m at a time.

The PROBBNML function can compute binomial probabilities. The syntax of the PROBBNML function is:

PROBBNML (p, n, m)

where:

p is the probability of success in a single trial where $0 \leq p \leq 1$,
n is a (whole) number of independent trials, $n > 0$, and
m is a (whole) number of successes in n trials, with $0 \leq m \leq n$.

The order of the arguments in the parenthesis is important. The PROBBNML function returns the cumulative probability of m or fewer successes in n trials, that is,

$$P(Y \le m \mid n) = \sum_{y=0}^{m} P(Y = y \mid n) = \sum_{y=0}^{m} \binom{n}{y} p^{y}(1-p)^{n-y}.$$

The appropriate syntax for this SAS function is similar to SAS functions introduced in Section 4.1. Within a DATA step, the SAS statement is:

```
newvariable = PROBBNML(p, n, m);
```

where p, n, and m are specified arguments within the PROBBNML function or are previously defined variables in the same DATA step.

OBJECTIVE 13.1: Given that the probability of success on a single trial is 0.3 and that there are 8 trials:

a. Find the probability of 5 or less successes in 8 trials. That is, compute $P(Y \le 5 \mid n = 8)$.

b. Find the probability of exactly 5 successes in 8 trials. That is, compute $P(Y = 5 \mid n = 8)$.

c. Find the probability of 4, 5, or 6 successes in 8 trials. That is, compute $P(4 \le Y \le 6 \mid n = 8)$.

d. Find the probability of more than 3 successes in 8 trials. That is, compute $P(Y > 3 \mid n = 8)$.

Use a temporary SAS data set and print the probability values.

```
DATA one;
a = PROBBNML(0.3, 8, 5);                                              *P(Y <= 5);
b = PROBBNML(0.3, 8, 5) - PROBBNML(0.3, 8, 4);                         *P(Y = 5);
c = PROBBNML(0.3, 8, 6) - PROBBNML(0.3, 8, 3);                    *P(4 <= Y <= 6);
d = 1 - PROBBNML(0.3, 8, 3);          *P(Y > 3) = P(Y >= 4) = 1 - P(Y <= 3);
PROC PRINT DATA=one;
TITLE 'Objective 13.1 - Binomial Probability Distribution';
RUN;
QUIT;
```

Different from previous examples of the DATA step, there are no INPUT, INFILE, SET, or MERGE statements in this objective. For ease of interpretation each new variable (a, b, c, d) corresponds to each of the parts a, b, c, and d of this objective. Thus four variables are created, and there is only one observation. Single-line comments (Section 20.2) are used to clarify lines of the program. The variables a, b, c, and d could have more definitive LABELs created in the DATA Step (Section 8.2).

Since PROBBNML computes cumulative probabilities, one has to subtract the probabilities to compute items b and c. Item d utilizes the concept of complementary probability.

Objective 13.1 – Binomial Probability Distribution

Obs	a	b	c	d
1	0.98871	0.046675	0.19281	0.19410

There are elementary statistical methods books that provide reference tables for either cumulative probabilities such as the PROBBNML function computes, or probabilities for single values of Y. These tables typically are for a limited number of values of p and n. This function is advantageous in that it can compute binomial probabilities for any values of p between 0 and 1 or for large values of n. Computing all probabilities for single values of Y from 0 to n can be a lengthy process for large values of n. In Objective 15.4, a DO loop will show an efficient way of computing probabilities for many or all values of Y in a binomial distribution.

13.1.2 The Poisson Distribution

The Poisson distribution is a discrete probability distribution typically used to describe count data. The values of a Poisson random variable Y are 0, 1, 2, 3, … (no upper limit). To compute the Poisson probability for a given value of Y = n the formula is:

$$P(Y = n) = \frac{e^{-m} m^n}{n!}$$

where the parameter m is the mean of the distribution. The syntax of the POISSON function is:

POISSON (m, n)

where m is the mean, m > 0, and n is the value of the random variable Y, and n = 0, 1, 2, 3, …

Here again, the order of the arguments in the parenthesis is important. The POISSON function computes the probability of n or fewer occurrences with a population mean of m, that is, $P(Y \le n) = \sum_{y=0}^{n} P(Y = y) = \sum_{y=0}^{n} \frac{e^{-m} m^y}{y!}$. Within a DATA step, the appropriate syntax for the SAS statement is:

```
newvariable = POISSON(m, n);
```

where m and n are specified arguments within the POISSON function or are previously defined variables in the same DATA step.

OBJECTIVE 13.2: Given that the mean of a Poisson random variable is 5.2:

a. Find the probability of 7 or less. That is, compute $P(Y \leq 7)$.
b. Find the probability of exactly 7. That is, compute $P(Y = 7)$.
c. Find the probability of at least 1. That is, compute $P(Y > 0) = P(Y \geq 1)$.

Use a temporary SAS data set and print the probability values.

```
DATA two;
a = POISSON(5.2, 7);                                          *P(Y <= 7);
b = POISSON(5.2, 7) - POISSON(5.2, 6);                        *P(Y = 7);
c = 1 - POISSON(5.2, 0);                        *P(Y >= 1)= 1 - P(Y = 0);

PROC PRINT DATA=two;
TITLE 'Objective 13.2 - Poisson Probability Distribution';
RUN;
QUIT;
```

Objective 13.2 – Poisson Probability Distribution

Obs	a	b	c
1	0.84492	0.11253	0.99448

There are also PDF and CDF functions in the SAS Data Step that can compute discrete probabilities and cumulative probabilities for these and other discrete distributions. SAS 9.4 Functions and CALL Routines: Reference in SAS Help and Documentation or SAS Institute (2016a) can be searched for syntax information for these functions. The PDF and CDF functions are applicable to the following discrete distributions: Bernoulli, Binomial, Geometric, Hypergeometric, and Negative Binomial.

13.2 Continuous Probability Distributions

There are also many continuous probability distributions available as a SAS function in a DATA step. The focus in this section will be on the distributions associated with the hypotheses testing and confidence interval topics discussed in the earlier chapters of the book. Using the normal distribution, a review of probabilities and quantiles will be done first. The function for a probability curve is referred to as probability density function (pdf). When graphed, the pdf function is zero or greater, and the total area under the curve (and above the horizontal axis) is always one. For continuous random variables it is important to remember that the probabilities are areas under the pdf curve. Quantiles are the values on the horizontal axis that determine a specified area under the curve. The median and the 95th percentile are examples of quantiles.

13.2.1 The Normal Distribution

The standard normal distribution has a mean of zero and a variance of one. This distribution of the standard normal random variable, Z, is symmetric about the zero mean, and thus the areas under the curve to either side of zero are 0.5. That is, $P(Z < 0) = P(Z > 0) = 0.5$. For continuous distributions, the interest typically is the area under the curve and the quantiles but not the y-coordinate (value of the pdf). Figure 13.1 is a plot of the standard normal pdf with two areas labeled a and b. The SAS probability function for the normal distribution is used to compute a and b.

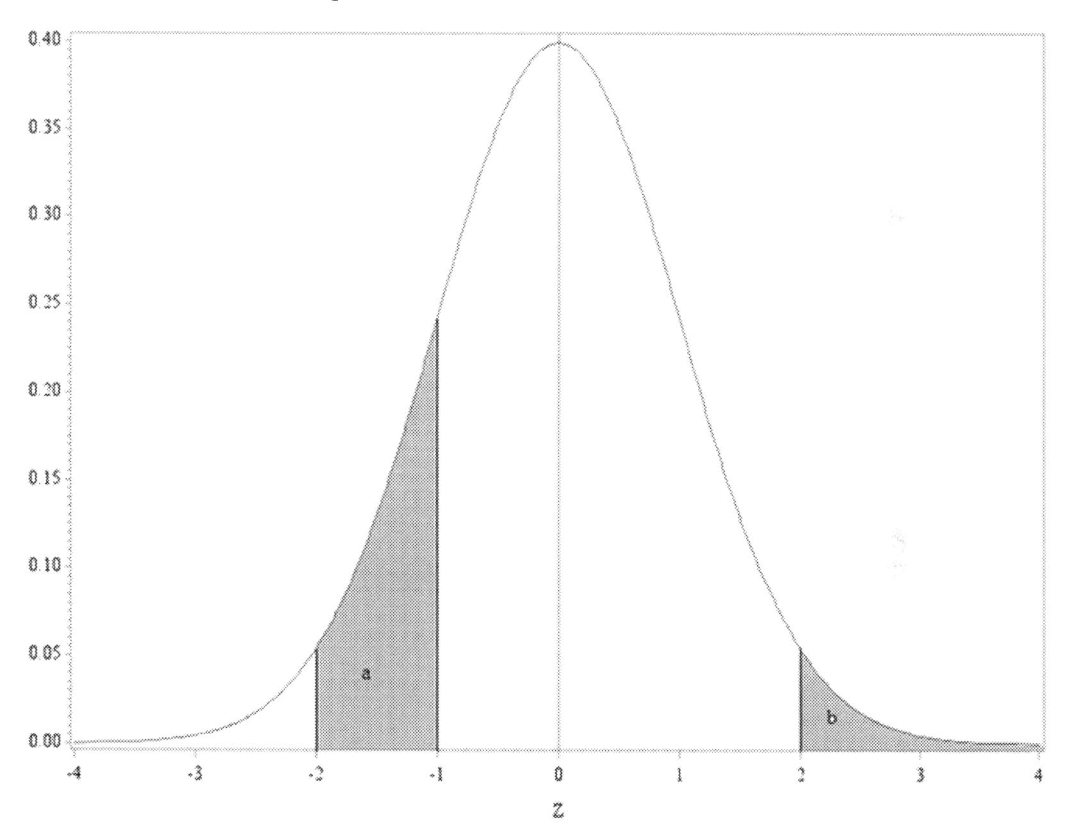

FIGURE 13.1
The standard normal probability distribution.

For the standard normal distribution, the PROBNORM function is used to compute probability, and the PROBIT function is used to determine quantiles.

The syntax of the PROBNORM function is:

PROBNORM (x)

where x is a specified numeric value of the standard normal random variable. The PROBNORM function returns the probability that an observation from the standard normal distribution is less than or equal to x, that is, $P(Z \leq x)$. Additionally, a function of values or variables can be specified as the argument, x, in this function. Within a DATA step, the appropriate syntax for this SAS statement is:

```
newvariable = PROBNORM(x);
```

The PROBIT function is the inverse of PROBNORM. The syntax of the PROBIT function is simple:

PROBIT (p)

where p is a numeric probability, with $0 < p < 1$. The PROBIT function returns the 100pth quantile from the standard normal distribution. The probability that an observation from the standard normal distribution is less than or equal to the returned quantile is p. Within a DATA step, the appropriate syntax for this SAS statement is:

```
newvariable = PROBIT(p);
```

OBJECTIVE 13.3: Given a standard normal distribution, compute the following:

a. Compute the area indicated by "a" in Figure 13.1. That is, compute $P(-2 < Z < -1)$.
b. Compute the area indicated by "b" in Figure 13.1. That is, compute $P(Z > 2)$.
c. Compute the standard normal random variable that has a cumulative probability (or left-side area) of 0.90. That is, find the 90th percentile of the standard normal distribution.
d. Compute the probability of a standard normal random variable below 1.645.

Use a temporary SAS data set and print the results.

```
DATA three;
a = PROBNORM(-1) - PROBNORM(-2);                *P(Z < -1) - P(Z < -2);
b = 1 - PROBNORM(2);                          *P(Z > 2) = 1 - P(Z < 2);
c = PROBIT(0.90);                                    *0.90 = P(Z < c);
d = PROBNORM(1.645);                                    *P(Z < 1.645);
;
PROC PRINT DATA=three ;
TITLE 'Objective 13.3';
RUN;
QUIT;
```

It is important to remember that the probability functions compute cumulative probability for a given value, so a difference between these cumulative probabilities is necessary for part a. In part b, the probability must be subtracted from 1 since the area to the right of 2 is needed. In the PROBIT function in part c, the argument must be between 0 and 1 and is the "left-side" area. From a basic statistics course, a z-score or standard normal score of 1.645 is often used in examples where the area above 1.645 is approximately 0.05; thus, for part d, the area below 1.645 is $1 - 0.05 = 0.95$.

Objective 13.3

Obs	a	b	c	d
1	0.13591	0.022750	1.28155	0.95002

More often a standard normal distribution is not the case, but a more general normal distribution with a mean μ and variance σ^2, $N(\mu, \sigma^2)$. To find probabilities associated with such a distribution, the transformation of the normal random variable Y to a standard normal random variable Z is done by this formula: $Z = \frac{Y - \mu}{\sigma}$. This transformation to a standard normal score (Z) can be written as the argument in the PROBNORM function. For example, suppose Y has a normal distribution with mean 12.8 and variance 6, $Y \sim N(12.8, 6)$, and the probability of a Y value in excess of 10 is to be computed. This can be done in the following ways:

```
DATA one;
P = 1-PROBNORM( (10-12.8)/SQRT(6) );
RUN;
```

```
DATA two;
Mu = 12.8;
Sigma2 = 6;
Y=10;
Z = (y-mu)/sqrt(sigma2);
P = 1 - PROBNORM(Z);
RUN;
```

On the left in SAS data set *one*, the transformation is completely calculated within the PROBNORM function. On the right in SAS data set *two*, each value is first entered as a SAS variable, and then transformed before entering the PROBNORM function. Both compute the same correct probability. There are other correct variations of this DATA step.

Likewise, the PROBIT function can assist in computing a quantile for a more general normal distribution. So, if $Y \sim N(12.8, 6)$ and the 75th percentile is to be computed, the line in a DATA step that would do this is:

```
Y75 = 12.8 + PROBIT(0.75)*SQRT(6);
```

This is an algebraic rearrangement of the transformation formula; that is, $Y = \mu + z\sigma$ where z is the 75th quantile of the standard normal distribution computed by the PROBIT function.

13.2.2 The t-Distribution

The t-distribution (or Student's t) has been used in Chapters 3 and 6 in hypothesis testing about one or two population means. The p-values or observed significance levels are probabilities determined by the area under the t-distribution function and more extreme than the computed test statistic, and the critical values of t, denoted $t_{\alpha, df}$, in the rejection regions (Tables 3.1, 6.2, and 6.4) are quantiles of the t-distribution.

To find a probability under the t-distribution curve, the PROBT function can be used. The syntax is:

PROBT (x, df <,nc>)

where:

x is the numeric random variable, or the value of the t-statistic on the horizontal or x-axis,
df is the degrees of freedom, and df > 0, and
nc is the non-centrality parameter, and nc ≥ 0. Omitting the comma and value for nc after ddf is the equivalent of nc = 0.

NOTE: Non-centrality parameter is a more advanced concept. In introductory statistical methods classes, the reference tables have a non-centrality parameter of zero. Non-centrality parameters are used in the evaluation of the power of a statistical test. Non-centrality parameters can be specified for the t- distribution here, and for the upcoming χ^2 and F distributions. For the hypothesis testing and confidence interval methods in this book, nc = 0.

The PROBT function computes the cumulative probability for a value from a Student's t distribution, with degrees of freedom df and non-centrality parameter nc; that is, $P(t_{df} \le x)$. When nc = 0, this distribution is called the central t-distribution. The df must be a positive number. Degrees of freedom measures are typically whole numbers but do not have to be. The Satterthwaite df* for the application in Table 6.4 is likely not a whole number.

The TINV function is the inverse of the PROBT function; that is, TINV finds the pth quantile (or critical t-value) given a cumulative probability p, df, and nc. The syntax of the TINV function is:

```
TINV (p, df <,nc>)
```

where:

p is a numeric probability, with $0 < p < 1$, and is the cumulative (or left side) probability,
df is the degrees of freedom, and df > 0, and
nc is the non-centrality parameter, and nc ≥ 0. For the applications in this book, nc = 0.

13.2.3 The χ^2-Distribution

The χ^2-distribution is an asymmetric distribution where the values of χ^2 are greater than or equal to zero, and the total area under the pdf is one. A χ^2-statistic was computed in Sections 7.3 and 9.1, and the degrees of freedom, df, depend on the application. Probabilities associated with the χ^2-distribution can be computed in the DATA step using the PROBCHI function. The syntax of the PROBCHI function is:

```
PROBCHI (x, df <, nc>)
```

where:

x is the numeric random variable (x ≥ 0), or the value of the χ^2-statistic on the horizontal or x-axis,
df is the degrees of freedom, and df > 0, and
nc is the non-centrality parameter, and nc ≥ 0. For the applications in this book, nc = 0.

The PROBCHI function computes the cumulative probability for a value, x, from a chi-square distribution, with degrees of freedom df and non-centrality parameter nc; that is, $P(\chi^2_{df} \le x)$. Like the t-distribution in the previous section, applications in the book are limited to nc = 0, or the central χ^2-distribution. Omitting the comma and value for nc after ddf is the equivalent of nc = 0. The CINV function is the inverse of the PROBCHI function, and its syntax is:

```
CINV (p, df <,nc>)
```

where:

p is a numeric probability, with $0 < p < 1$, and is the cumulative (or left side) probability,
df is the degrees of freedom, and df > 0, and
nc is the non-centrality parameter, and nc ≥ 0. For the applications in this book, nc = 0.

The CINV function computes the pth quantile from the chi-square distribution with degrees of freedom df and a non-centrality parameter nc.

13.2.4 The F-distribution

Like the χ^2-distribution the F-distribution is an asymmetric distribution where the values of F are greater than or equal to zero, and the total area under the pdf is one. An F-statistic was computed in Sections 6.4, 7.1, and 11.1. The F-statistic has numerator degrees of freedom, ndf, and denominator degrees of freedom, ddf. The shape of the F-distribution changes as ndf or ddf or both change. These degrees of freedom measures depend on the application. Probabilities associated with the F-distribution can be computed in the DATA step using the PROBF function. The syntax of the PROBF function is:

```
PROBF (x, ndf, ddf <, nc>)
```

where:

x is the numeric random variable ($x \geq 0$), or the value of the F-statistic on the horizontal or x-axis,
ndf is the numerator degrees of freedom, and ndf > 0,
ddf is the denominator degrees of freedom, and ddf > 0, and
nc is the non-centrality parameter, and nc ≥ 0. For the applications in this book, nc = 0.

The PROBF function computes the cumulative probability for a value, x, from an F distribution, with numerator degrees of freedom ndf, denominator degrees of freedom ddf, and non-centrality parameter nc; that is, $P(F_{ndf,\, ddf} \leq x)$. Omitting the comma and value for nc after ddf is the equivalent of nc = 0.

The FINV function is the inverse of the PROBF function.

```
FINV (p, ndf, ddf <,nc>)
```

where:

p is a numeric probability, with $0 < p < 1$, and is the cumulative (or left side) probability, df is the degrees of freedom, and df > 0,
ndf is a numeric numerator degrees of freedom parameter, with ndf > 0,
ddf is a numeric denominator degrees of freedom parameter, with ddf > 0, and
nc is an optional numeric non-centrality parameter, with nc ≥ 0.

The FINV function computes the pth quantile from the F distribution with numerator degrees of freedom ndf, denominator degrees of freedom ddf, and non-centrality parameter, nc.

Like the discrete probability distributions there are also PDF and CDF functions in the SAS Data Step for continuous distributions. Again, SAS 9.4 Functions and CALL Routines: Reference in SAS Help and Documentation or SAS Institute (2016a) can be searched for syntax information for these functions. Because event probabilities for continuous random variables are determined by the area under the curve, the CDF function must be used and not the PDF function.

The inverse of the probability functions (PROBIT, TINV, CINV, FINV) can produce the critical values one is accustomed to reading from a reference table for selected values of the Type I error rate, α, and much more.

OBJECTIVE 13.4: Use the SAS functions for t, χ^2, and F to compute the following. Use a temporary SAS data set and print the results.

a. Find the value of t when the right-hand area is 0.12 and the df = 14.
b. What is the probability of a t-statistic larger than 2.104 with 20 df?
c. What is the value of the critical value $\chi^2_{0.04,\ 14}$? (Find the χ^2 value with 14 df that determines a right-hand area of 0.04.)
d. Find the p-value (observed significance level) of the χ^2 statistic 17.04 with 6 df for the hypothesis test H_0: $\sigma^2 = 42$ versus H_1: $\sigma^2 \neq 42$.
e. Find the p-value associated with the ANOVA F-statistic of $F_{4,\ 16} = 7.83$.
f. Find the critical F-value determined by $\alpha = 0.03$ for the ANOVA in e.

```
DATA four;
a = TINV(0.88, 14);                    * P(t > a)=0.12 or 0.88=P(t < a), df=14;
b = 1 - PROBT(2.104, 20);                            *P(t > 2.104, df=20)=b;
c = CINV(0.96, 14);                       *P(X2 > c)=0.04 or P(X2 < c)=0.96;
d = 2*(1 - PROBCHI(17.04, 6) );                        *2P(X2 > 17.04)=d, df=6;
e = 1 - PROBF(7.83, 4,16) ;                  *1 - P(F>7.83)=e, ndf=4, ddf=16;
f = FINV(0.97, 4, 16) ;                      *0.03 = P(F > f), ndf=4, ddf=16;
PROC PRINT DATA=four;
TITLE 'Objective 13.4';
RUN;
QUIT;
```

Objective 13.4

Obs	a	b	c	d	e	f
1	1.22716	0.024115	24.4855	0.018275	.001075471	3.53420

13.3 Chapter Summary

The probabilities of events defined by discrete or continuous random variables can be computed using SAS functions in a DATA step. There are many families of distributions for which there are SAS functions available. For the continuous random variable probability functions, there are also inverse probability functions which compute quantiles of the distributions. This chapter presents the two discrete distributions, binomial and Poisson, and the continuous distributions associated with hypothesis test statistics overviewed in this book, standard normal, t, χ^2, and F. For those with a stronger background in statistics, these functions can be used to conduct power studies for a proposed experiment. A knowledge of non-centrality parameters is typically needed for these types of power computations.

14

Reading and Writing Data Files

The ability for one software product to be able to read or write files that are compatible with other software products is desirable. For the SAS user, reading data files created by another software product is certainly an attractive feature. In Chapter 8, the INFILE statement of the DATA Step was introduced and used to read data from an external file. These files are generally some variation of a text file, typically space, tab, or comma delimited. Data sets are often created using popular spreadsheet software or other statistical software. Can the data from one of these spreadsheet sources be "read" into SAS? Conversely, can SAS write or output SAS data sets to an external *non-SAS* file? The answer to both of these questions is, "Yes!" To do this, Microsoft® Excel (Microsoft Corporation, 2016) will be used, but SAS is not limited to Microsoft® Excel for these actions.

14.1 The Import Wizard

The IMPORT procedure is a procedure in SAS that can read many forms of spreadsheet files, text files, data base files, and data files of other forms. Thus, it is possible to read an Excel file directly, that is, it does not have to be saved as a text file in an intermediate step. Keeping current with the available options for the IMPORT procedure can be challenging. The Import Wizard is a point-and-click approach to reading external data files. To demonstrate the Import Wizard, the data in Table 14.1 is entered in Microsoft® Excel 2016 exactly as it is shown in the table beginning in line 1 (below the table title) and Excel column A with TRT. This file is saved as data1.xlsx in the folder G:\CLGoad\.

Whenever one takes a file developed in one software product for use in another software product, care should always be taken. Specifically, one should be mindful of how missing observations are handled and examine both character and numeric values.

While there is syntax that one can learn for this procedure, it is presented here first using the Import Wizard. The Import Wizard is a menu-driven procedure for importing data files. Readers are reminded that menu items and other selections made by point-and-click appear in **boldface** in these objectives and instructions.

During a SAS session **File – Import Data** is selected from the pull-down menus. This sequence will open an Import Wizard, as in Figure 14.1. Within this dialog window one can select from a drop-down list the type of data file to import into SAS. Among these file types are (space) Delimited File (*.*), Tab Delimited File (*.txt), and Comma Separated Values (*.csv) which were file types that INFILE could also manage. Additionally, there are several other file types in this list that INFILE methods will not recognize. For this example **Microsoft Excel Workbook (*.xls *.xlsb *.xlsm *.xlsx)** is selected, and then **Next**.

The next window in the dialog is **Connect to MS Excel**. That is, where is the file located? G:\CLGoad\data1.xlsx is entered, or one can **Browse** and select the directory:\folder and filename from the lists available. **OK**

TABLE 14.1

Microsoft® Excel file: data1.xlsx for the Import Wizard

TRT	X	Y	Z
A	5	15	15
A	3	21	44
A	5	11	79
A	7	55	46
A	1	22	16
A	6	23	54
A	8	.	52
B	4	44	32
B	1	11	53
B	6	77	51
B	.	88	24
B	1	55	26
B	8		59
B	9	11	54
B	4	33	21

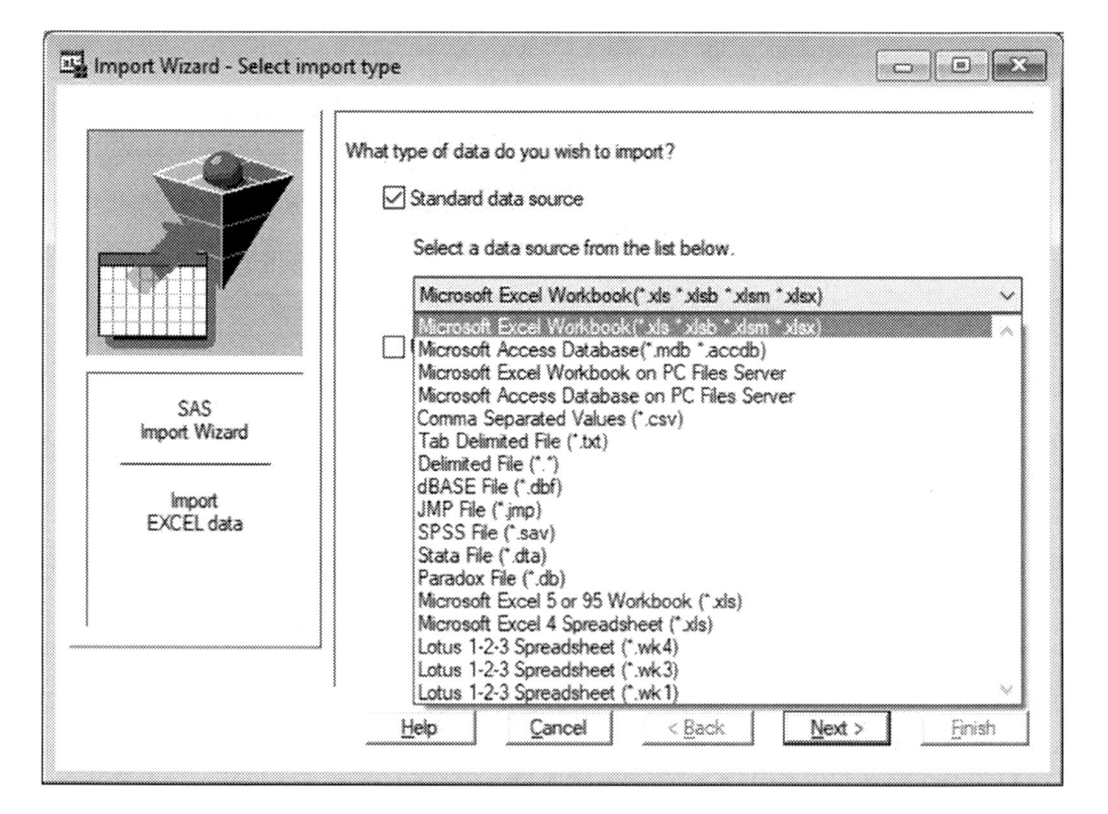

FIGURE 14.1
Opening dialog of the Import Wizard.

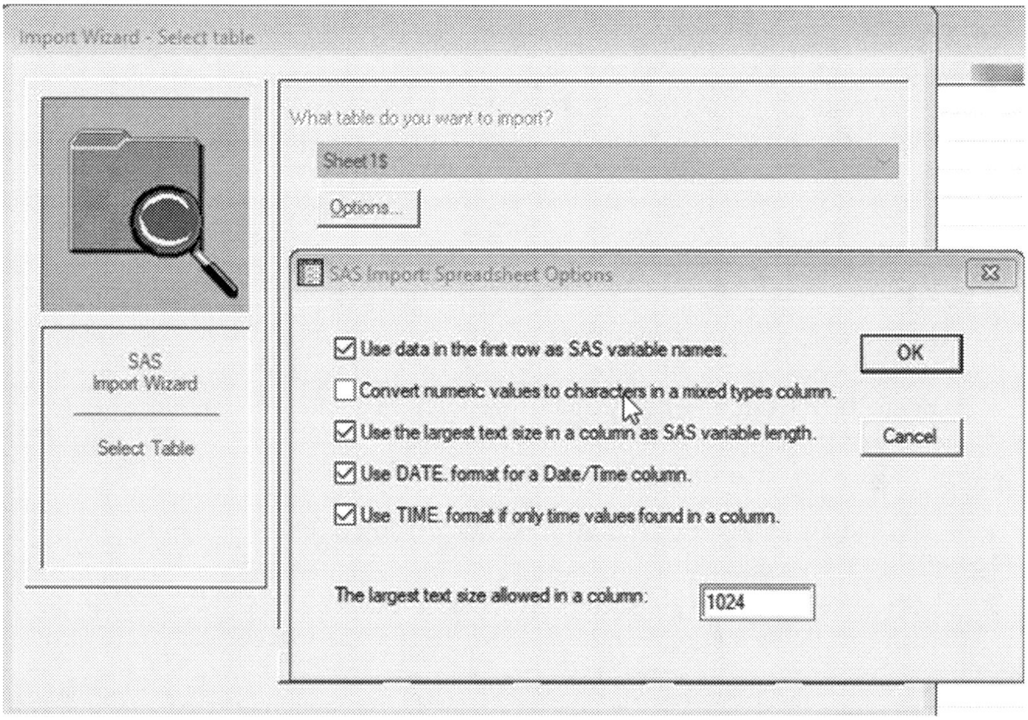

FIGURE 14.2
The default options for the data file to be imported.

What table do you want to import? (Figure 14.2) If the file is a multi-sheet spreadsheet, and the data are not on the first sheet, the available worksheet names in the targeted Excel file are listed when the down arrow is used to expand the list. The default worksheet names are "Sheet1$", "Sheet2$", and so on if the worksheets were not renamed in the data file. If the worksheets have been renamed, the Import dialog will identify the names. Worksheet names should be less than 30 characters. Longer worksheet names will not be recognized by the Import Wizard. It may be necessary to shorten the sheet name(s) in Excel and resave the file before the IMPORT procedure may work. Once the sheet or page of the spreadsheet has been identified, the **Options** button is then selected to identify any special case situations. Generally the default settings in this dialog are adequate.

Available options are:

Use data in the first row as SAS variable names – As in Chapter 8, it is strongly recommended that the first line of an external data file contain column headings to be used as variable names. For this simple data file the column headings were kept to one word. Later, using ViewTable, longer column headings in Excel will be examined. The box should be checked if one line for column headings has been allowed. Using the Import Wizard there is a limit of one line for headings. No headings on the worksheet? Then the option should be deselected (unchecking the box).

Convert numeric values to characters in a mixed types column – SAS will determine whether the data entries are character or numeric based on the values in the first lines of data (below headings). This item should be checked if the column entries could be either numeric or character. For example, suppose there is a column header "Location", and some of the values in the column are: 56, 84, 29N, 29S. If 56 is the first entry in the "Location" column, then it is presumed that all of the data in the column are numeric. When the values "29N" and "29S" are encountered, these will be read in as missing values since they are character strings. Likewise, if a character string is the first entry, all values in the column are presumed to be character strings. If there is missing data in a column, a period is not necessary in that cell of the Excel sheet. The blank will be recognized as missing. If the column variable is numeric and contains missing values and if a period is used to denote a missing value, then the mixed types column should not be used. Otherwise, the column variable will be identified as character due to the use of the period denoting a missing value.

Use the largest text size in a column as SAS variable length – If the data file contains multiple word or lengthy text expressions in a single column, this option should be selected if the text strings should not be truncated.

DATE and TIME formats are not covered in this book. The default settings are generally recommended for the beginning SAS programmer.

After selecting or verifying the **Spreadsheet Options**, then **OK – Next.**

Select library and member – In Figure 14.3, the name of an existing SAS library and the name of the SAS data set are entered in two blanks. The down arrow lists the available libraries. If the desired library does not appear in this list,

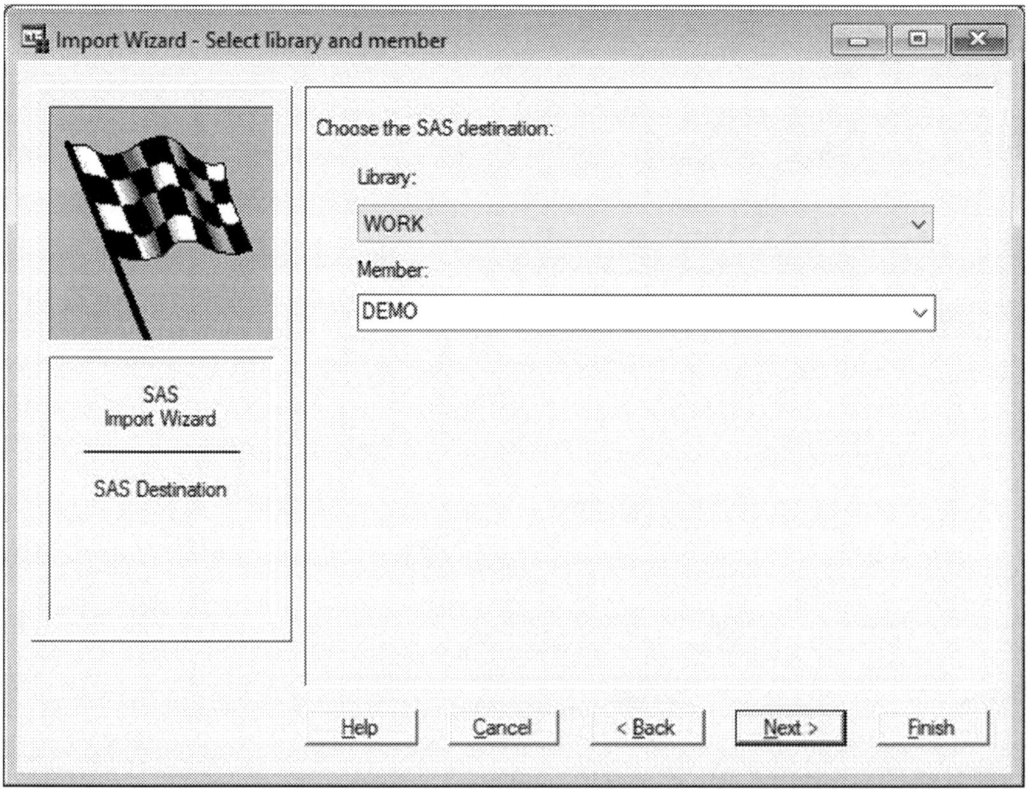

FIGURE 14.3
Selecting the Library and SAS data set name.

a new library will need to be created before this IMPORT can be completed (Section 12.2). Thus, the IMPORT procedure can create either a permanent SAS data set or a temporary one. For this demonstration, the *WORK* library is used, and the SAS data set is *DEMO*. That is, this two blank dialog is used to create the two-level SAS data set name.

At this point the **Finish** button can be selected. Confirmation of the successful completion of the IMPORT procedure will appear in the log window:

```
NOTE: WORK.DEMO was successfully created.
```

Or a message indicating cancellation or other errors with importation may appear in the log window.

This menu-driven process can produce the SAS code for the IMPORT procedure if **Next** instead of **Finish** is selected as the last step in Figure 14.3. An additional dialog window opens when **Next** is selected and is shown in Figure 14.4. In **Create SAS Statements** the complete path and filename is input for the IMPORT procedure code to be recovered. **Replace file if it exists** should be checked if that is appropriate. **Finish** will complete this importation of data. Using this additional step in the Import Wizard will also result in feedback in the log window as indicated before, and a SAS program, IMPORT DEMO.SAS (in Figure 14.4) which can later be opened in the Editor is produced.

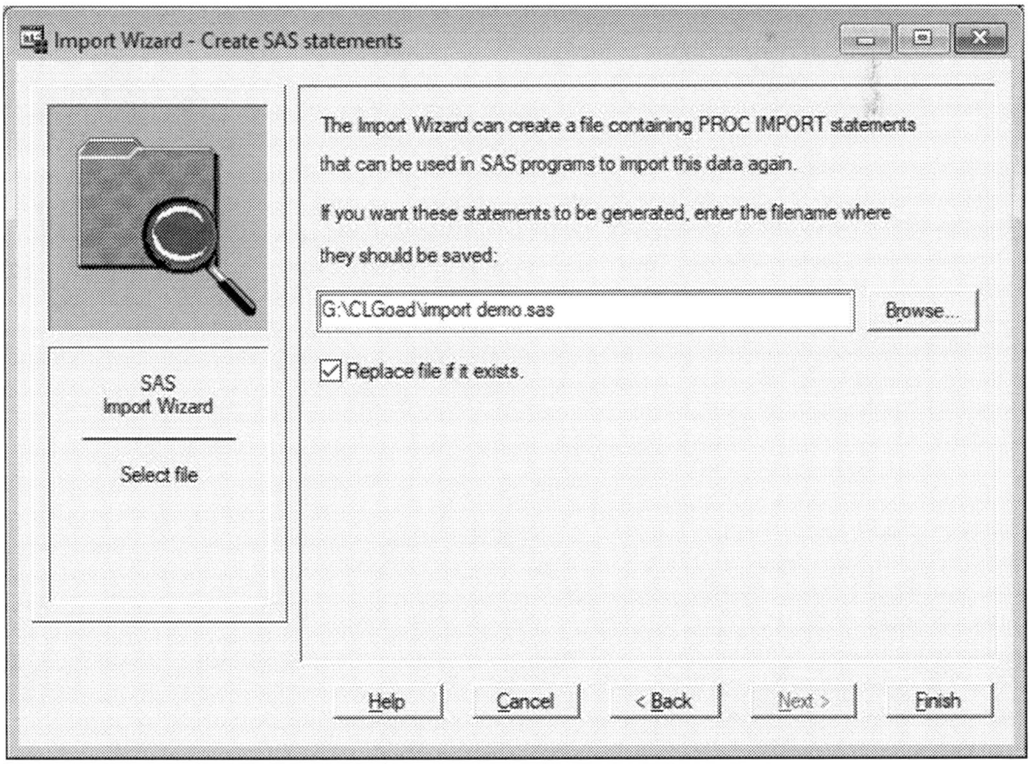

FIGURE 14.4
Recover the PROC IMPORT code using the Import Wizard.

VIEWTABLE: Work.Demo

	TRT	X	Y	Z
1	A	5	15	15
2	A	3	21	44
3	A	5	11	79
4	A	7	55	46
5	A	1	22	16
6	A	6	23	54
7	A	8	.	52
8	B	4	44	32
9	B	1	11	53
10	B	6	77	51
11	B	.	88	24
12	B	1	55	26
13	B	8	.	59
14	B	9	11	54
15	B	4	33	21

FIGURE 14.5
Verifying data using ViewTable.

If the Excel sheet will be needed again in SAS, saving the SAS code for the IMPORT procedure is recommended so that the Import Wizard dialog does not need to be repeated. Submitting the simple IMPORT code produced by the Import Wizard is all that is necessary. Additionally, recovering the IMPORT code would be advantageous if the Excel sheet has revisions or additions. In this way, when the Excel file is updated, all that is necessary is to submit the Import procedure code produced at the end of the previous Import Wizard session.

After the completion of the Import Wizard, the resulting SAS data set *demo* is in the *WORK* library. In the SAS Explorer window, the sequence **Libraries – WORK – DEMO** is selected to open the SAS data set using ViewTable, shown in Figure 14.5. The missing numeric values denoted by the "." or the space in initial Excel spreadsheet are both correctly recorded as missing in SAS. One should always verify how missing observations are managed when reading data files. As indicated earlier, caution is advised when enabling the "mixed types column" option. The "." in the place of the missing values in a column of numeric values will convert the variable to character.

If there were no column headers in the external file, SAS will assign default variable names F1, F2, F3, and so on for the variable names. These default variable names can be changed in a new DATA step using the RENAME statement (Section 8.4).

14.2 The IMPORT Procedure

The IMPORT procedure code requested in the Import Wizard can be opened in the **Editor**. From the demonstration in Section 14.1, **File – Open G:\CLGoad\import demo.sas** opens the following program in the Editor:

```
PROC IMPORT OUT= WORK.demo
            DATAFILE= "G:\CLGoad\data1.xlsx"
            DBMS=EXCEL REPLACE;
     SHEET="Sheet1$";
     GETNAMES=YES;
     MIXED=NO;
     SCANTEXT=YES;
     USEDATE=YES;
     SCANTIME=YES;
RUN;
```

The selections made in the dialog windows shown in Figures 14.1 through 14.3 can be identified in the syntax options on the PROC IMPORT statement and in the following statements. To learn the IMPORT procedure, it is this author's opinion that the Import Wizard should be utilized and recover the PROC IMPORT code. Options in the PROC IMPORT statement and the statements that follow it are dependent upon the platform in which SAS is operating (here, Windows) and the type of file that is being accessed. The demonstration here is using Microsoft® Excel 2016 and SAS 9.4. Older versions of Excel have additional statements in the IMPORT procedure are not applicable to the 2016 version. For example, if the data values do not appear in the second row but, say, in row 10, older versions of Excel have a STARTROW statement that would identify that; STARTROW=10. This is just one example of how the programming code can be affected. One could expect different dialog windows and hence different options and statements when importing data from different sources, such as SPSS® (IBM Corp., 2017), Stata® (StataCorp., 2019), or Microsoft® Access (Microsoft Corporation, 2016).

Keeping the SAS code for importing the data file is also useful if the external file may be updated periodically thus requiring the SAS data set creation and analysis in SAS to be updated.

The FILENAME statement (Section 8.1) can be used in conjunction with IMPORT procedure code. The FILENAME assigns a *fileref* to an external file, and that *fileref* can be used in the IMPORT procedure. Using the drive:\folder specification in the previous demonstration, the SAS code is modified as:

```
FILENAME fileref 'G:\CLGoad\data1.xlsx';
PROC IMPORT OUT= WORK.demo
            DATAFILE= fileref
            DBMS=EXCEL REPLACE;
⋮
RUN;
```

When multiple pages or worksheets are to be read from the same Excel file, the Import Procedure code can be copied, and the SHEET must be modified for the additional sheet(s).

If the external Excel file has simple column headers such as those in *data1.xlsx*, variable labels can be added in the PROC IMPORT block of statements using the LABEL statement introduced in Section 8.2. The LABEL statement is positioned before the RUN statement for the procedure. Such as:

```
PROC IMPORT OUT= WORK.demo
            DATAFILE= "G:\CLGoad\data1.xlsx"
            DBMS=EXCEL REPLACE;
     SHEET="Sheet1$";
     GETNAMES=YES;
```

```
      MIXED=NO;
      SCANTEXT=YES;
      USEDATE=YES;
      SCANTIME=YES;
LABEL x = "Text for Variable X"
          Y = "Text for Variable Y"
          Z = "Text for Variable Z";
RUN;
```

The Import Wizard does not allow for the creation of labels.

No other variable assignment statements or other DATA Step commands can be used in the IMPORT Procedure. The next demonstration will examine variable names and labels in greater detail.

In Chapter 11, Table 11.2 contained measured responses on 56 head of beef steers. If this data were originally created in an Excel spreadsheet, the data could be imported from Excel using the methods of this chapter. An Excel file *Chapter 14 Import data.xlsx* has been created. The first sheet in the file contains the Chapter 11 data from 56 steers supplied by Producer 1, the second sheet is from Producer 2, and the third sheet is from Producer 3. While Producer 1 recorded data for 56 steers, Producers 2 and 3 have data from both heifers (H) and steers (S), and this is represented on their respective sheets in the file with the addition of a column for the variable Sex. Figure 14.6 shows the first ten lines of data on each sheet of the Excel file. The columns are not exactly in the same order, the column names differ somewhat, and the sheet names have different spacing as well as different text.

Chapter 14 Import data.xlsx can be imported using three different point-and-click sequences, or using PROC IMPORT code. The effects of these different column headings and different sheet names on coding and the SAS data sets will be examined.

The PROC IMPORT code is:

```
FILENAME beef 'G:\CLGoad\Chapter 14 Import data.xlsx';
PROC IMPORT OUT= WORK.prod1
            DATAFILE= beef
            DBMS=EXCEL REPLACE;
      SHEET="'Producer1 Steer1$'";
      GETNAMES=YES;      MIXED=NO;
      SCANTEXT=YES;      USEDATE=YES;
      SCANTIME=YES;
RUN;

PROC IMPORT OUT= WORK.prod2
            DATAFILE= beef
            DBMS=EXCEL REPLACE;
      SHEET="Sheet2$";
      GETNAMES=YES;      MIXED=NO;
      SCANTEXT=YES;      USEDATE=YES;
      SCANTIME=YES;
RUN;

PROC IMPORT OUT= WORK.prod3
            DATAFILE= beef
            DBMS=EXCEL REPLACE;
      SHEET="'Producer 3'";
      GETNAMES=YES;      MIXED=NO;
      SCANTEXT=YES;      USEDATE=YES;
      SCANTIME=YES;
RUN;
```

	A	B	C	D	E	F
1	DMI	ADG	REA	CWT	BackFat	
2	17.009	3.05	13.3821	819	0.54571	
3	17.0183	2.87	13.7779	857	0.53974	
4	17.4826	3.12	13.5262	826	0.46	
5	18.0043	2.73	.	831.256	.	
6	16.2731	2.83	12.8589	799	0.54622	
7	17.5451	2.98	13.8188	850	0.5055	
8	16.6	2.78	13.48	790	0.5028	
9	17.1309	2.8	12.646	805	0.512	
10	17.8689	2.85	13.4662	814	0.49434	
11	17.8994	3.03	13.2404	838	0.53535	

Producer1 Steer | Sheet2 | Producer 3 | She

	A	B	C	D	E	F
1	Sex	CWT	REA	Back Fat	DMI	ADG
2	H	769	13.2774	0.52667	16.844	2.91
3	H	800	13.5977	0.57591	16.8912	2.89
4	H	743.466	.	.	17.4981	2.8
5	H	787.154	.	.	18.5111	2.88
6	H	793	13.2446	0.63815	17.5085	3.03
7	H	756	13.2446	0.63815	17.5085	3.03
8	H	759	13.1011	0.53111	17.6471	2.87
9	H	787.694	13.1682	0.59482	18.0717	3.01
10	H	802	12.9752	0.59759	18.7439	3.14
11	H	805	13.8792	0.59583	18.5986	3.01

Producer1 Steer | **Sheet2** | Producer 3 | She

	A	B	C	D	E	F
1	Sex	DMI(lb)	ADG(lb)	CWT, #	REA	BackFat
2	H	16.2967	2.56	726	12.9626	0.51354
3	H	16.9471	2.75	755	13.333	0.49787
4	H	16.9732	2.76	760	13.4218	0.49425
5	H	18.169	2.72	743.278	.	.
6	H	19.8894	2.99	754.236	9.2	0.28
7	H	16.3229	2.68	729	12.5832	0.51032
8	H	16.7013	2.57	770	12.6817	0.58839
9	H	17.9253	2.8	778.888	13.1624	0.55775
10	H	17.7497	2.85	714.831	.	.
11	H	19.6525	3.05	753.93	.	.
12	H	16.1886	2.5	714	12.2161	0.55586

Producer1 Steer | Sheet2 | **Producer 3** | She

FIGURE 14.6
The first ten observations from each producer. a. First lines of the data file for Producer 1; data are for steers only. b. First lines of the data file for Producer 2; data are for heifers (H) and steers (S). c. First lines of the data file for Producer 3; data are for heifers (H) and steers (S).

The following notes appear in the Log window:

```
NOTE: The data set WORK.PROD1 has 56 observations and 5 variables.
NOTE: The data set WORK.PROD2 has 171 observations and 6 variables.
NOTE: The data set WORK.PROD3 has 101 observations and 6 variables.
```

First the usage of the FILENAME statement should be noted. It is not necessary if the complete file path is specified in the DATAFILE= option in the PROC IMPORT statement. For lengthy file paths in repetitive IMPORT code, it does help with efficiency.

Next is the effect of spacing in the sheet names. For Producer 1, the sheet name is "Producerl Steer". Because it is a multiple word sheet name, the syntax uses single quotes inside double quotes in the SHEET statement. That is, `SHEET = "'Producerl Steer1$'";` Producer 2 has data on the second sheet, and the default sheet name, "Sheet2", has been used. The SHEET statement does not require both types of quotation marks since there are no spaces in the sheet name. Thus, `SHEET="Sheet2$";` And finally, Producer 3 uses both types of quotation marks since there is a space in the sheet name. That is, `SHEET="'Producer 3'";` These very subtle differences in the SHEET statement are correctly detected when one uses the Import Wizard. For many sheets in a single Excel file, renaming the sheets for clarity is recommended.

Each of these three SAS data sets for the beef data were created successfully as indicated in the aforementioned notes from the SAS Log window. In addition to the different spacing and text on the sheet name tabs, the column headers from one sheet to the next are not the same. Producer 1 information has the simple variable names with no spaces, Producer 2 has "Back Fat" as two words, and Producer 3 has units indicated on three of the column headers. One of the objectives is to form one SAS data set with common variable names and add the variable "Producer" to the combined data set. To concatenate the three SAS data sets, the SET command (Section 4.4) is needed. But before the three SAS data sets can be concatenated, the variables from one data set to another should be the same. To view what happens to the variable names when importing data from an external file, ViewTable (Section 8.3) will be used.

From the **Explorer** window on the left side of the SAS screen, the Work library is selected and then each of the SAS data sets Prod1, Prod2, and Prod3 can be examined in ViewTable. The column headers initially displayed are the actual text from the first row of the Excel file unless a LABEL statement was included in the PROC IMPORT code. If no LABEL statement was used, the actual text from the first row determines the labels. As in Section 8.3, one can switch to a view where the column headers are the variable names rather than the variable labels by selecting **View – Column Names** from the pull-down menu while in ViewTable. Figure 14.7 shows the **Column Labels** on the left and **Column Names** on the right for each of the three producers' data sets.

The two *Prod1* screen captures at the top of Figure 14.7 show that the labels and the columns names are exactly the same. This is because in the Excel file, SAS naming conventions for variables were met with these choices of column variable names. In the middle row of Figure 14.7 *Prod2* shows "Back Fat" in a label, and the space is read as an underscore (_) in the variable names on the right. And for *Prod3*, any special characters or spaces in the header row of Excel are represented in the label but are underscores in the variable name in the SAS data set. So while GETNAMES=YES is recommended since column headers are always encouraged in external data files, it must be remembered that special characters and spaces in these column names are converted to an underscore

Column Headers are Labels

VIEWTABLE: Work.Prod1

	DMI	ADG	REA	CWT	BackFat
1	17.009	3.05	13.3821	819	0.54571
2	17.0183	2.87	13.7779	857	0.53974
3	17.4826	3.12	13.5262	826	0.46
4	18.0043	2.73	.	831.256	.
5	16.2731	2.83	12.8589	799	0.54622
6	17.5451	2.98	13.8188	850	0.5055
7	16.6	2.78	13.48	790	0.5028
8	17.1309	2.8	12.646	805	0.512
9	17.8689	2.85	13.4662	814	0.49434
10	17.8994	3.03	13.2404	838	0.53535

VIEWTABLE: Work.Prod2

	Sex	CWT	REA	Back Fat	DMI	ADG
1	H	769	13.2774	0.52667	16.844	2.91
2	H	800	13.5977	0.57591	16.8912	2.89
3	H	743.466	.	.	17.4981	2.8
4	H	787.154	.	.	18.5111	2.88
5	H	793	13.2446	0.63815	17.5085	3.03
6	H	756	13.2446	0.63815	17.5085	3.03
7	H	759	13.1011	0.53111	17.6471	2.87
8	H	787.694	13.1682	0.59482	18.0717	3.01
9	H	802	12.9752	0.59759	18.7439	3.14
10	H	805	13.8792	0.59583	18.5986	3.01

VIEWTABLE: Work.Prod3

	Sex	DMI(lb)	ADG(lb)	CWT, #	REA	BackFat
1	H	16.2967	2.56	726	12.9626	0.51354
2	H	16.9471	2.75	755	13.333	0.49787
3	H	16.9732	2.76	760	13.4218	0.49425
4	H	18.169	2.72	743.278	.	.
5	H	19.8894	2.99	754.236	9.2	0.28
6	H	16.3229	2.68	729	12.5832	0.51032
7	H	16.7013	2.57	770	12.6817	0.58839
8	H	17.9253	2.8	778.888	13.1624	0.55775
9	H	17.7497	2.85	714.831	.	.
10	H	19.6525	3.05	753.93	.	.

Column Headers are Variable Names

VIEWTABLE: Work.Prod1

	DMI	ADG	REA	CWT	BackFat
1	17.009	3.05	13.3821	819	0.54571
2	17.0183	2.87	13.7779	857	0.53974
3	17.4826	3.12	13.5262	826	0.46
4	18.0043	2.73	.	831.256	.
5	16.2731	2.83	12.8589	799	0.54622
6	17.5451	2.98	13.8188	850	0.5055
7	16.6	2.78	13.48	790	0.5028
8	17.1309	2.8	12.646	805	0.512
9	17.8689	2.85	13.4662	814	0.49434
10	17.8994	3.03	13.2404	838	0.53535

VIEWTABLE: Work.Prod2

	Sex	CWT	REA	Back_Fat	DMI	ADG
1	H	769	13.2774	0.52667	16.844	2.91
2	H	800	13.5977	0.57591	16.8912	2.89
3	H	743.466	.	.	17.4981	2.8
4	H	787.154	.	.	18.5111	2.88
5	H	793	13.2446	0.63815	17.5085	3.03
6	H	756	13.2446	0.63815	17.5085	3.03
7	H	759	13.1011	0.53111	17.6471	2.87
8	H	787.694	13.1682	0.59482	18.0717	3.01
9	H	802	12.9752	0.59759	18.7439	3.14
10	H	805	13.8792	0.59583	18.5986	3.01

VIEWTABLE: Work.Prod3

	Sex	DMI_lb_	ADG_lb_	CWT___	REA	BackFat
1	H	16.2967	2.56	726	12.9626	0.51354
2	H	16.9471	2.75	755	13.333	0.49787
3	H	16.9732	2.76	760	13.4218	0.49425
4	H	18.169	2.72	743.278	.	.
5	H	19.8894	2.99	754.236	9.2	0.28
6	H	16.3229	2.68	729	12.5832	0.51032
7	H	16.7013	2.57	770	12.6817	0.58839
8	H	17.9253	2.8	778.888	13.1624	0.55775
9	H	17.7497	2.85	714.831	.	.
10	H	19.6525	3.05	753.93	.	.

FIGURE 14.7
Column labels (left) versus column names (right) for each of the three SAS data sets in ViewTable.

when SAS imports the data. This can be challenging at times. For example, in SAS data set *Prod3* the carcass weight variable is read in as CWT___. It may be puzzling as to the number of underscores in the variable name. There are three. An extra space was entered after # in the Excel file.

OBJECTIVE 14.1: Given the SAS data sets *Prod1, Prod2,* and *Prod3* in the Work library created using the previous SAS program. Concatenate these three data sets after adding a Producer variable to each and create uniform variable names and labels. Use the RENAME statement where needed (Section 8.4). Name this new data set *beef3* in the Work library.

```
DATA prod1;
SET prod1;
Producer = 1;
Sex = "S";          *This variable was not in the data set for Producer 1;
                     *All observations from this producer were steers (S);
LABEL cwt  = 'Carcass Weight, lb'
      adg  = 'Average Daily Gain, lb'
      dmi  = 'Dry Matter Intake, lb'
      rea  = 'Rib Eye Area'
   Backfat= 'Back Fat Thickness';

DATA prod2;
SET prod2;
Producer = 2;
RENAME back_fat = backfat;

DATA prod3;
SET prod3;
Producer = 3;
RENAME dmi_lb_ = dmi       adg_lb_ = adg       cwt___ = cwt;

DATA beef3;
SET prod1 prod2 prod3;
RUN;
QUIT;
```

The only objective of this short SAS program is to create a SAS data set that contains all of the data from these three producers. Variables in common among the SAS data sets to be concatenated must have the same variable name, thus the use of the RENAME statements. The labels could be entered for the first SAS data set that is included in the SET command (of the fourth DATA step) combining all three of the SAS data sets or in the fourth DATA step creating *beef3*. The case sensitivity of the variable names is also determined by the first SAS data set in the SET command.

No PRINT procedure has been included since there are over 300 lines of data. SAS data set *beef3* can be more efficiently examined in ViewTable. The order of the column variables is determined by the first SAS data set in the SET statement, and this is shown in Figure 14.8. Producer and Sex were the last two variables created in SAS data set *prod1* and are the last two variables in *work.beef3*.

VIEWTABLE: Work.Beef3

	Dry Matter Intake, lb	Average Daily Gain, lb	Rib Eye Area	Carcass Weight, lb	Back Fat Thickness	Producer	Sex
1	17.009	3.05	13.3821	819	0.54571	1	S
2	17.0183	2.87	13.7779	857	0.53974	1	S
3	17.4826	3.12	13.5262	826	0.46	1	S
4	18.0043	2.73	.	831.256	.	1	S
5	16.2731	2.83	12.8589	799	0.54622	1	S
6	17.5451	2.98	13.8188	850	0.5055	1	S
7	16.6	2.78	13.48	790	0.5028	1	S
8	17.1309	2.8	12.646	805	0.512	1	S
9	17.8689	2.85	13.4662	814	0.49434	1	S
10	17.8994	3.03	13.2404	838	0.53535	1	S

FIGURE 14.8
ViewTable of WORK.BEEF3 with column labels (first ten observations)

14.3 The Export Wizard

Like the Import Wizard, there is an Export Wizard (**File – Export Data** selected from the pull-down menu) that can write a SAS data set to an external file in formats compatible with various software. As with the Import Wizard, the Export Wizard can generate SAS code that accompanies this action.

Completing the dialog in the windows of the Export Wizard is much like the dialog for the Import Wizard. Since importing data from Microsoft Excel was done in the early part of this chapter, this chapter will finish with exporting a SAS data set to Microsoft Excel.

OBJECTIVE 14.2: Compute the means of each of the variables ADG, DMI, CWT, REA, and BackFat and other default summary statistics for each producer and sex of animal. Recover these values in a SAS data set called *beef3means*. Suppress the printing of the procedure results. Export these summary values to a fourth sheet in the Microsoft Excel file *Chapter 14 Import data.xlsx*.

```
PROC SORT DATA=beef3;
BY producer sex;
PROC MEANS DATA=beef3 NOPRINT;
BY producer sex;
VAR adg dmi cwt rea backfat;
OUTPUT OUT=beef3means;
RUN;
```

The UNIVARIATE procedure could also have been used as it also has a NOPRINT option and can output summary statistics to a SAS data set. The Output Delivery System can also be used to suppress printed results. This will be covered in Chapter 19. The OUTPUT statement in this MEANS Procedure does differ from the form presented in Chapter 3. Without specifying which statistics should be output to *beef3means* SAS selects a default set of statistics of which includes the mean.

Or the following could have been submitted to produce only the means:

```
PROC MEANS DATA=beef3 NOPRINT;
BY producer sex;
VAR adg dmi cwt rea backfat;
OUTPUT OUT=beef3means MEAN=ADG_mn DMI_mn CWT_mn REA_mn BackFat_mn;
RUN;
```

The *beef3means* created by the first PROC MEANS code will be exported to Excel so the reader can observe the default summary statistics information included.

Once a SAS data set has been created, it can be exported using the Export Wizard. When the Export Wizard is initiated, similar dialog as the Import Wizard appears.

File – Export Data – Choose the source SAS data set: WORK.BEEF3MEANS

This SAS data set will be exported to Excel. The list of software products to which one can export a SAS data set is the same as that shown in Figure 14.1. Microsoft Excel should be chosen from the list. This is similar to Figure 14.2, then **Browse** to an existing Excel file or enter the path and filename of a new Excel file to be created. When selecting an existing file, there will be a prompt in the **Save As**, "This file already exists". Since a new sheet is to be added to the existing file, **Append** should be selected in Figure 14.9.

After confirming the Append action, a new sheet name in the Excel file (less than 30 characters) must be defined. In Figure 14.10, "Producer Means" is entered in the blank. This will become the name of the sheet in Excel. As with the Import Wizard, one can **Finish** the exportation of a SAS data set, or select **Next** to recover the SAS code generated by the Export Wizard. If multiple SAS data sets are going to be exported, or if one is learning PROC EXPORT code, then selecting **Next** is recommended. Enter a path and file name for the SAS code to export the SAS data set. Additionally, the destination file must be closed when running the Export Wizard. If not, an error message in the log window will result.

FIGURE 14.9
Append creates a new Excel sheet.

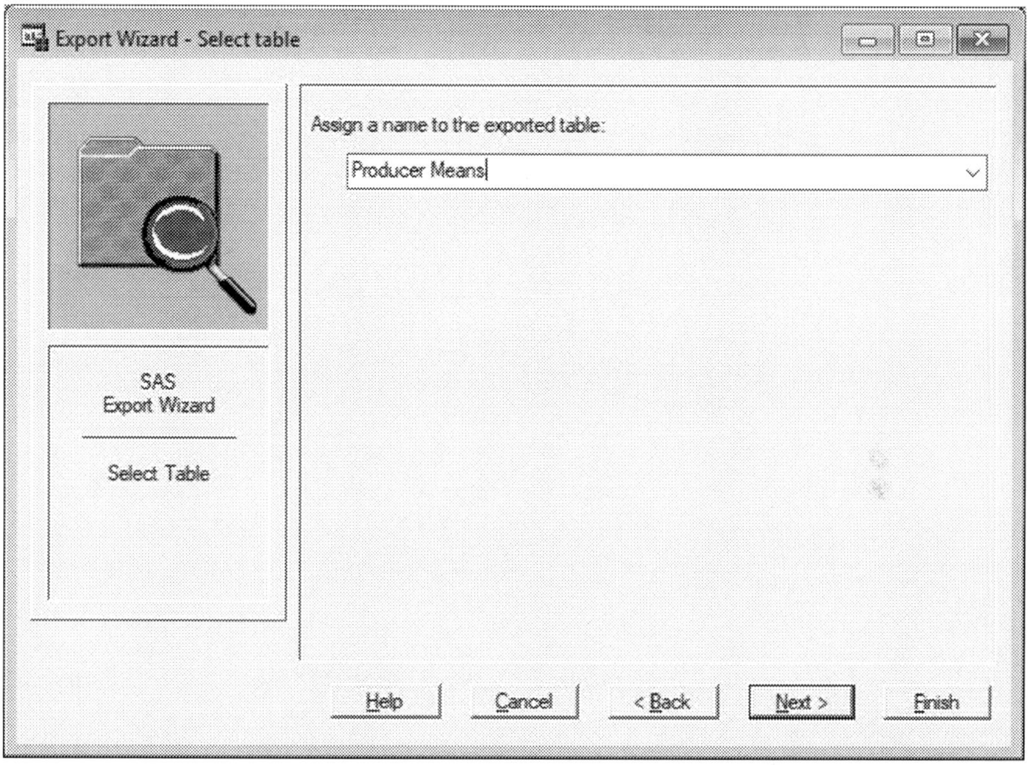

FIGURE 14.10
Naming the Excel sheet in the Export dialog.

Once the export has completed, the Excel file can be opened. Figure 14.11 shows the newly created sheet in the Excel file. "Producer_Means" is the newly created sheet. The content of this sheet is the default output mentioned after the Objective 14.2 SAS code.

14.4 The EXPORT Procedure

Recovering the EXPORT code from the Export Wizard is recommended especially when one is just learning the procedure. Some of the options or statements may be software specific, and the Export Wizard will generate the correct syntax for use. To export to a new software file for the first time the Export Wizard is recommended. For the export of *beef3means* to the original *Chapter 14 Import data.xlsx* file, the code generated is:

```
PROC EXPORT DATA= WORK.BEEF3MEANS
            OUTFILE= "'G:\CLGoad\\Chapter 14 Import data.xlsx"
            DBMS=EXCEL REPLACE;
     SHEET="Producer Means";
RUN;
```

	A	B	C	D	E	F	G	H	I	J
1	Producer	Sex	_TYPE_	_FREQ_	_STAT_	ADG	DMI	CWT	REA	BackFat
2	1	S	0	56	N	56	56	56	54	54
3	1	S	0	56	MIN	1.53	16.2731	701	12.3437	0.39456
4	1	S	0	56	MAX	3.84	24.5569	965	15.9731	0.5839
5	1	S	0	56	MEAN	3.070536	18.85129	837.3976	13.74598	0.509169
6	1	S	0	56	STD	0.355832	2.108601	39.8766	0.701741	0.038078
7	2	H	0	84	N	84	84	84	72	72
8	2	H	0	84	MIN	2.27	15.0786	701	12.181	0.45267
9	2	H	0	84	MAX	3.94	25.2228	897	15.1254	0.68
10	2	H	0	84	MEAN	3.152976	20.38602	797.7571	13.54374	0.580104
11	2	H	0	84	STD	0.396851	2.580068	41.30939	0.531257	0.055513
12	2	S	0	87	N	87	87	87	71	71
13	2	S	0	87	MIN	2.58	16.0269	762	12.7222	0.44434
14	2	S	0	87	MAX	4.33	25.8127	958.203	16.1842	0.68219
15	2	S	0	87	MEAN	3.346437	20.54329	862.0739	14.06278	0.560301
16	2	S	0	87	STD	0.394824	2.336091	39.96613	0.662429	0.048657
17	3	H	0	31	N	31	31	31	27	27
18	3	H	0	31	MIN	2.34	15.4203	704	9.2	0.28
19	3	H	0	31	MAX	3.47	22.9539	835	14.7	0.58839
20	3	H	0	31	MEAN	2.887419	18.91238	762.0696	13.3055	0.505068
21	3	H	0	31	STD	0.33667	2.299663	36.07723	1.051248	0.059884
22	3	S	0	70	N	70	70	70	60	60
23	3	S	0	70	MIN	2.38	16.3187	766.713	13.0286	0.42606
24	3	S	0	70	MAX	4.14	26.1766	918	15.5428	0.61285
25	3	S	0	70	MEAN	3.044857	19.32714	834.8593	13.88054	0.493276
26	3	S	0	70	STD	0.433462	2.319371	33.11232	0.582566	0.047037
27										

Producer1 Steer | Sheet2 | Producer 3 | Producer_Means

FIGURE 14.11
The new Excel sheet produced by the EXPORT procedure.

14.5 Chapter Summary

As software has evolved, it has become more important for different software products to work together compatibly. Sharing data files among researchers is an example of the necessity of this. The ability to read data files produced by other software products into SAS is a huge asset. In this chapter, the Import Wizard was used to read in data from a Microsoft Excel file. SAS is not limited to reading Excel files though. Data files produced by other statistical and data management software can also be read. Because of the variations in the types of files that can be imported and the variations in the IMPORT procedure syntax needed for those file types, the point-and-click method of the Import Wizard is encouraged. Once a data file is imported as a SAS data set, all of the DATA step manipulations and SAS procedures can be used to analyze the data. Similarly, the EXPORT procedure is a wonderful convenience for creating files compatible with another software. The point-and-click Export Wizard is recommended for this task also.

15

DATA Step Information 5 – DO Loops, ARRAY, and Random Number Generators

Using the DATA step, data can be entered directly using an INPUT statement, read in from an external file using an INFILE statement, data can be transformed, or excluded. This does not cover all the DATA step skills in this book, but it has been shown that a wide variety of operations can be achieved in a DATA step. In this chapter, some new tools for performing repetitive actions more efficiently in a DATA step are presented. Repetitive actions can be accomplished using loops in the DATA step. Arrays of variables and random number generators are two concepts that are useful in these programming loops.

15.1 DO Loops – DO and END Statements

Within a DATA step operation, one or more DO loops can be included. The syntax of a DO loop:

```
DO < options >;
     SAS statements
END;
```

Each DO statement must have an END statement that "closes" the loop. DO loops can also be nested within each other, or can be encountered sequentially.

DO statements can have the following formats:

1. DO *index-variable* = *specification-1, specification-2, …, specification-n;*
   Example: DO i = 8;
   DO day = "Mon", "Wed", "Fri";
   DO month = 3, 6, 9, 12;
2. DO *index-variable* = *start* TO *stop;*
   Example: DO k = 1 TO 100;
3. DO *index-variable* = *start* TO *stop* BY *increment;*
   (If unspecified, the increment is 1, as in item 2.)
   Example: DO k = 1 to 100 BY 5;
   DO m = 75 to 50 BY -1;
4. DO WHILE (*expression*);
   Example: DO WHILE (n lt 5);
5. DO UNTIL (*expression*);
   Example: DO UNTIL (n >= 5);

15.2 The OUTPUT Statement

With many DO loops, an OUTPUT statement may be necessary. The OUTPUT statement tells SAS to write the current observation to the SAS data set immediately rather than at the end of the DATA step. The OUTPUT statement has other applications than in a DO loop.

The syntax of the OUTPUT statement can take one of the following forms:

1. OUTPUT;
   This statement records the current observation to the SAS data set(s) named in the DATA statement.
2. OUTPUT *SAS-data-set;*
   This OUTPUT statement directs the observations to a specifically named SAS data set.
3. IF *condition* THEN OUTPUT *SAS-data-set;*
   Previously in Chapter 4, following THEN was a variable assignment statement or DELETE. DELETE has been the only action seen here before. The result of an IF – THEN OUTPUT statement records the entire observation to the named SAS data set when the *condition* is true. Objective 15.3 will demonstrate this.

OBJECTIVE 15.1: Rather than enter all of the variables in a SAS data set, it may sometimes be appropriate to enter response variables only. When the values of response variable are in the appropriate order, DO loops can assign the treatment and replication number.

```
DATA loopex1;
DO trt=1 TO 6;
  DO rep = 1 TO 2;
    INPUT y @@;
    OUTPUT;
  END;
END;

DATALINES;
12 14 18 16 12 11 17 19 20 11 13 15
;
*The PRINT procedure can be omitted if ViewTable is used.;
PROC PRINT DATA=loopex1;
TITLE 'Objective 15.1 - With OUTPUT Statement';

RUN;
QUIT;
```

With this DO statement syntax the values of trt and rep advance to their respective limits without advancing the variables inside the loop(s). This is an example of nested DO loops. Each DO statement has an END statement. It is not necessary to indent the lines in the DATA step as it has been shown here, nor are the blank lines necessary. This has been done to assist in proofreading. The first DO statement is paired with the second END statement. That is the outside loop. The second DO statement and the first END statement are the inner loop and are similarly indented to visually identify the inner loop. So, when this DATA step starts, the variable trt is set to 1, and the second or inner loop runs completely while trt=1. For each value of rep the INPUT statement reads one variable, y. The OUTPUT statement records all variables trt, rep, and y in each iteration of the inner loop. When rep=2 is completed in the inner loop, the outer loops moves to trt=2, and rep resets at 1 in the inner loop. Each of the 6 values for trt have 2 values of rep, so 12 observations or lines in SAS data set *loopex1* are expected, and there are 12 values of y entered after the DATALINES statement. It should also be noted that the double trailing @ is used in the INPUT statement, though it is not required to do so. If not in use, each value of y should be entered on a new line for 12 lines of numbers after the DATALINES statement (Section 2.2).

The SAS log message confirms the number of variables and observations in the data set *WORK.LOOPEX1.*

```
NOTE: There were 12 observations read from the data set WORK.LOOPEX1.
```

Objective 15.1 – With OUTPUT Statement

Obs	trt	rep	y
1	1	1	12
2	1	2	14
3	2	1	18
4	2	2	16
5	3	1	12
6	3	2	11
7	4	1	17
8	4	2	19
9	5	1	20
10	5	2	11
11	6	1	13
12	6	2	15

As indicated by the line comment in the program (Section 20.2), one could check the data for correctness using ViewTable. If so, closing ViewTable after proofreading should be always done. Submitting the program a second time after correcting for errors requires that the SAS data set not be open in ViewTable if amendments to the SAS data set are needed.

OBJECTIVE 15.2: Run the above Objective 15.1 program without the OUTPUT statement and note the difference in the log and results.

From the SAS log:

```
NOTE: There were 1 observations read from the data set WORK.LOOPEX1.
```

Clearly a severe error has occurred.

And in the Results Viewer

Objective 15.2 – Without OUTPUT Statement

Obs	trt	rep	y
1	7	3	15

What happened? The DO loops did execute. Since there was no OUTPUT statement in the inner loop, recording an observation in each iteration was not done. Only the final values of the variables **after** the loops execute are recorded.

Data transformations, new variable creation, and IF – THEN statements covered in Chapter 4 can also be included when using DO loops. Previously, the guideline was that these types of statements occur after the INPUT statement and before the DATALINES statement. That is still the case here. Any of those types of modifications should occur after the INPUT statement. Since there is an OUTPUT statement directing that observations be recorded to the SAS data set in each iteration, then those types of modification statements should typically occur before the OUTPUT statement.

OBJECTIVE 15.3: Create two SAS data sets. The first data set, *loopex2*, contains only information from the first three treatments. The second data set, *loopex3*, contains only information from treatment levels 4, 5, and 6. Log transform the y variable also.

Modify the previous program to read as follows:

```
DATA loopex2 loopex3;
DO trt=1 TO 6;
  DO rep = 1 TO 2;
    INPUT y @@;
    y_log = log(y);
    IF trt <= 3 THEN OUTPUT loopex2;
    ELSE OUTPUT loopex3;
  END;
END;

DATALINES;
12 14 18 16 12 11 17 19 20 11 13 15
;
PROC PRINT DATA=loopex2;
TITLE 'Objective 15.3 - Trt 1, 2, 3 Data';
PROC PRINT DATA=loopex3;
TITLE 'Objective 15.3 - Trt 4, 5, 6 Data';

RUN;
QUIT;
```

There are a few new lessons in this program. First, more than one SAS data set name is specified in the DATA statement. This capability has not been previously used in this book. If the DATA step is creating multiple SAS data sets, all of them must be named in the DATA statement. The SAS data sets that are to be created can be either temporary or permanent. If permanent, SAS libraries need to be defined prior to the submission of the DATA step. Second, the new variable assignment *y_log* and IF – THEN occurs inside the inner loop and before OUTPUT.

In the SAS Log, the two SAS data sets were created and appear to be the correct size.

```
NOTE: The data set WORK.LOOPEX2 has 6 observations and 4 variables.
NOTE: The data set WORK.LOOPEX3 has 6 observations and 4 variables.
```

Objective 15.3 – Trt 1, 2, 3 Data

Obs	trt	rep	Y	y_log
1	1	1	12	2.48491
2	1	2	14	2.63906
3	2	1	18	2.89037
4	2	2	16	2.77259
5	3	1	12	2.48491
6	3	2	11	2.39790

Objective 15.3 – Trt 4, 5, 6 Data

Obs	trt	rep	Y	y_log
1	4	1	17	2.83321
2	4	2	19	2.94444
3	5	1	20	2.99573
4	5	2	11	2.39790
5	6	1	13	2.56495
6	6	2	15	2.70805

Inspection of the each of the printed SAS data sets verifies the correct data sets.

OBJECTIVE 15.4: Use PROBBNML function (Section 13.1.1) to generate cumulative probability and individual probability distributions for the discrete binomial distribution with n=6 trials and three different probabilities of success, p = 0.4, 0.45, 0.5. Print a table with columns y, probability of y, and the cumulative probability of y for each value of p. Include LABELs for these printed columns.

Since n = 6, the values of Y are 0, 1, 2, 3, 4, 5, 6. There is more than one way to accomplish this objective.

```
DATA four;
DO p = 0.4, 0.45, 0.5;
            DO y = 0 to 6;
            cp = PROBBNML(p, 6, y);   *P(Y <= y);
                   IF y=0 THEN prob=cp;
                   ELSE prob = PROBBNML(p, 6, y) - PROBBNML(p, 6, y-1);
                   *P(Y = y);
                   OUTPUT;
             END;
END;
LABEL y ="Y successes in 6 trials"
            cp = 'Cumulative Probability, P(Y <= y)'
            prob = 'Probability, P(Y = y)';
PROC SORT DATA=four; BY p y ;
PROC PRINT DATA=four LABEL NOOBS; BY p;
```

```
VAR y prob cp;
TITLE 'Objective 15.4';
RUN;
```

Programming note: The arguments in the PROBBNML function can be variables (such as p, y), numbers (such as 6), or arithmetic expressions (such as y–1), or a combination of these.

Objective 15.4

p=0.4

Y successes in 6 trials	Probability, P(Y = y)	Cumulative Probability, P(Y <= y)
0	0.04666	0.04666
1	0.18662	0.23328
2	0.31104	0.54432
3	0.27648	0.82080
4	0.13824	0.95904
5	0.03686	0.99590
6	0.00410	1.00000

p=0.45

Y successes in 6 trials	Probability, P(Y = y)	Cumulative Probability, P(Y <= y)
0	0.02768	0.02768
1	0.13589	0.16357
2	0.27795	0.44152
3	0.30322	0.74474
4	0.18607	0.93080
5	0.06089	0.99170
6	0.00830	1.00000

p=0.5

Y successes in 6 trials	Probability, P(Y = y)	Cumulative Probability, P(Y <= y)
0	0.01563	0.01563
1	0.09375	0.10938
2	0.23438	0.34375
3	0.31250	0.65625
4	0.23438	0.89063
5	0.09375	0.98438
6	0.01563	1.00000

Reference tables in textbooks exist for binomial probabilities and cumulative probabilities for select values of p and n, but generating these tables for any values of n and p are quite simple to do. Any of the families of discrete distributions (Section 13.1) can similarly be computed using one or more DO loops.

The last two syntax options for the DO statement, DO WHILE and DO UNTIL, will be demonstrated with a couple of very simple objectives to demonstrate how these loops flow.

OBJECTIVE 15.5: Run the following program with and without the OUTPUT statement and compare the results. Then run the program a third time using the OUTPUT statement and changing "r<20" to "r<=20". Note the differences.

In this first option, the "counter" variable r must be initialized outside of the DO WHILE loop. The DO WHILE syntax does not advance the value of r, thus the r+1 step is needed inside the loop. If r+1 is not included inside the loop, the result is an infinite loop that one must "break" using the exclamation point button on the SAS tool bar or the Break key on the keyboard.

```
DATA a;
r = 16;
DO WHILE (r<20);
        r+1;
        OUTPUT;
END;
PROC PRINT DATA=a;
TITLE 'Objective 15.5';
RUN;
QUIT;
```

Obs	r
1	17
2	18
3	19
4	20

Omitting the OUTPUT statement in the DO WHILE loop, results in the single observation where r = 20.

```
DATA a;
r = 16;
DO WHILE (r<20);
        r+1;
END;
```

Obs	r
1	20

Changing the condition on r in the DO WHILE statement affects the range of the values of r.

```
DATA a;
r = 16;
DO WHILE (r<=20);
        r+1;
        OUTPUT;
END;
```

Obs	r
1	17
2	18
3	19
4	20
5	21

The increment on the counter variable can be changed. Positive and negative values can be used for this increment.

```
DATA a;
r = 16;
DO WHILE (r<20);
        r+2;
        OUTPUT;
END;
```

Obs	r
1	18
2	20

OBJECTIVE 15.6: Change the syntax of the DO WHILE loop to DO UNTIL and observe the effects. The DO UNTIL loop values evaluates the condition at the bottom of the loop, and the DO WHILE evaluates the condition at the top of the loop.

The DO UNTIL loop also requires that a counter variable be initialized in the same DATA step prior to the loop, and the counter variable needs to be advanced in the loop.

```
DATA a;
r = 16;;
DO UNTIL (r=20);
        r+1;
        OUTPUT;
END;
PROC PRINT DATA=a;
RUN;
QUIT;
```

Obs	r
1	17
2	18
3	19
4	20
5	21

In the DO UNTIL statement, (r > 20) has the same effect as (r = 20) in this scenario, whereas the condition (r < 20) will not work because r starts at a value less than 20.

```
DATA a;
r = 16;;
DO UNTIL (r>20);
        r+1;
        OUTPUT;
END;
```

Obs	r
1	17
2	18
3	19
4	20
5	21

The Objectives 15.5 and 15.6 were minimal code to demonstrate the control of the counter or index variable one has when using the loops. The repetitive action(s) or function(s) can be placed inside the loop before the OUTPUT statement just as done in the initial DO loops of Objectives 15.1. through 15.4.

There is also a conditional DO loop that can be included in a DATA step. IF – THEN statements specify a condition that must be satisfied in order for an action to take place. That is,

```
IF condition THEN action ;
```

Syntax for *condition* is covered in Section 4.2. The *action* can be a variable assignment statement or transformation. Actions include DELETE (Section 4.2), OUTPUT (Section 15.2), and now a DO action. That is,

```
IF condition THEN DO;
            DO loop syntax for the repetitive task
            END;
```

This conditional DO loop will be demonstrated in Objective 15.7.

15.3 The ARRAY Statement

The ARRAY statement defines a selected set of variables as elements of an array. Arrays can be either one-dimensional or two-dimensional. In this book only the one-dimensional array will be introduced. The ARRAY statement is used only in DATA step operations. The simple syntax of the ARRAY statement is:

ARRAY *array-name {subscript}* <$> *<array elements or variables>*;

ARRAY statements are positioned after an INPUT statement (and before DATALINES), or after SET and MERGE statements (Sections 4.1, 4.4, and 4.5). In this first example, four variables are read into the SAS data set using an INPUT statement. The ARRAY statement defines *v* as an array of three variables, and those variables are x1, x2, and x3. The value in { } in the ARRAY statement sets the number of variables or columns to be grouped into array *v*.

```
DATA a;
INPUT y x1 x2 x3;
ARRAY v{3} x1 x2 x3;
```

```
.
.
.
DATALINES;
```

Consider a second ARRAY statement: ARRAY w{3} x1 x3 x2; The order of the variables in array *w* is different from the order of the variables in array *v*. One has to be mindful of the order of the variables in the named arrays when programming. Additionally, if the number in { } does not match the number of variables listed, an error message will result when the DATA step is submitted as shown here:

```
ERROR: Too few variables defined for the dimension(s) specified for the
array v.
```

or

```
ERROR: Too many variables defined for the dimension(s) specified for the
array v.
```

The dimension of the array or the variable list in the ARRAY statement will need to be corrected if either of these errors are observed.

In the following SAS data set *b*, the use of the "double dash" in the ARRAY statement is illustrated.

```
DATA b;
INPUT response month1 $ month2 $ month3 $ month4 $ month5 $ month6 $;
ARRAY mo{5} month1 -- month4 month6;
.
.
.
DATALINES;
```

The double dash "- -" can be used in the ARRAY statement only if all of the variables in between the month1 and month4 are to be included. The double dash executes successfully because month2 and month3 have been previously defined. See Section 20.1 for more information on the usage of the double dash "- -". The effect of this ARRAY statement is a five variable, one-dimensional array, and month5 has not been included.

The benefit of using an array is to efficiently perform a repetitive action on variables in the array. This can be done in a DO loop. Consider the *beef* SAS data set from Chapter 11. The five columns in *beef* were DMI, ADG, CWT, BackFat, REA (in that order). DMI, ADG, and CWT were measured in pounds. Suppose they were to be transformed to kilograms (kg). For only three variables, that is simple enough to do in a DATA step without the use of an ARRAY. That is,

```
DATA beef2;
SET beef;
ADG_kg = adg/2.2046;
DMI_kg = dmi/2.2046;
CWT_kg = cwt/2.2046;
RUN;
```

With the use of a PRINT procedure or ViewTable, the order of the columns in *beef2* are: DMI, ADG, CWT, BackFat, REA, ADG_kg, DMI_kg, and CWT_kg since the last three variable assignment statements occurred in that order. Two one-dimensional arrays in a DO loop can also accomplish this. The creation of *beef2* will be rewritten to demonstrate the use of arrays. An array of the variables to be transformed and an array of the new variable names must also be defined in the DATA step.

```
DATA beef2;
SET beef;
ARRAY b{3} dmi adg cwt;
ARRAY c{3} DMI_kg ADG_kg CWT_kg;        *create the new variables in beef2;
DO i = 1 to 3;
          c{i} = b{i}/2.2046;
END;
RUN;
```

Uppercase letters were used in the second ARRAY statement for the new variable names because the case of the variable names when they are initialized is what appears in the output (Section 2.2). Previously defined variable names are shown in lowercase. Clearly, the use of ARRAY statements might not appear advantageous for only three algebraic conversions. However, for larger numbers of conversions, the usage of ARRAY statements can reduce the number of repetitive operations in a DATA step.

Submitting the DATA step with ARRAY statements from the Editor, SAS data set *beef2* is created. In the SAS log window:

```
NOTE: There were 56 observations read from the data set WORK.BEEF.
NOTE: The data set WORK.BEEF2 has 56 observations and 9 variables.
```

The number of observations was not increased. The SAS data set *beef* had five column variables, and *beef2* has nine column variables. The expected number of variables was eight since only three new variables were named in the second array. What is the ninth variable?

Figure 15.1 shows the first seven observations of *beef2* in ViewTable. From ViewTable, the ninth variable is the indexing variable, *i*, and its value is 4 since i = 4 when the loop finishes. No OUTPUT statement was included in this loop. If an OUTPUT statement were included, then after each variable was converted to kilograms, SAS would have moved to a new line before computing the kilogram value for the next column variable. The result would be that *beef2* would have 3 × 56 = 168 observations rather than 56.

	DMI	ADG	CWT	BackFat	REA	DMI_kg	ADG_kg	CWT_kg	i
1	17.009	3.05	819	0.54571	13.3821	7.7152317881	1.3834709244	371.49596299	4
2	17.0183	2.87	857	0.53974	13.7779	7.7194502404	1.30182346	388.73264991	4
3	17.4826	3.12	826	0.46	13.5262	7.9300553388	1.4152227161	374.67114216	4
4	18.0043	2.73	831.256	.	.	8.1666969065	1.2383198766	377.05524812	4
5	16.2731	2.83	799	0.54622	12.8589	7.3814297378	1.2836795791	362.4240225	4
6	17.5451	2.98	850	0.5055	13.8188	7.9584051529	1.3517191327	385.55747074	4
7	16.6	2.78	790	0.5028	13.48	7.5297106051	1.2609997278	358.34164928	4

FIGURE 15.1
ViewTable of SAS Data Set beef2 (partial).

The creation of the new kilogram variables using arrays could have been done in the original DATA step that created SAS data set *beef* in the following way:

```
DATA beef;
INPUT DMI ADG CWT BackFat REA;
ARRAY b{3} dmi adg cwt;
ARRAY c{3} DMI_kg ADG_kg CWT_kg;                        *create the new variables;
DO i = 1 to 3;
            c{i} = b{i}/2.2046;
END;
DATALINES;
17.0090     3.05     819.000     0.54571     13.3821
17.0183     2.87     857.000     0.53974     13.7779
17.4826     3.12     826.000     0.46000     13.5262
⋮
```

The double dash cannot be used in an ARRAY statement defining new variables even if the new variables have an implied indexing or order. For example, consider the creation of a SAS data set *demo*.

```
DATA demo ;
INPUT p q r s t u v x y z;
ARRAY b{8} p--x;          *defines an array of 8 variables in the data set;
ARRAY c{8} d e f g h i j k;                           *define 8 new variables;
DO m=1 to 8;
       Variable assignment statements for c
END;
DATALINES;
```

The ten variables in the INPUT statement are initialized in that statement, therefore the double dash works in defining array *b*. However, variables d e f g h i j k must each be defined in the definition of array *c* since those variables do not already exist in SAS data set *demo*. Thus, the statement

```
ARRAY c{8} d -- k;
```

is meaningless in the DATA step even though the implication to the reader may be all of the letters in the alphabet ranging from d to k.

15.4 Random Number Generators

There are many different types of random number generators. Typically, these are used to generate a list or sample of random observations from a population with a specified distribution. A few of these random number generators will be examined in this section. Additionally, the use of DO loops can facilitate the selection of a random sample. Random number generators are programmed in a DATA step. For each of these generators, a seed value should be specified. The seed value for any of the random number generators must be an integer between 0 and $2^{31} - 1$.

There are two types of random number generators in the SAS DATA step. First there are functions, and a second type is a SAS statement referred to as a CALL routine. CALL routines give the user greater control over the seed values. Seed values can be initialized prior to a random generator function or be written as an argument in the function. However, in a CALL routine, seed values must be initialized in a variable assignment statement prior to the CALL routine.

Some researchers use the computer's current clock value as the seed value. (This is seed value 0.) However, this is not generally recommended. Usually when one is generating a random sample, the results need to be reproducible. Keeping a record of the starting seed value achieves this reproducibility.

15.4.1 Seed Values and Standard Normal Random Numbers

Many statistical applications assume that the data is from a normally distributed population. There are two methods of generating a standard normal random number. First is the RANNOR function. The syntax of the RANNOR function is:

RANNOR(seed) – generates a random variable from a standard normal distribution.

Example: `y = RANNOR(2816);`

where 2816 is the seed value chosen for the function, and y is the randomly generated value from the standard normal distribution, N(0, 1).

Second is the CALL routine which differs from a SAS function. The syntax of this CALL RANNOR routine is:

```
CALL RANNOR(seed, x);
```

where the seed is initialized prior to this statement, and x is the randomly generated value from the standard normal distribution, N(0, 1).

OBJECTIVE 15.7: The following demonstrates the control one can exercise over the seed values used in a random number generation program. Note all seed values start out the same and progress the same until i=6. Used in a function, the seed values do not change when the value of the seed is specified outside the interval. (Note: Seed values typically do not all have to start at the same value. This is done here to illustrate the differences between functions and CALL routines using seed values declared outside of the DO loop and within the function.)

```
DATA seven;
seed1 = 2120;
seed2 = 2120;
seed3 = 2120;
DO i=1 TO 10;
        CALL RANNOR(seed1, x1);
        CALL RANNOR(seed2, x2);
        y1 = RANNOR(seed3);
        y2 = RANNOR(2120);
        IF i=6 THEN DO;
                seed2=17;
                seed3=17;
                   END;
         OUTPUT;
END;
```

```
PROC PRINT DATA=seven;
TITLE 'Objective 15.7';
RUN;
QUIT;
```

Objective 15.7

Obs	seed1	seed2	seed3	i	x1	x2	y1	y2
1	1262886128	1262886128	2120	1	-1.74966	-1.74966	-1.74966	0.86050
2	1986697677	1986697677	2120	2	0.86050	0.86050	0.45429	0.80717
3	1863976022	1863976022	2120	3	0.45429	0.45429	0.77593	-0.23818
4	400403431	400403431	2120	4	0.80717	0.80717	-1.63992	0.45672
5	260037493	260037493	2120	5	0.77593	0.77593	-0.21056	0.25768
6	1300313190	17	17	6	-0.23818	-0.23818	2.19918	1.74418
7	926215726	1055505749	17	7	-1.63992	-1.96464	-0.61851	-1.63069
8	109959750	879989345	17	8	0.45672	-0.56825	-0.26321	2.03344
9	1397620086	1025885034	17	9	-0.21056	-0.89075	-0.18213	1.71000
10	420776480	83234937	17	10	0.25768	1.66228	-2.88848	-0.58530

Though all of the seed values start out the same, what happens to them when operated on by a CALL routine or a function differs. The values in columns x1, x2, y1, and y2 are each a different random sample from a N(0,1) population. In the case of x1 and x2, these columns are the same until observation 6 when a change in seed2 was initiated. (Note the use of the DO and END statements with the IF – THEN statement.) The main point here is that with an initialized seed variable, three of the four methods start with exactly the same value in observation 1. Putting the numeric seed value in the function has a different outcome. For most readers, extensively controlling the seed value is not a priority. Choosing any one of these methods is adequate for most programmers. Introducing CALL routines is not necessary here, but a brief introduction to the concept of CALL routines may be beneficial to the reader in other applications.

Random number generation can be done for continuous distributions and for discrete distributions. In addition to the seed values for the random number generation, there may be additional syntax for parameters of the distribution.

15.4.2 Continuous Distributions

The normal and continuous uniform distributions are, perhaps, the easiest of the continuous distributions to introduce. For the each of the following pairs of statements, the value of x is the randomly generated value using the given statement in a SAS DATA step.

Normal Distribution

```
x = RANNOR(seed);
CALL RANNOR(seed, x);
```

For this normal distribution x is a numeric variable from a normal distribution with mean 0 and variance 1. Recall that all or almost all of the standard normal distribution is between –3 and 3. Values beyond those limits occur with a very small probability.

Uniform Distribution

```
x = RANUNI(seed);
CALL RANUNI(seed, x);
```

For the uniform distribution x is a variable that has the value chosen from the uniform distribution on the interval (0,1). There are other uniform distributions that can be described using a more general interval (a,b).

Cauchy, Exponential, and Gamma random number generators are also available. They are RANCAU(seed), RANEXP(seed), and RANGAM(seed, shape_parameter), respectively.

Random number generation from other distributions in the same family, say $N(\mu, \sigma^2)$ where μ and σ^2 are specified, can be generated fairly simply. Some general rules apply when multiplying the random number x by a constant or adding a constant.

If X is the randomly generated number, it could be modified in a DATA step by:

$Y = X + \beta$; adds a constant β to each value X generated. This would change the mean of the distribution by the constant β. Adding a constant to the generated random variable translates the distribution by β units.

$W = a^*X$; multiplies each randomly generated X by a. This would change the variance of the distribution of X by a factor of a^2. Multiplying the generated random variable by α 'stretches' the distribution if $|\alpha| > 1$, or 'shrinks' the distribution if $|\alpha| < 1$.

$U = a^*X + \beta$; or $V = a^*(X + \beta)$; changes both the mean and variance of the randomly generated X.

Each of these modifications, Y, W, U, and V, can be used to achieve more general $N(\mu, \sigma^2)$ and Uniform(a,b) distributions.

OBJECTIVE 15.8: Using a SAS function, generate four random samples of size 10 from three populations: N(0,1), N(5, 1), N(0, 6), and N(8, 10). Recall the transformation from a standard normal score to $N(\mu, \sigma^2)$ variable is: $y = \mu + z\sigma$ where $z \sim N(0,1)$. Use one DATA step to do this. PRINT the resulting SAS data set suppressing observations numbers and including the total number of observations in the data set.

```
DATA eight;
DO i = 1 to 10;
   X = RANNOR(28374);                            * X ~ N(0, 1);
   Y = 5 + RANNOR(39587209);                     * Y ~ N(5, 1);
   W = SQRT(6) * RANNOR(659363);                 * W ~ N(0, 6);
   U = 8 + SQRT(10) * RANNOR(494703);            * U ~ N(8, 10);
   OUTPUT;
END;
RUN;
PROC PRINT DATA=eight NOOBS N;
TITLE 'Objective 15.8';
RUN;
QUIT;
```

X, Y, W, and U are independent of each other since they each are generated separately.

Objective 15.8

i	X	Y	W	U
1	−0.59283	5.43577	−0.42667	8.6847
2	0.03273	4.53032	−1.21178	5.1910
3	2.03762	5.71119	0.50189	7.7227
4	−1.32452	5.15949	−3.31791	5.5878
5	0.99156	2.75990	−0.86996	7.0926
6	−0.37300	4.44577	−0.37013	8.8623
7	−0.04682	4.92706	3.47000	11.9147
8	−1.25877	4.89920	−3.38246	3.9745
9	−1.26162	5.31555	−1.97299	9.0753
10	−1.88935	4.99068	−1.30302	11.3208
		N = 10		

Dependent samples could have been generated. The following SAS Data step (*eight*) does this. A CALL routine for a standard normal random number is used, but the RANNOR function could also have been used. These are dependent samples since the standard normal random value is generated first, and Y, W, and U are arithmetic modifications of the random number. The seed value must be initialized outside of the loop.

```
DATA eight;
Seed = 6474983
DO i = 1 to 10 ;
   CALL RANNOR(seed, x);                * X ~ N(0, 1);
   Y = 5 + x;                           * Y ~ N(5, 1);
   W = SQRT(6) * x;                     * W ~ N(0, 6);
   U = 8 + SQRT(10) * x;                * U ~ N(8, 10);
   OUTPUT;
END;
RUN;
```

It is left as an exercise for the reader to replace RANNOR with RANUNI in the previous DATA steps.

15.4.3 Discrete Distributions

Similarly, random samples from discrete distributions can also be generated. The binomial and Poisson distributions were first overviewed in Section 13.1. Additionally, a more general discrete distribution with a finite number, n, of responses is included.

Binomial Distribution

```
x = RANBIN(seed, n, p);
CALL RANBIN(seed, n, p, x);
```

where n is the number of independent trials, p is the probability of success on a single trial, $0 < p < 1$, and x is the number of successes randomly selected; that is, x = 0, 1, 2, 3, ..., n.

Poisson Distribution

```
RANPOI(seed, m)
CALL RANPOI(seed, m, x);
```

where m is the mean of the Poisson distribution, $m > 0$, and x is the randomly selected number. Recall, x = 0, 1, 2, 3, ...

Tabled Probability Distribution

```
x = RANTBL(seed, p1, p2, …, pn)
CALL RANTBL(seed, p1, p2, …, pn, x)
```

There are n possible values for x each with probability p_i where i = 1, ..., n, $p_1+p_2+\ldots+p_n = 1$, and x is the random value generated.

OBJECTIVE 15.9: The number of calls received at the technology support desk of a local computer store is described by a Poisson distribution with an average of 14 calls per day. Generate two random samples of size 5 days for a random variable C = # of daily calls. Compute the average value of C for each of the random samples using the MEANS Procedure (Section 3.5).

```
DATA nine;
DO Sample = 1 TO 2;
     DO Day = 1 to 5;
              C = RANPOI(739284, 14);        * RANPOI(seed, mean);
              OUTPUT;
     END;
END;

PROC MEANS DATA=nine;
CLASS sample;
VAR c;
TITLE 'Objective 15.9 - Option 1';
RUN;
QUIT;
```

Objective 15.9 – Option 1

The MEANS Procedure

Analysis Variable : C

Sample	N Obs	N	Mean	Std Dev	Minimum	Maximum
1	5	5	12.6000000	4.3358967	8.0000000	18.0000000
2	5	5	14.8000000	3.8987177	11.0000000	20.0000000

Checking the SAS Log, one can verify that ten observations are created in *WORK.NINE*.

```
NOTE: The data set WORK.NINE has 10 observations and 3 variables.
```

Variables in *WORK.NINE* are: Sample, Day, and C.
Programming Note: The MEANS Procedure could also have been programmed with a BY statement. That is,

```
PROC MEANS DATA=nine;
BY sample;
VAR c;
TITLE 'Objective 15.9 - Option 2';
RUN;
```

Objective 15.9 – Option 2

The MEANS Procedure

Sample=1

Analysis Variable : C

N	Mean	Std Dev	Minimum	Maximum
5	12.6000000	4.3358967	8.0000000	18.0000000

Sample=2

Analysis Variable : C

N	Mean	Std Dev	Minimum	Maximum
5	14.8000000	3.8987177	11.0000000	20.0000000

The caution when using the BY statement in Option 2 regards the order of the BY variable *sample*. If the order of two DO statements in the DATA step for SAS data set *nine* were switched, then the resulting SAS data set *nine* is not ordered according to Sample but by Day, and a SORT Procedure BY *sample* is needed before the Option 2 MEANS Procedure. If there are any other data reordering steps before the Option 2 MEANS Procedure, a SORT procedure will be necessary.

For a larger number of samples, the output in Option 1 is more concise. If the results are to be output to a new SAS data set using an OUTPUT statement, either option can be chosen, but the content of the output SAS data sets will differ slightly as shown in Section 3.5.

15.5 Chapter Summary

One or more DO loops in a DATA step can facilitate a repetitive action in a DATA step. Each DO statement requires an END statement to close the loop. In this chapter, one-dimensional arrays and random number generators were paired with the DO loop to demonstrate its utility. ViewTable or the PRINT Procedure in addition to checking the feedback in the SAS Log are essential tools when assessing the DATA step for errors in DO loops, ARRAY statements, and random number generation. For random number generation, graphing procedures in Chapters 5, 16, and 17 can also be useful summary methods when generating large random samples.

16

Statistical Graphics Procedures

Throughout this book, ODS Graphics for procedures have been observed as a part of the output of some statistical procedures. These ODS Graphics are diagnostic tools or support for the statistical results produced. Very little code was required to produce these quality graphics because SAS has defined the templates to be used in these graphics. In addition to the ODS Graphics that are options for statistical procedures, ODS Graphics include Statistical Graphical procedures which are a component of Base SAS. By default, these procedures produce image files, and no understanding of graphics catalogs is needed. So, with very little syntax, readers will be able to produce presentation quality graphics using these procedures. Statistical Graphics Procedures are SGPLOT, SGSCATTER, SGPANEL, SGDESIGN, and SGRENDER. With any of these procedures, there are many, many options that can be included, and only some of the basics for enhancing colors or symbols will be covered here. SGDESIGN and SGRENDER are more advanced topics and are not covered in this book.

The procedures in this chapter require that ODS Graphics are enabled during the SAS session which it typically is by default at the start of the SAS session. If it has been disabled, Section 19.1 has information about enabling ODS Graphics.

For the Statistical Graphics or "SG" procedures, TITLE, FOOTNOTE, and BY statements (Chapter 2), WHERE statement (Chapter 4), LABEL statement (Chapter 8), and FORMAT statement (Chapter 18) can be included in all of these procedures.

16.1 The SGPLOT Procedure

There are many chart statements that can be used in the SGPLOT procedure. Some of the introductory statements or options are presented here. Basic SGPLOT procedure syntax is:

PROC SGPLOT DATA=*SAS-data-set;*
Chart statement(s)
RUN;

At most two chart statements can be chosen in the SGPLOT procedure. Types of charts in the SGPLOT procedure are listed in Table 16.1. When two chart statements are included, the images are overlaid onto the same axes if the chart statements are compatible. If the chart statements are not compatible, an error results. This can be addressed by placing each of the chart statements in separate SGPLOT procedures.

There are many options for each of the chart or plot types. These options control colors, symbols, size of bars, size of symbols, types of density curves, and more. The objectives in this chapter will demonstrate some of these options. To see a complete list of options, the reader should consult SAS Help and Documentation or SAS Institute Inc., 2013a.

TABLE 16.1

Chart Statements for the SGPLOT Procedure

Type of Chart or Plot	SGPLOT statements	Example
Bar chart	HBAR VBAR	HBAR *discrete-variable </options>;*
Histogram	HISTOGRAM	HISTOGRAM *continuous-variable </options>;*
Density curve	DENSITY	DENSITY *continuous-variable </options>;*
Box plot	HBOX VBOX	HBOX *variable </options>;*
(X,Y) plots	SCATTER SERIES	SCATTER Y=*vertical-axis-variable* X = *horizontal-axis-variable* *</options>;*

16.1.1 Horizontal and Vertical Bar Charts

Bar charts can be used to represent the frequency of a discrete variable in the data set. Bar lengths can also be chosen to represent the mean response of a second variable for each bar in the chart. The mean response may be just one of the sample statistics illustrated. A bar chart for one discrete or categorical variable can be superimposed on another if the bar lengths are compatible.

HBAR creates a horizontal bar chart, and VBAR creates a vertical bar chart. A category variable is summarized by these types of charts. The size, color, and transparency of the bars in the chart can be controlled.

```
HBAR category-variable </options> ;
VBAR category-variable </options> ;
```

The default chart for either of these statements is a frequency bar chart for the specified category variable. The options on these statements can refine these charts. Introductory options for these chart statements are given in Table 16.2.

OBJECTIVE 16.1: For the *beef3* SAS data set created in Chapter 14, produce the default horizontal bar chart for the category variable Sex.

```
PROC SGPLOT DATA=beef3;
HBAR sex;
TITLE 'Objective 16.1 - Default Horizontal Bar Chart';
RUN;
```

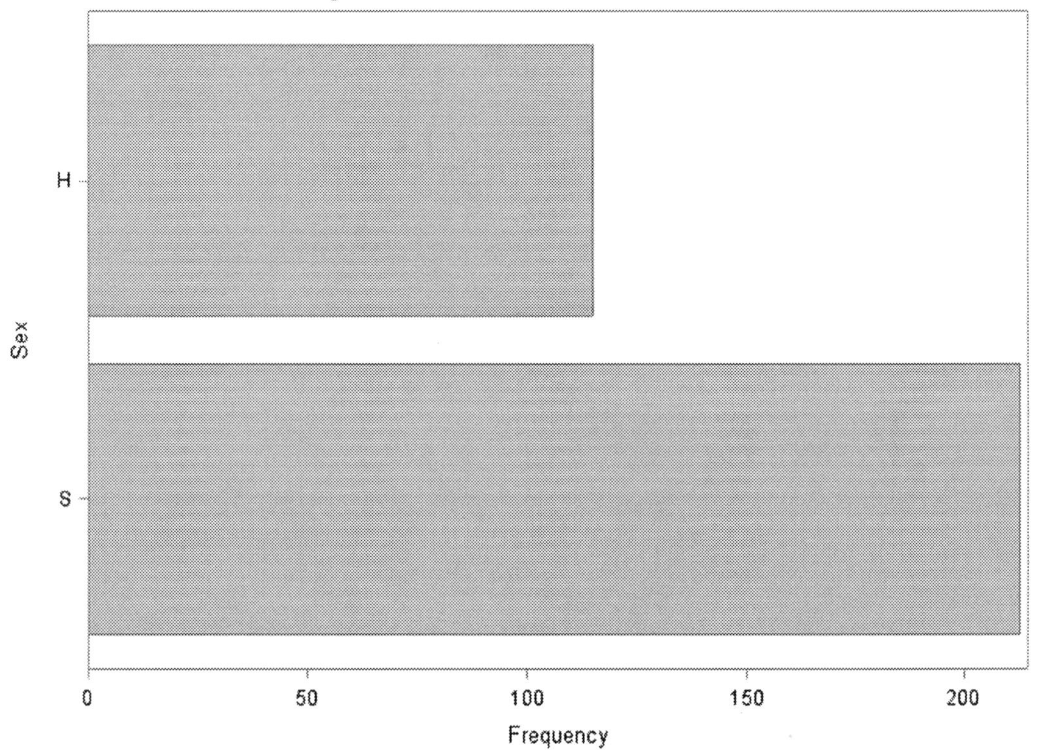

TABLE 16.2

Options for the HBAR and VBAR Statements

Options	Explanation
BARWIDTH = *b*	0 < *b* < 1; 1 is the default setting. Values smaller than 1 reduce the width of the bars.
FILL | NOFILL	Pick one. FILL is the default, producing solid color bars. NOFILL produces only the outline of the bars.
GROUP = *category-variable*	The GROUP option will divide each of the bars determined by the chart variable into sections attributable to the values of the GROUP variable. GROUP is applicable to FREQ bars only. See also Section 17.2.1.
LIMITS=BOTH | UPPER | LOWER ALPHA = *p*	Pick one limit. Specify an interval (BOTH), upper bound (UPPER), or lower bound (LOWER) and set ALPHA = p where 0 < *p* < 1 for (1–*p*)100% confidence limit(s).
RESPONSE = *summary variable* STAT = FREQ | MEAN | SUM | MEDIAN	Both statements are needed when another variable is to be summarized. Pick one statistic, STAT. The bar length is the value of the specified statistic for the summary variable. When no summary variable is specified, the frequency of the chart variable determines the bar length.
TRANSPARENCY = *t* FILLATTRS = (COLOR= *color*)	0 < *t* < 1; The solid bar color saturation can be controlled by the TRANSPARENCY option. The default is 1 for a solid color bar. Transparency values less than one are recommended when overlaying bar charts. The bar color is controlled by *color*. Keep it simple when starting out: BLACK, BLUE, RED, GREEN, ORANGE, PURPLE, GRAY, and so on. There is a more extensive color palette (Section 17.1), but simple choices are best initially.

The default chart is a frequency bar chart with solid bars. SAS has default colors selected for these graphs. Later objectives in this chapter will demonstrate options to change the colors.

OBJECTIVE 16.2: For each producer in *beef3*, construct a horizontal frequency bar chart for the category variable Sex.

```
PROC SORT DATA=beef3; BY Producer;
PROC SGPLOT DATA=beef3; BY Producer;
HBAR sex;
TITLE 'Objective 16.2 - BY Producer';
RUN;
```

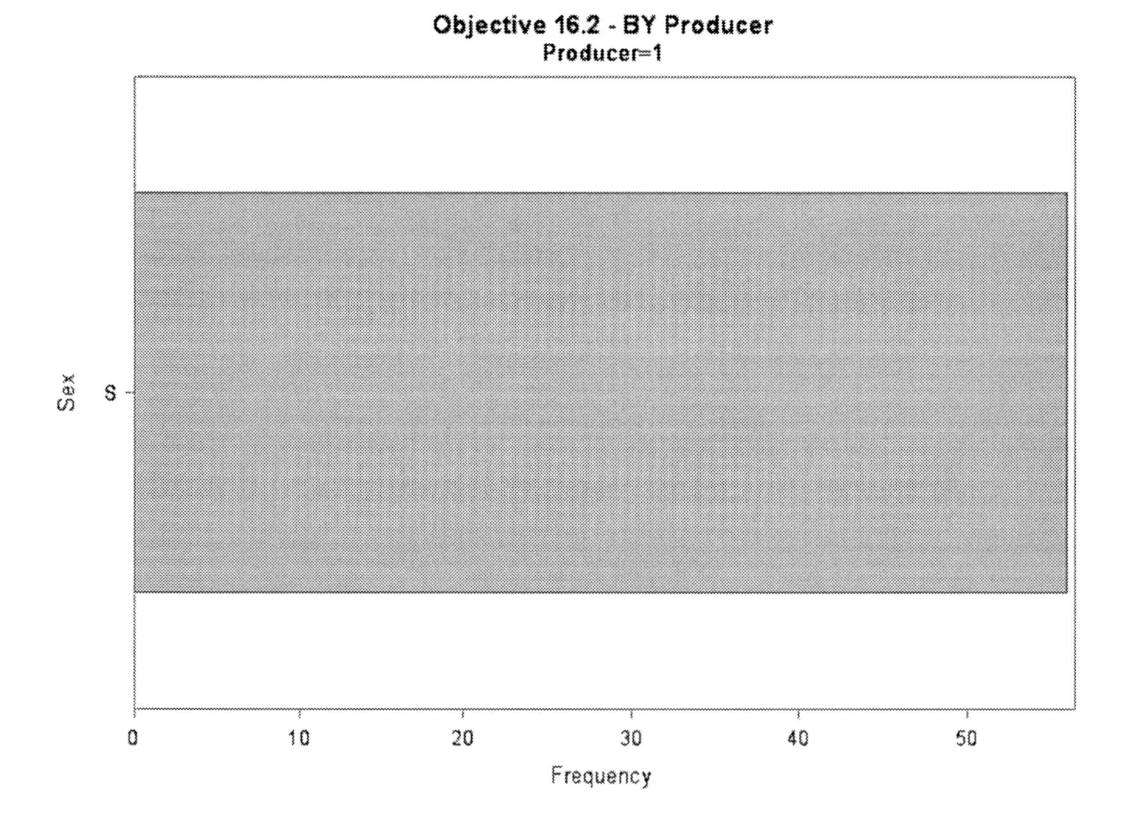

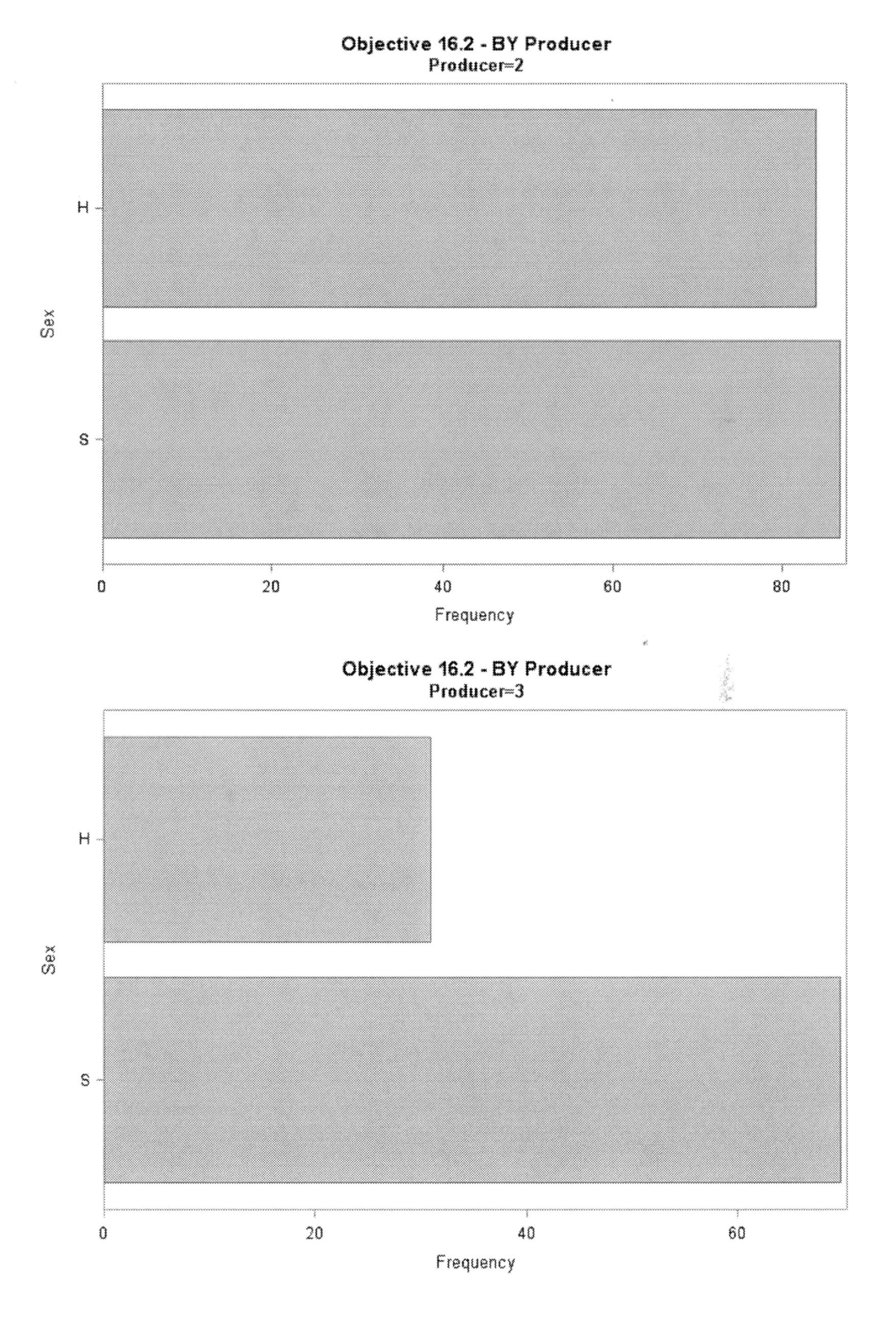
Objective 16.2 - BY Producer
Producer=2
H
S
Sex
0
20
40
60
80
Frequency
Objective 16.2 - BY Producer
Producer=3
H
S
Sex
0
20
40
60
Frequency

Three different bar charts were produced. There is a better way to include this BY variable so that a single-graph image results. See Section 16.3. There is also a GROUP option (GROUP = producer) that could have been used on the HBAR statement. It does produce a single chart, and each bar is subdivided into the lengths (each a different color) corresponding to the frequencies for each producer.

OBJECTIVE 16.3: Using the *beef3* SAS data set, produce a bar chart for Producer 3 for the Sex variable. Let the length of each bar represent the mean average daily gain (ADG). Include a 90% confidence interval for the mean ADG.

```
PROC SGPLOT DATA=beef3;
WHERE Producer = 3;
HBAR sex / RESPONSE = adg STAT=MEAN ALPHA=0.10 LIMITS=BOTH;
TITLE 'Objective 16.3 - Bar Length is ADG mean';
TITLE2 '90% CI for ADG Mean - Producer 3';
RUN;
```

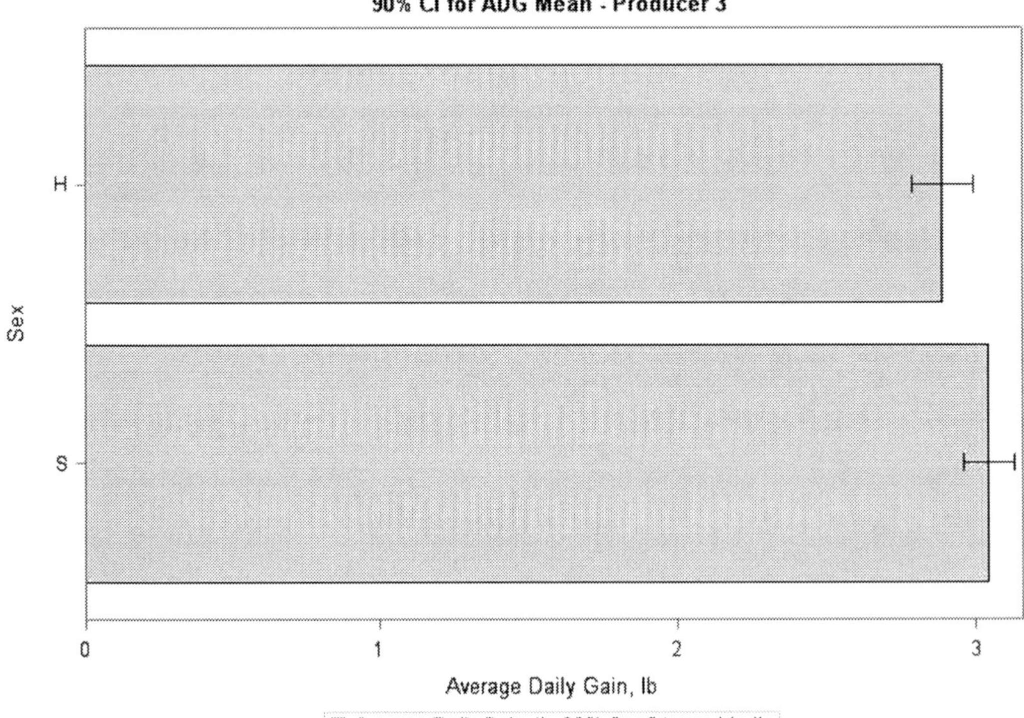

Up to this point, only a single chart or graph statement has been included in the SGPLOT Procedure. Both the HBAR and VBAR statements will overlay bar charts. One must make sure the charts to be overlaid are measured on similar axes before overlaying them. To do this, the two charts to be overlaid are included in the same SGPLOT procedure. When overlaying charts, bar widths, color selections, and transparency options can assist in producing a legible bar chart.

OBJECTIVE 16.4: Produce two horizontal bar charts for the Sex variable for Producer 3. Let the bar length be the mean of ADG in the first chart and the mean of DMI in the second bar chart. Overlay these two charts.

```
PROC SGPLOT DATA=beef3;
WHERE Producer = 3;
HBAR sex / RESPONSE = adg STAT=MEAN FILLATTRS=(COLOR=BLACK);
HBAR sex / RESPONSE = dmi STAT=MEAN BARWIDTH=0.70 TRANSPARENCY=0.50
FILLATTRS=(COLOR=black);
TITLE 'Objective 16.4 - Overlaying Bar Charts - Producer 3';
RUN;
```

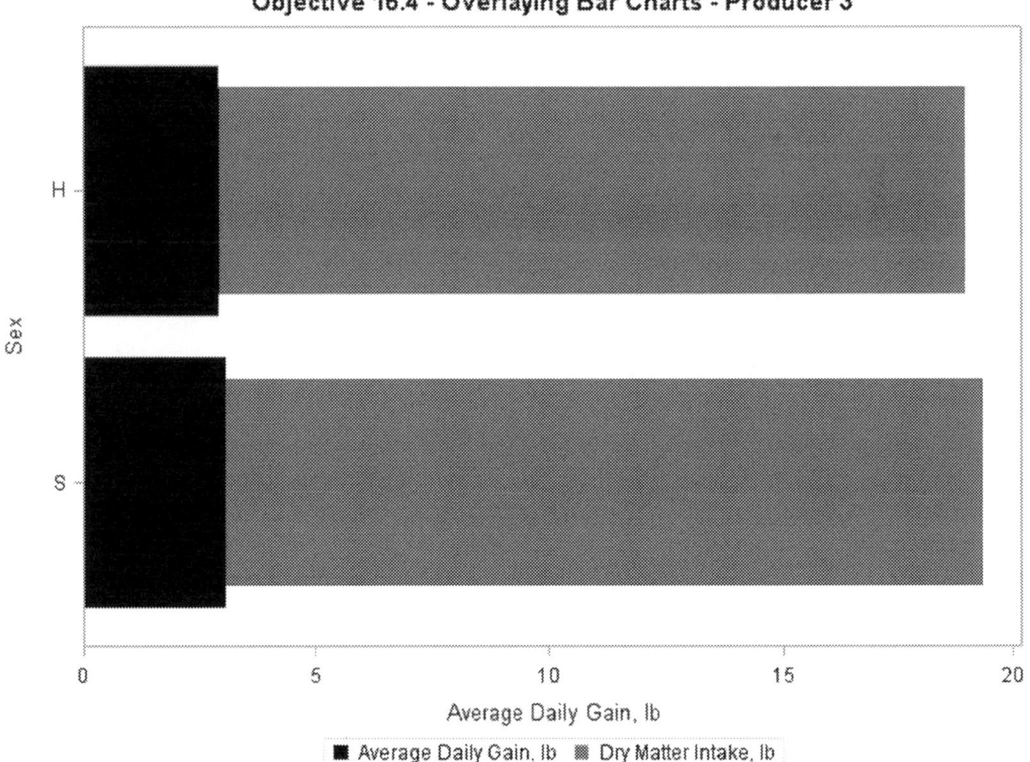

ADG and DMI are both measured in pounds. And though ADG is much smaller than DMI, one can still compare ADG between steers and heifers, and likewise examine DMI. The first HBAR statement is the first chart, and the second HBAR statement is overlaid on the first. If the bar widths here were the same and the DMI bar chart was produced first, the ADG bars would be completely hidden behind the longer DMI bars. When the bar lengths for the two response variables are close in length or there are more bars for the levels of the category variable, the choices of bar color, bar width, and transparency can aid in producing a more legible bar chart. Here the choice of the color black and black with a transparency option may be one approach to produce an image for black and white or grayscale media.

16.1.2 Histograms and Density Curves

For a continuous random variable, a frequency histogram can be produced. These histograms are vertical in the SGPLOT procedure. Color, fill, and transparency options for the histogram are the same as for the HBAR and VBAR statements. For a histogram, the number of bars or bins, as SAS calls them, can be specified, or the width of the bars or bins can be specified. The syntax of the HISTOGRAM statement is:

```
HISTOGRAM continuous-random-variable </options>;
```

A few of the introductory or basic options for the HISTOGRAM statement are in Table 16.3.

Often when histograms are produced, they are used to investigate the continuous distribution of the data by superimposing a curve on the histogram. A density curve is estimated with the DENSITY statement in the SGPLOT procedure. The syntax of the DENSITY statement is:

TABLE 16.3

HISTOGRAM and DENSITY Options

HISTOGRAM *continuous-random variable </options>*;

Option	Explanation
BINWIDTH=*numeric-value*	Specifies the width of the bars or bins. SAS will determine the number of bins. If NBINS= is specified, this option has no effect.
NBINS=*numeric-value*	Requests the number of bars or bins. SAS will determine the width of the bins.
SCALE= COUNT | PERCENT | PROPORTION	Pick one. The selected option specifies the bar heights. Percent is the default. Count is the observed frequency.
	See also the color, fill, and transparency options in Table 16.2.

DENSITY *response-variable </options>*;

Option	Explanation
TYPE = NORMAL	Creates a normal density curve with the sample mean and sample standard deviation as the population parameters.
TYPE = NORMAL (MU = *numeric-value* SIGMA = *numeric-value*)	Creates a normal density curve with the specified population parameters. If either MU or SIGMA is not specified, the sample values for mean and standard deviation will be used.

```
DENSITY response-variable </options>;
```

There are normal and other kernel density estimators as an option, but only the normal density will be included here. Normal DENSITY options are also in Table 16.3. The DENSITY statement can only appear as a second statement to produce an overlaid density curve when the first statement is HISTOGRAM or DENISTY in the SGPLOT procedure.

OBJECTIVE 16.5: Produce a frequency histogram for the CWT variable in SAS data set *beef3* for all producers. Observe the default number of bins (bars) and the bin width.

```
PROC SGPLOT DATA=beef3;
HISTOGRAM cwt;
TITLE 'Objective 16.5 - Default Histogram';
RUN;
```

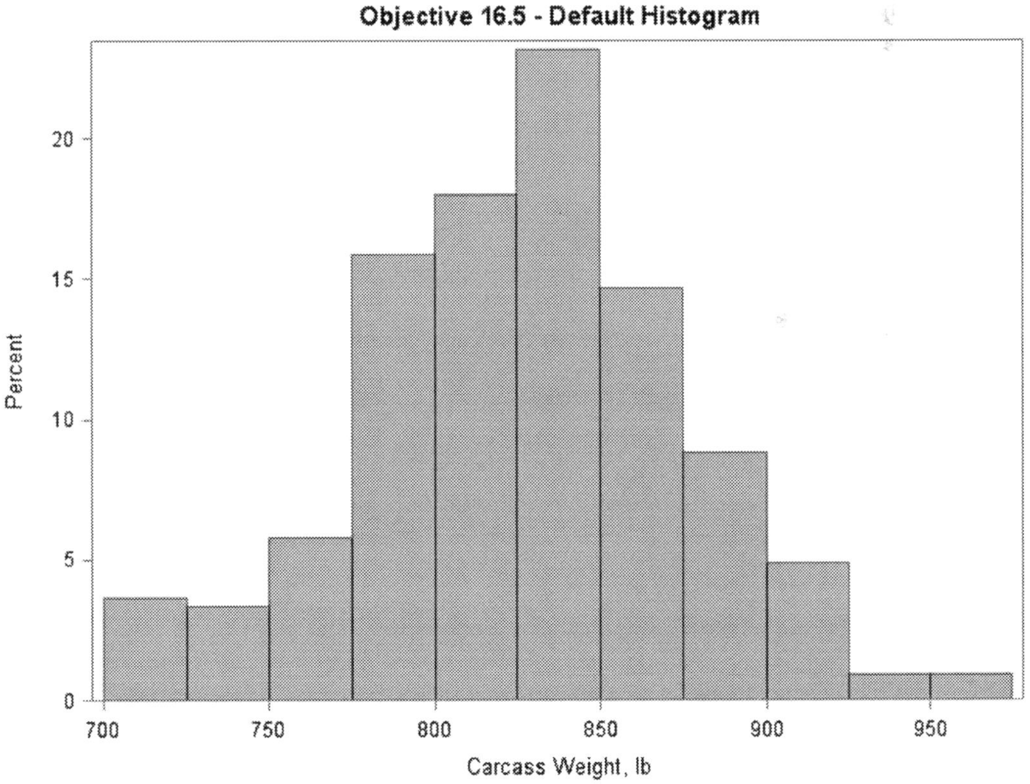

Objective 16.5 - Default Histogram

Percent

0 5 10 15 20

700 750 800 850 900 950

Carcass Weight, lb

In this instance SAS selected eleven bars or bins, and a bin width of 25.

OBJECTIVE 16.6: Modify the histogram in Objective 16.5 by overlaying a normal density curve with the sample statistics as the normal population parameters.

```
PROC SGPLOT DATA=beef3;
HISTOGRAM cwt;
DENSITY cwt / TYPE=NORMAL;
TITLE 'Objective 16.6';
RUN;
```

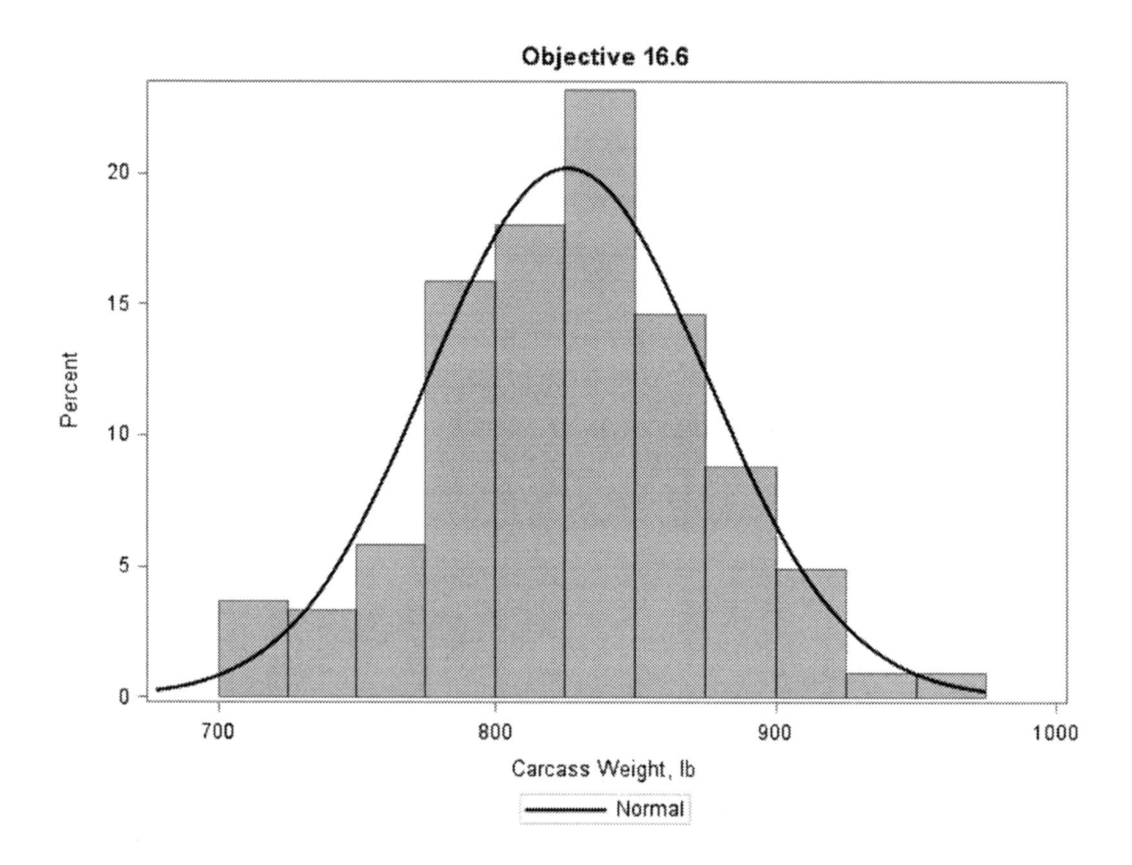

In conjunction with the DENSITY statement, the horizontal axis changed slightly in this example to capture the symmetric normal density requested.

16.1.3 Box Plots

Box plots are a simple graphic that illustrate the mean and/or median of the data as well as multiple measures of variation (range, interquartile range = IQR, selected percentiles). The SGPLOT procedures produce both horizontal and vertical box plots with the HBOX and VBOX statements, respectively. The syntax of these statements is:

```
HBOX response-variable </options>;
VBOX response-variable </options>;
```

Table 16.4 lists a few of the elementary options to get started.

OBJECTIVE 16.7: For SAS data set *beef3* produce vertical box plots for the BackFat response variable for each producer on the same set of axes. Connect the means of the three producers.

```
PROC SGPLOT DATA=beef3;
VBOX backfat/CATEGORY=producer CONNECT=MEAN;
TITLE 'Objective 16.7';
RUN;
```

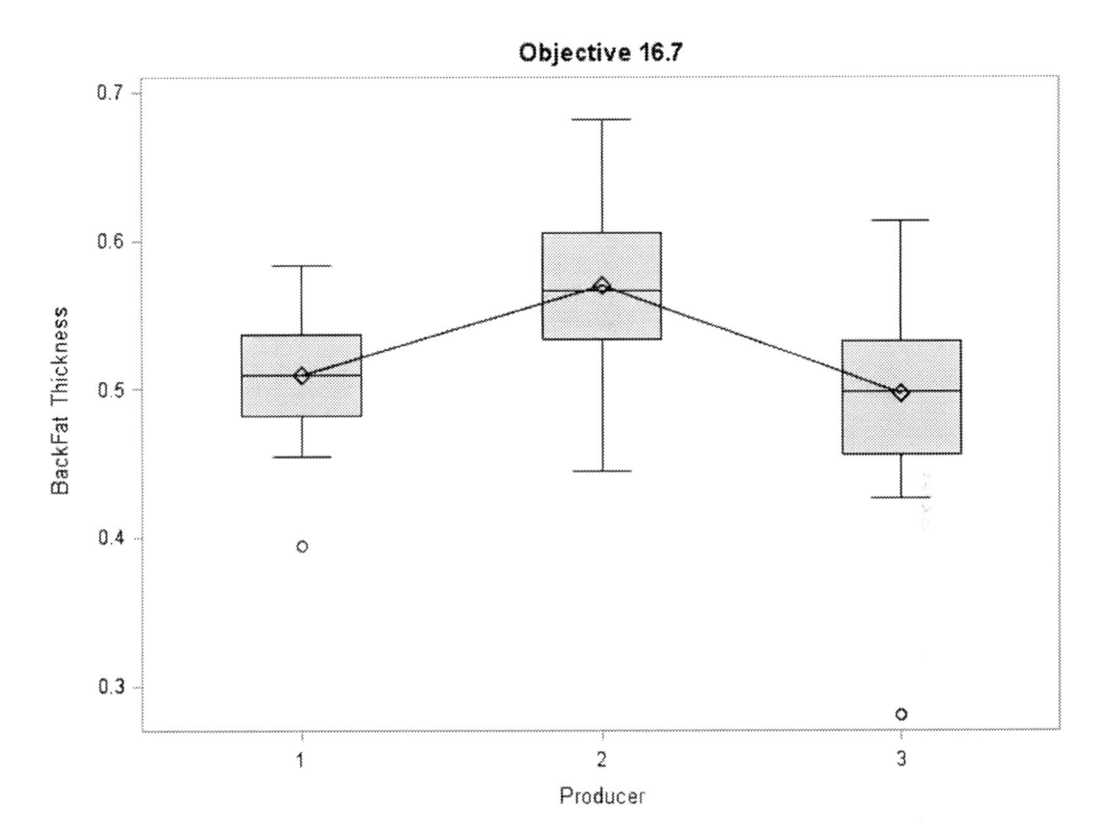

TABLE 16.4

Introductory Options for the HBOX and VBOX Statements

Options	Explanation
BOXWIDTH = *n*	0 < n < 1; default is 0.4
CATEGORY = *categorical-variable*	A box plot for the response variable is produced for each level of the categorical variable on one set of axes.
CONNECT= MEAN | MEDIAN | Q1 | Q3 | MIN | MAX	Pick one. Can only be used with the CATEGORY option. A line connects the indicated statistic across the levels of the category variable.
FILL | NOFILL	Pick one. FILL is the default. The fill color can be set in the FILLATTRS option. NOFILL produces an outline of the box plot.
FILLATTRS=(COLOR=*color*)	The box color is controlled by *color*. Keep it simple when starting out: BLACK, BLUE, RED, GREEN, ORANGE, PURPLE, GRAY, and so on.
MEANATTRS=*style-element*	Style-element is the shape of the symbol used to identify the mean on the box plots. Shape outlines or filled shapes are available. Shape outlines are: CIRCLE, DIAMOND, HOMEDOWN, SQUARE, STAR, and TRIANGLE. Filled shapes are the previous shapes followed by "FILLED", such as CIRCLEFILLED (no spaces). There are other style elements available. DIAMOND is the default symbol.
NOMEAN NOMEDIAN NOOUTLIERS	Suppress the mean, median, or outliers from being included in the box plot.

16.1.4 Scatter Plots and Series Plots

Plotting ordered pairs of variables on an X-Y coordinate system can be done using either the SCATTER statement or the SERIES statement. The SCATTER statement plots the ordered pairs as points on the two-dimensional axes. The SERIES statement connects the points as they are plotted. The syntax of the two statements is:

```
SCATTER X=horizontal-variable Y=vertical-variable </options>;
SERIES X=horizontal-variable Y=vertical-variable </options>;
```

A few of the basic options for controlling plotting symbols, colors, line styles, and more are listed in Table 16.5.

OBJECTIVE 16.8: Using the SAS data set *beef3* produce a scatter plot of CWT versus DMI for Producer 3 only. Plot this data with three plotting options.
Option 1: Use black-filled circles as the plotting symbols.
Option 2: Mark the plotting symbol with S or H indicating the sex of the animal.
Option 3: Use only S or H as the plotting symbol.

```
PROC SGPLOT DATA=beef3;
WHERE producer = 3;
SCATTER Y=cwt X=dmi / MARKERATTRS=(COLOR=BLACK SYMBOL=CIRCLEFILLED);
TITLE 'Objective 16.8 - Option 1';
RUN;
PROC SGPLOT DATA=beef3;
WHERE producer = 3;
SCATTER Y=cwt X=dmi / DATALABEL=sex;
TITLE 'Objective 16.8 - Option 2';
RUN;
PROC SGPLOT DATA=beef3;
WHERE producer = 3;
SCATTER Y=cwt X=dmi / MARKERCHAR=sex;
TITLE 'Objective 16.8 - Option 3';
RUN;
```

TABLE 16.5

Introductory Options for the SCATTER and SERIES Statements

Options	Explanation
DATALABEL= <*variable*>	The value of the specified variable is to be used as a label in the SCATTER. If no variable is specified, the value of Y is the label.
GROUP = *category-variable*	For the SCATTER statement – this produces different color plotting symbols for each level of the category variable. For the SERIES statement – the plotted points are connected by a line for each level of the category variable.
LINEATTRS=(COLOR=*color* PATTERN=*pattern*)	For the SERIES statement – this produces different color lines using different line patterns. For the colors, keep it simple when starting out: BLACK, BLUE, RED, GREEN, ORANGE, PURPLE, GRAY, and so on. More information on color selection is given in Section 17.1. The choices for *pattern* are: 1, 2, . . . , 42. 1 is a solid line, and the remaining are patterns of dots and dashes.
MARKERATTRS=(COLOR=*color* SYMBOL=*symbol*)	For the SCATTER statement – plotting symbols are called markers. The marker color is controlled by *color*, and color choices are as in LINEATTRS. Symbols are outlines or filled shapes. Shape outlines are: CIRCLE, DIAMOND, HOMEDOWN, SQUARE, STAR, and TRIANGLE. Filled shapes are the previous shapes followed by "FILLED", such as CIRCLEFILLED (no spaces).
MARKERCHAR= *variable*	The value of the specified variable is the plotting symbol.

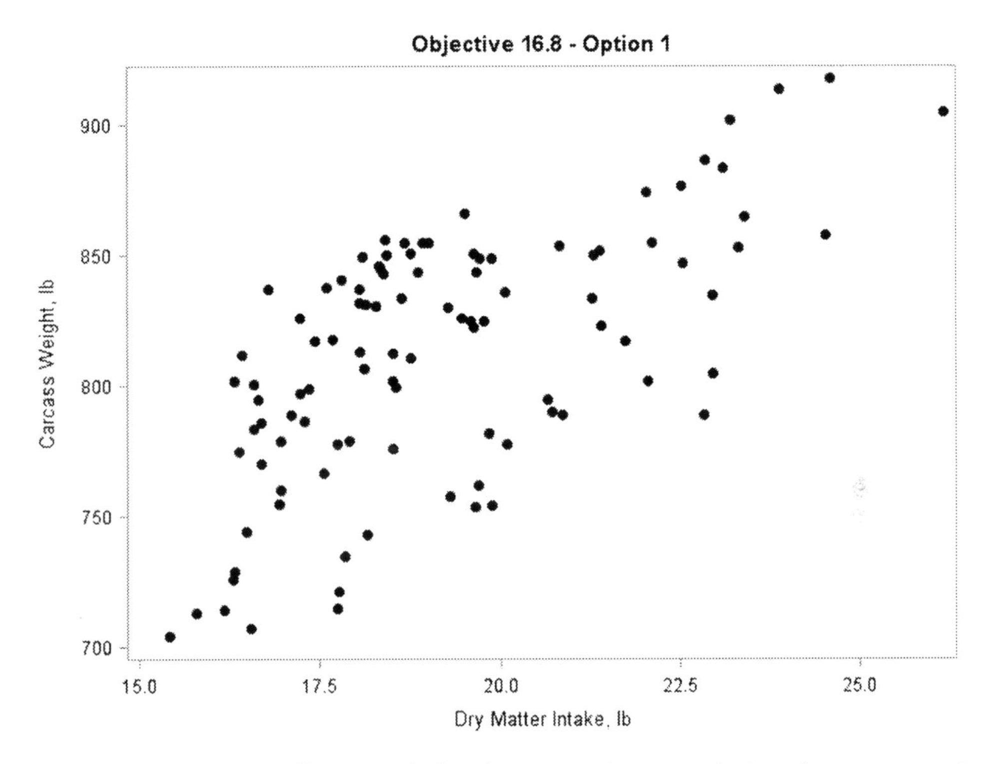

A scatter plot is an excellent graph for demonstrating associations between a pair of numeric variables as seen in Chapter 11. Using Option 1 the strong positive association between CWT and DMI is evident. Additionally, selecting the plotting symbol in this simple X-Y graph in SGPLOT is fairly easy to do.

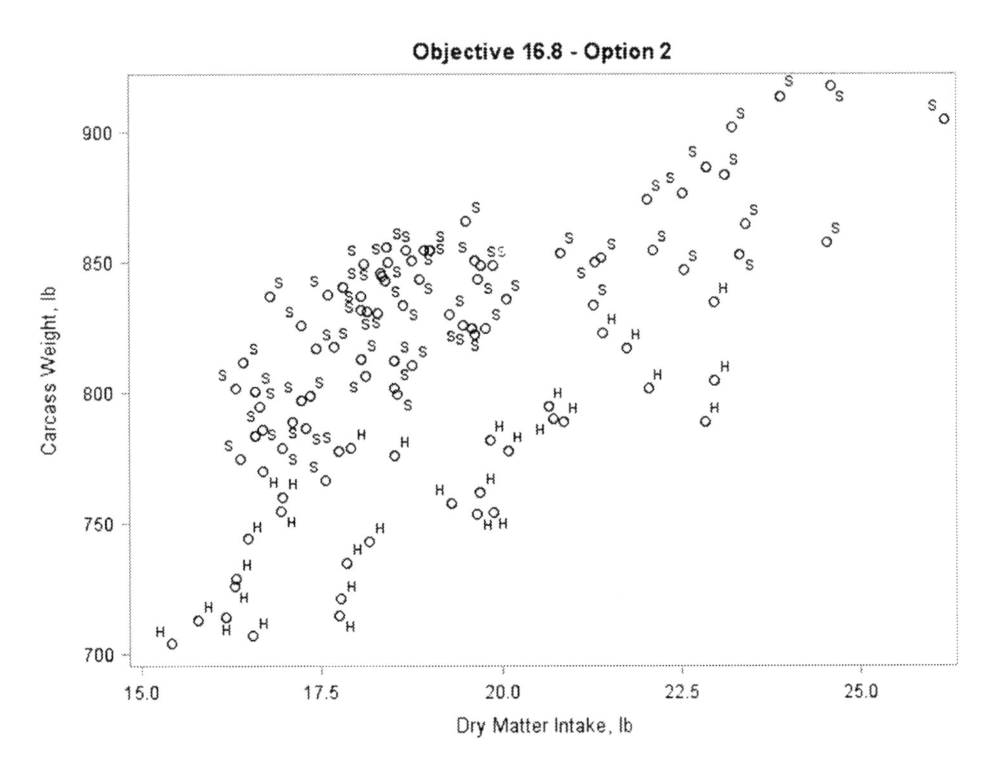

The open circle plotting symbols in Option 2 are the default plotting symbol for the SCATTER statement. The DATALABEL option places the value of specified variable next to the plotting symbol. If the values of the DATALABEL variable were lengthier text, that text would appear next to the plotting symbol. Since the single letters S and H adequately identify the sex of the animals, longer expressions in the SAS data set are not needed. However, it may be desirable to use the full words "Steer" and "Heifer" in some procedures (not just graphics). The FORMAT procedure in Chapter 18 can be used to achieve this effect without changing the contents of the SAS data set.

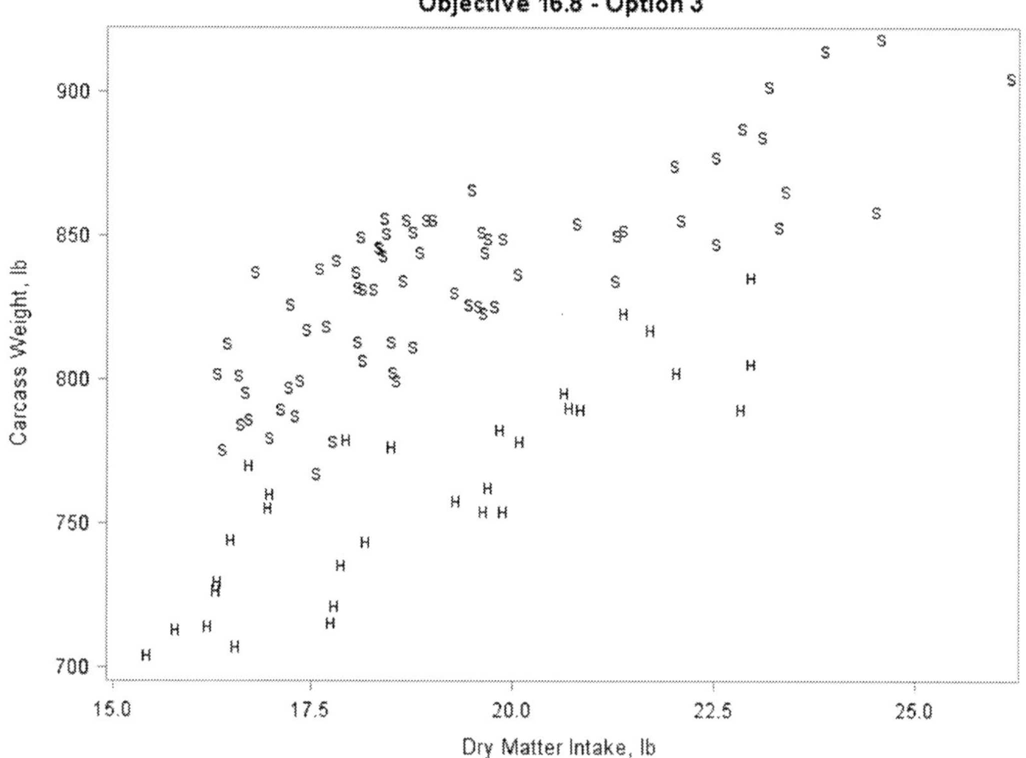

In Option 3, the MARKERCHAR option instructs the SCATTER statement to use the value of the named variable as the plotting symbol. When scatter plots become somewhat congested using Option 2, this option may result in a less congested scatter plot.

OBJECTIVE 16.9: For the SAS data set *beef3means* from Section 14.3 plot the mean CWT versus Producer identifying the variable Sex as the group variable. Use a line or series plot. Use a scatter plot with the values of "H" and "S" to identify the group variable in the graph. This is typically referred to as an interaction plot for the response variable CWT and the two classification variables Producer and Sex.

```
PROC SGPLOT DATA=beef3means;
WHERE _STAT_="MEAN";
SERIES Y=cwt X=producer / GROUP=sex LINEATTRS=(COLOR=BLACK PATTERN=3);
SCATTER Y=cwt X=producer / MARKERCHAR=sex;
TITLE 'Objective 16.9';
RUN;
```

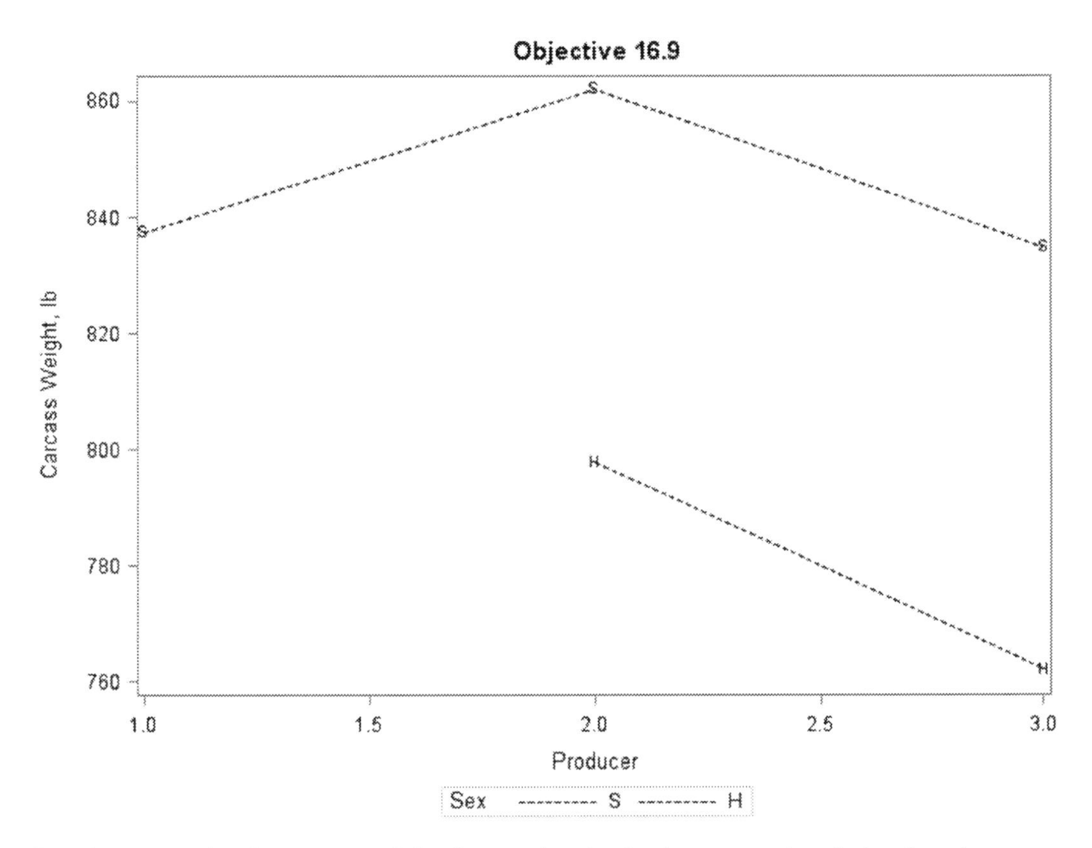

For this example, the context of the data assists in the interpretation. In beef cattle, steers tend to be heavier than heifers. For the *beef3* SAS data set used to compute the *beef3means* one should recall that Producer 1 had only steers (S) in the data set. Thus, when examining the series plot for this objective, the top jointed segment in the plot represents the steer means, and the segment from Producer 2 to 3 represents the heifer means.

When the GROUP option is used on the SCATTER and SERIES statements, SGPLOT assigns different colors to each of the groups. Since this book is limited to grayscale, line and marker attributes are chosen to accommodate viewing in grayscale.

16.2 The SGSCATTER Procedure

The SGSCATTER Procedure does differ from the scatter plots obtained in the SGPLOT procedure. In the SGSCATTER procedure, scatter plots can be arranged in a matrix or other paneled graphs typically involving more than one Y variable or more than one X variable. For a simple scatterplot, both the SGSCATTER and SGPLOT procedure produce an adequate presentation quality image. When multiple plots or comparative scatter plots are needed, the SGSCATTER procedure has the templates to create these panels.

The syntax of the SGSCATTER procedure is:

```
PROC SGSCATTER DATA=SAS-data-set <options>;
COMPARE X=variablelist  Y=variablelist </options>;
MATRIX variablelist </options>;
```

For the SGSCATTER procedure, either the COMPARE statement or the MATRIX statement should be chosen, but not both in the same procedure.

```
COMPARE X=variablelist   Y=variablelist </options>;
```

To compare more than one X or more than one Y variable, the list of variable names should each be enclosed in parenthesis, such as X= (*variable-1 variable-2 … variable-n*).

Options for the COMPARE statement include the DATALABEL, MARKERATTRS, and TRANSPARENCY that have been defined for statements in the SGPLOT procedure, see Table 16.5. To compare plots, the GROUP = *variable* option is needed. The GRID option will draw grid lines in the background of the scatterplots at each value of the tick marks on both the X- and Y-axes.

```
MATRIX variablelist </options>;
```

All variables in the list must be numeric but they do not need to be enclosed in parenthesis as in the COMPARE statement. Specifying n variables in the matrix statement will result in an n × n scatter plot matrix. The MATRIX statement has the options listed for the COMPARE statement.

OBJECTIVE 16.10: Use the *beef3* SAS data set to construct scatter plots of each of the following dependent variables, CWT, REA, and BackFat, versus the independent variables, DMI and ADG. Add the GRID option to see its effect on the scatter plots.

```
PROC SGSCATTER DATA=beef3;
COMPARE Y= (cwt rea backfat) X=(dmi adg) / GRID
MARKERATTRS=(SYMBOL=CIRCLE COLOR=BLACK);
TITLE 'Objective 16.10';
RUN;
```

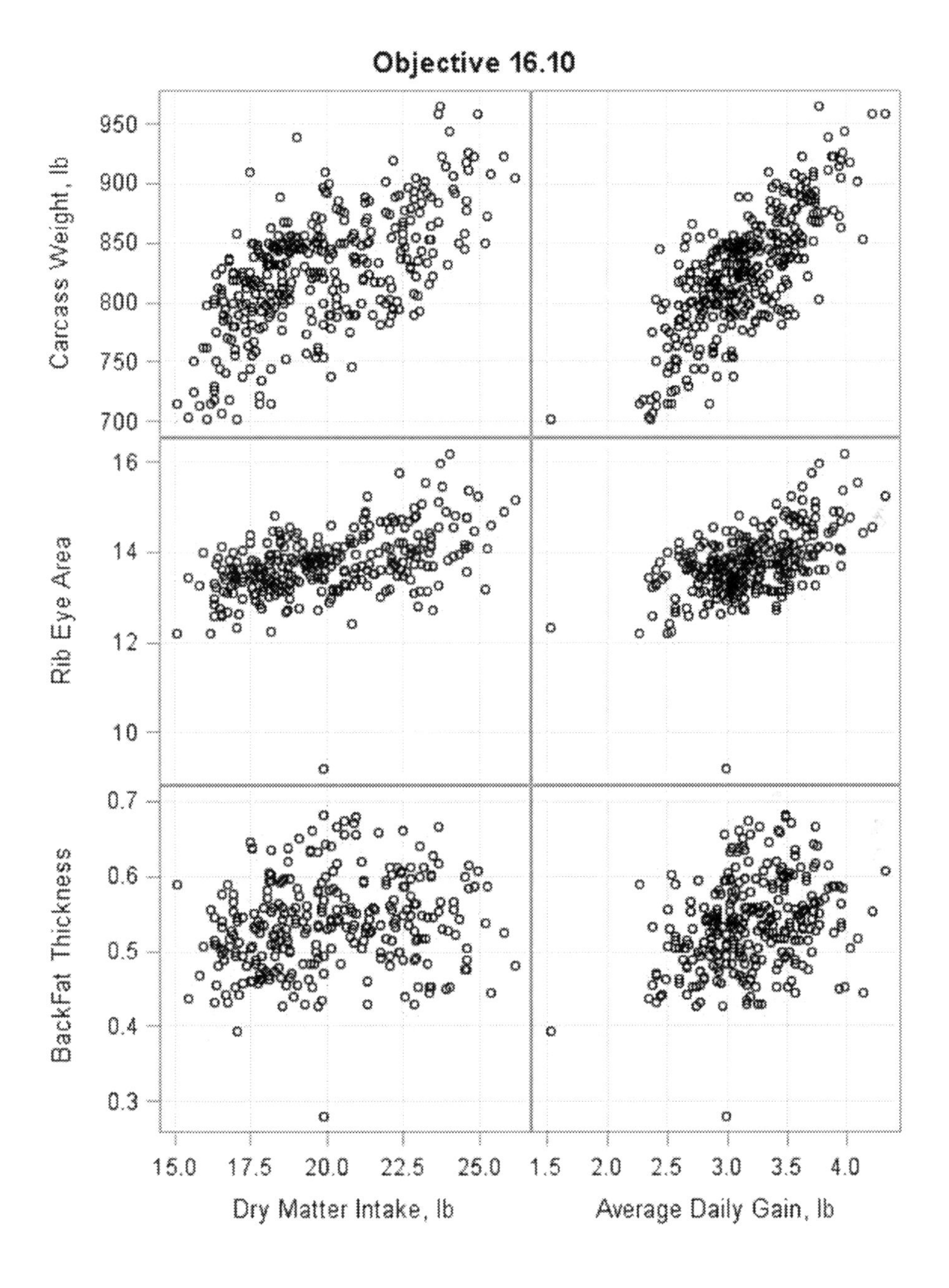
Objective 16.10
Carcass Weight, lb
Rib Eye Area
BackFat Thickness
950
900
850
800
750
700
16
14
12
10
0.7
0.6
0.5
0.4
0.3
15.0 17.5 20.0 22.5 25.0
1.5 2.0 2.5 3.0 3.5 4.0
Dry Matter Intake, lb
Average Daily Gain, lb

In this panel of scatter plots, the multiple choices for the Y variable result in the number of rows and the two variables chosen for the X variable determine the number of columns. Either X= or Y= can be specified first in the COMPARE statement. Specifying only one Y variable and one X variable will result in a single scatterplot as shown in the SGPLOT procedure. The labels created with the SAS data set *beef3* in Chapter 14 are used on the axes in this panel graph. SAS picked the tick marks for each axis, and the GRID option draws the reference lines in the background. Marker attributes available for the SCATTER statement are applicable here.

OBJECTIVE 16.11: Using the *beef3* SAS data set, produce a 5 × 5 matrix of scatter plots for all five of the response variables: CWT, REA, BackFat, DMI, ADG.

```
PROC SGSCATTER DATA=beef3;
MATRIX cwt rea backfat dmi adg /MARKERATTRS=(SYMBOL=CIRCLE COLOR=BLACK);
TITLE 'Objective 16.11';
RUN;
```

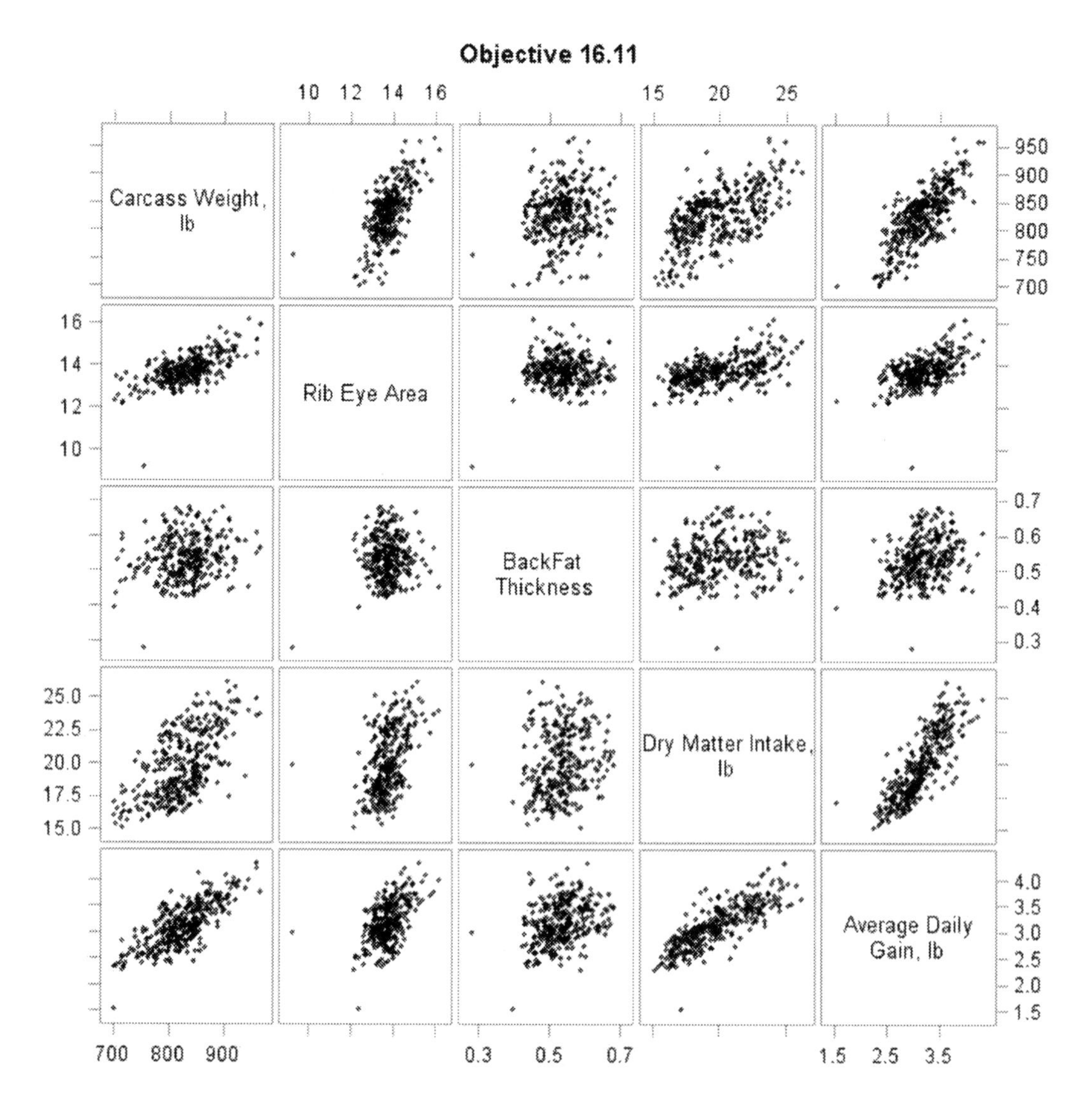

The order of the variables in the MATRIX statement determines the order of the plots in the scatter plot matrix. This is very similar to the panel or matrix generated by the CORR Procedure in Section 11.5.

16.3 The SGPANEL Procedure

The SGPANEL procedure creates multiple graphs in a row, column, or lattice (row by column) for one or more classification variables. Within the SGPANEL procedure many of the statements and options used in SGPLOT are available, and these images can be combined in a panel. The essential syntax of the SGPANEL procedure is:

PROC SGPANEL DATA=*SAS-data-set;*
PANELBY *variables </options>;*
chart statement(s)

The chart statements available are the same as those for SGPLOT listed in Table 16.1. There are other statements and options to enhance the appearance of the produced panel, but this overview will focus on the essentials. Once the chart statements for SGPLOT are learned, the essential statement in the SGPANEL procedure is the PANELBY statement as it controls the layout of the panel of graphs.

PANELBY is the **required** first statement of the SGPANEL procedure. This statement specifies the classification variable(s) for the panel and the type of layout. The options for the PANELBY statement appear in Table 16.6.

TABLE 16.6

Options for the PANELBY Statement in the SGPANEL Procedure

Options	Explanation
BORDER | NOBORDER	Pick one. Adds or suppresses cell borders within the panel. BORDER is the default.
COLHEADERPOS = TOP | BOTTOM | BOTH	Pick one. Columns in the panel are captioned with the variable name or label. Specify the location. TOP is the default. If PANEL=LAYOUT, then this option is not effective.
COLUMNS = c	Chooses the number of columns in the panel.
ROWHEADERPOS = LEFT | RIGHT | BOTH	Pick one. Rows in the panel are captioned with the variable name or label. Specify the location. RIGHT is the default. If PANEL=LAYOUT, then this option is not effective.
ROWS = r	Chooses the number of rows in the panel.
LAYOUT = LATTICE | PANEL | COLUMNLATTICE | ROWLATTICE	Pick one. LATTICE creates an r × c panel when two classification variables are specified in this statement. PANEL forms an r × c panel with row and column headings available. COLUMNLATTICE arranges the graph cells into a single row, and ROWLATTICE arranges the graph cells into a single column. The default setting is PANEL.

There are many other options in the SGPANEL procedure that control the legend and the horizontal and vertical axes, but those are not covered here.

To demonstrate the SGPANEL procedure, some of this chapter's earlier objectives for SGPLOT will be reconfigured for display in a panel graph.

OBJECTIVE 16.12: In Objective 16.2 a vertical bar chart for Sex was produced for each value of Producer. Arrange these three bar charts using SGPANEL with the default LAYOUT.

```
PROC SGPANEL DATA=beef3;
PANELBY producer;
VBAR sex / FILLATTRS=(COLOR=WHITE);
TITLE 'Objective 16.12';
RUN;
```

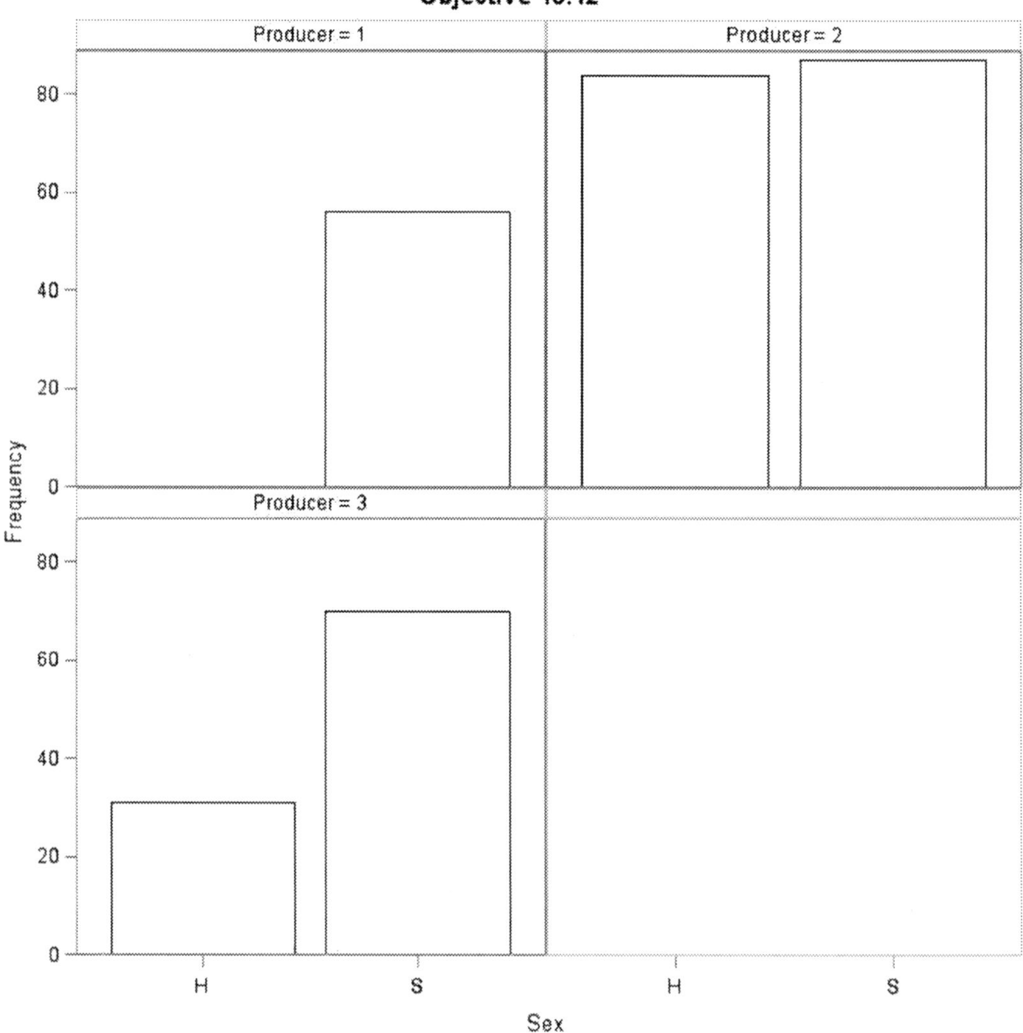

Programming Note: COLOR=WHITE produces the outline of the bars in the chart and is compatible with black and white or grayscale graphics.

Three larger images were the result in Objective 16.2. Here all three bar charts appear in a single graphic. With only a change in the name of the procedure and the PANELBY statement, a more concise graph is obtained.

OBJECTIVE 16.13: In Objective 16.3 a vertical bar chart illustrating the mean ADG for each sex for Producer 3 only was produced. Examine vertical bar charts for all producers using a column lattice. Omit the confidence intervals included in that objective.

```
PROC SGPANEL DATA=beef3;
PANELBY producer / LAYOUT=COLUMNLATTICE;
VBAR Sex / RESPONSE=adg STAT=MEAN FILLATTRS=(COLOR=LIGHTGRAY);
TITLE 'Objective 16.13';
RUN;
```

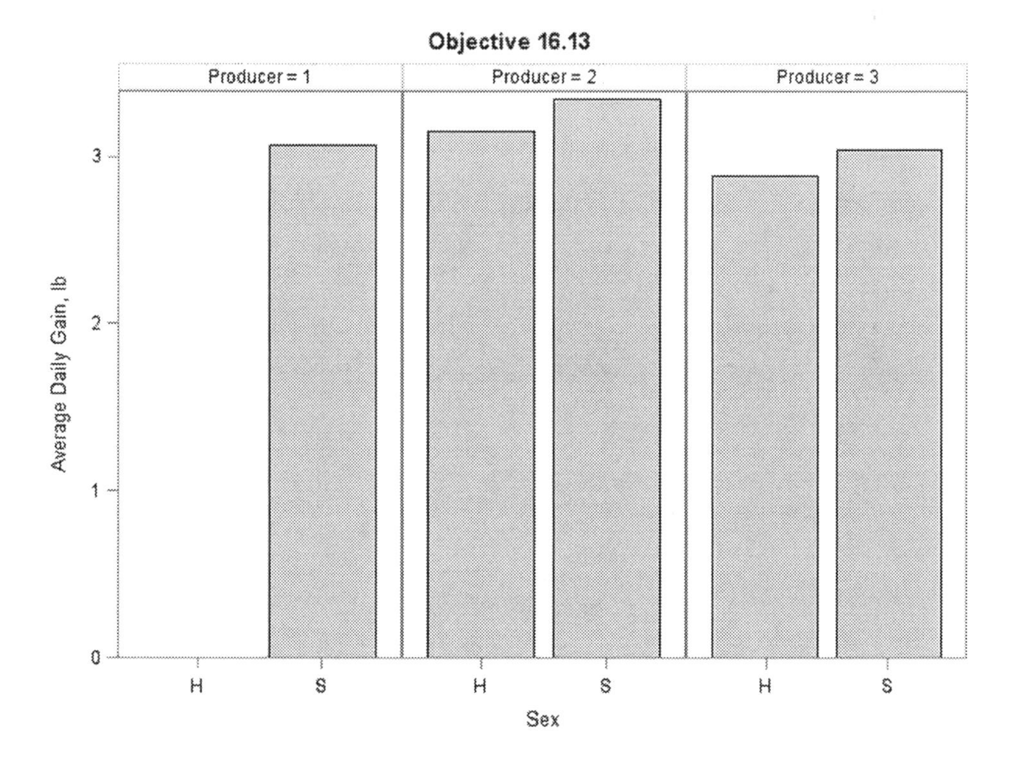

For this column lattice, there is a visual comparison of the mean ADG for each combination of the category variables. Additionally, the chart statement (VBAR) uses the same syntax as covered in the SGPLOT procedure. For a larger number of producers, a break in the column lattice may occur as it "wraps" to a new line. The COLUMNS = c option may be needed to control the number of columns.

OBJECTIVE 16.14: The SGPANEL procedure also allows for graphs to be overlaid in a panel plot. In Objective 16.6 a histogram for CWT with a normal density curve overlaid was produced. Produce a lattice plot for each combination of the producer and sex variables.

```
PROC SGPANEL DATA=beef3;
PANELBY producer sex / LAYOUT=LATTICE;
HISTOGRAM cwt;
DENSITY cwt / TYPE=NORMAL;
TITLE 'Objective 16.14';
RUN;
```

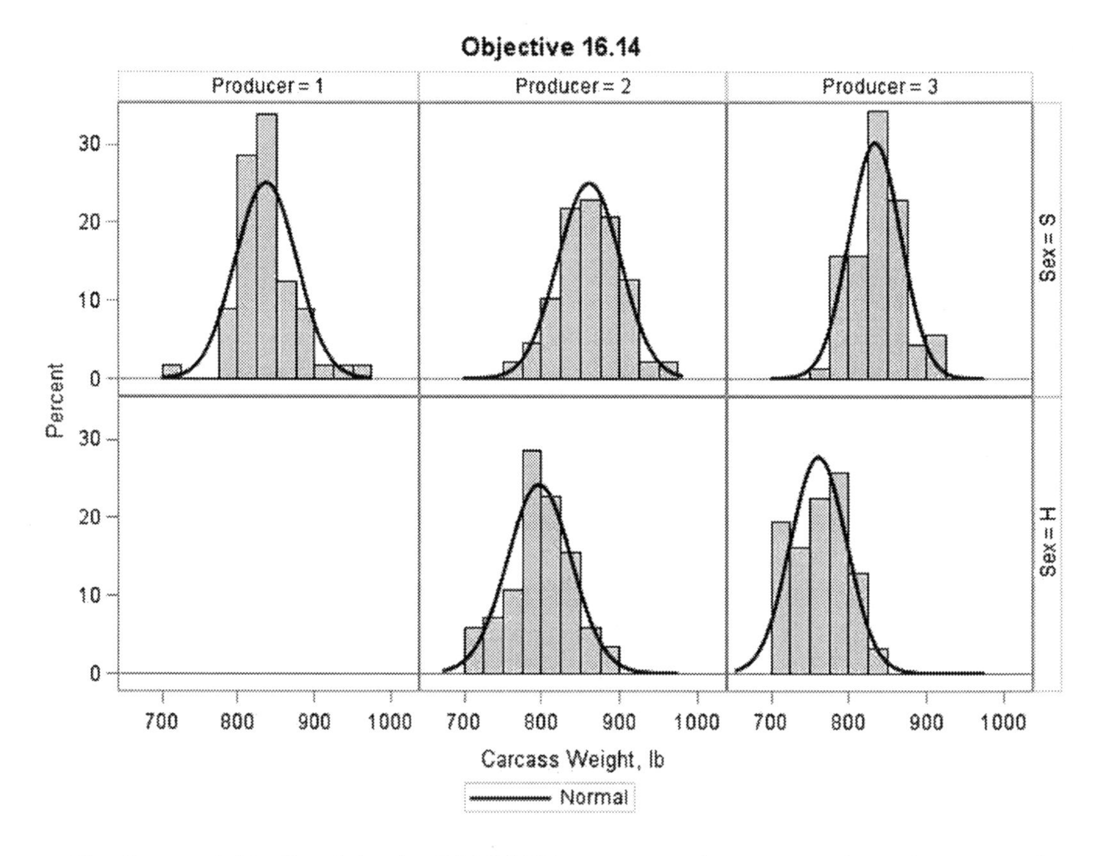

The first variable listed in the PANELBY statement determines the columns in the PANEL layout, and the second variable determined the rows. Switching the order of the variables in the PANELBY statement would result in three rows by two columns in the PANEL layout.

OBJECTIVE 16.15: In Objective 16.7 side-by-side vertical box plots for BackFat were produced for each producer, and the means were connected. Modify that objective by constructing a row lattice where one cell in the row is for heifers and the other is for steers. Place row headers on both the left and right sides of the row lattice.

```
PROC SGPANEL DATA=beef3;
PANELBY sex / LAYOUT=ROWLATTICE ROWHEADERPOS=BOTH;
VBOX backfat / CATEGORY=producer CONNECT=MEAN;
TITLE 'Objective 16.15';
RUN;
```

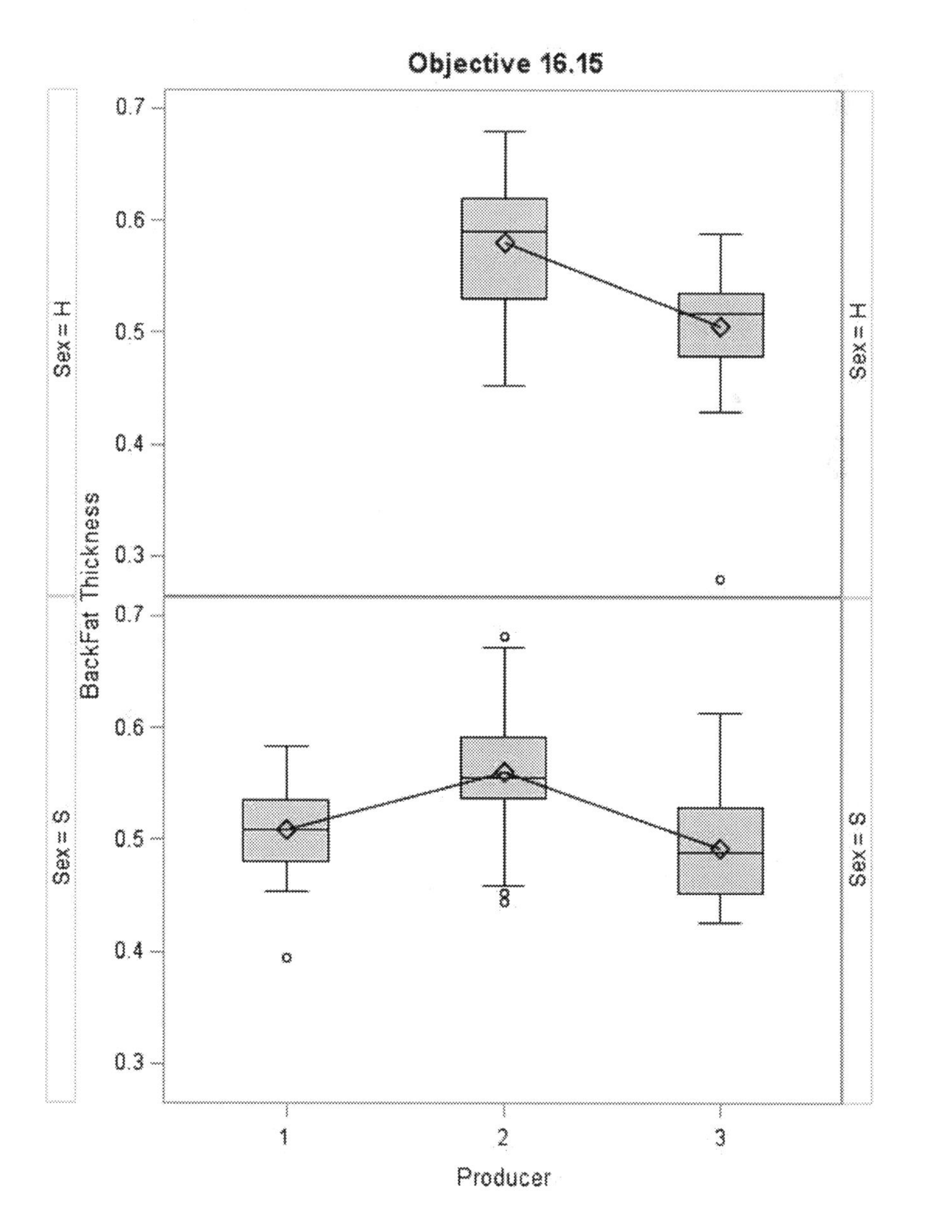

Like the column lattice, the row lattice may require the ROWS = r option when the PANELBY variable has a large number of levels.

There are two other Statistical Graphics procedures that require a stronger background, and they are SGDESIGN and SGRENDER. These procedures allow for greater customization of panel graphs.

16.4 Saving the Statistical Graphics Images

Statistical Graphics procedures produce image files with a .PNG extension by default. These PNG (Ping) files can be edited in an ODS Graphics Editor, but more familiarization with ODS Graphics Editor is recommended. From the Results Viewer, the PNG images can be selected by right clicking the image and selecting **Save picture as**. A **Save Picture** dialog window will open and a prompt for a File name with the default file type as PNG.

Using the Output Delivery System the file types can be changed and image files can be created in a destination folder identified in an ODS statement. Chapter 19 covers some features of the Output Delivery System.

16.5 Chapter Summary

The ODS Graphics available in many of the procedures in the earlier chapters of this book are excellent diagnostic tools. From the simple plots associated with a paired t-test to the panel of residuals in the REG procedure, these graphics are extremely helpful in data analysis. What made them even more appealing is that very little programming code was needed to produce them. This chapter introduced Statistical Graphics produres that also produce ODS Graphics. The Statistical Graphics procedures are procedures that have graphics templates built into them. These procedures can produce some well-designed graphics images with very little code in the Editor. As with any SAS procedure, the more statements and options one knows, the more sophisticated the produced graphic image can become.

17

SAS/GRAPH Procedures

The SAS software system has many products or components. Most of the procedures in this book are procedures in the Base SAS product. The GLM, NPAR1WAY, and REG procedures are in a group of procedures in the SAS/STAT product. SAS/GRAPH is another product within the SAS software system. SAS/GRAPH is a collection of procedures that are sometimes referred to as "legacy" graphics procedures since these are the graphics procedures have been available for a number of years in comparison to the Statistical Graphics procedures in Chapter 16. When producing graphics using current computing and printing capability, one does not give much thought to color, pattern, and quality until a color printer cartridge runs out. Some of the SAS/GRAPH syntax for these procedures control graphics catalogs and printer (or plotter) destinations, and these options were necessary in early versions of SAS. With current computing capability, this is not necessary but some of these options can be beneficial to the experienced SAS/GRAPH programmer. Like the Statistical Graphics procedures, there are default settings for the graphics produced by SAS/GRAPH procedures. Modifying those settings in SAS/GRAPH is no more challenging than modifying the default setting for Statistical Graphics procedures.

With SAS/GRAPH, presentation quality graphics can be produced. These images can be saved as graphics files for use in presentations and publications. These images can be produced in color or grayscale. SAS/GRAPH procedures begin with the letter G: GPLOT, GCHART, G3D, GCONTOUR, GMAP, GREPLAY, etc. There is also a GOPTIONS statement similar to the OPTIONS statement (Chapter 20). GOPTIONS can be specified to control image size, specify vertical or horizontal orientation, select fonts and font colors, and much more.

For the SAS/GRAPH procedures, TITLE, FOOTNOTE, BY statements (Chapter 2), WHERE statement (Chapter 4), LABEL statement (Chapter 8), and FORMAT statement (Chapter 18) can be included in these procedures.

17.1 GOPTIONS Statement

The GOPTIONS statement is a global statement. Thus, the selections made in a GOPTIONS statement are enabled until they are reset or replaced or the SAS session ends.

17.1.1 Selecting Colors for Graphics

Choosing the color for text used in graphics is just one of the options for the GOPTIONS statement. Since color choices will also be made in other SAS/GRAPH statements, it is a good idea to quickly overview color choices here.

There are numerous color choices that can be made. It is recommended that beginners stick with the basics, such as: RED, BLUE, ORANGE, BLACK, GREEN, GRAY, PURPLE, PINK, YELLOW, and BROWN when making a color selection. These simple name levels are from the color-naming scheme referred to as "SAS Color Names" in the SAS Registry. There are many, many SAS color names available. The following SAS code submitted from the Editor will produce the list of SAS Color Names in the Log window.

```
PROC REGISTRY LIST STARTAT="COLORNAMES";
RUN;
```

For the really creative SAS programmer "khaki", "limegreen", "turquoise", and many other colors can be found in this list. These SAS color names can also be used in the Statistical Graphics Procedures' attributes in Chapter 16.

There are eight different color-naming schemes in SAS/GRAPH which include: RGB, HLS, Gray scale, and others. If an alphanumeric or other alphabetic code is used to select a color in a SAS program, the programmer used one of the other SAS color-naming schemes. "SAS Color Names" is the color-naming scheme used in this book. SAS Color Names will appear in the GOPTIONS, PATTERN, and SYMBOL statements in this chapter.

17.1.2 A Few Basic Options

GOPTIONS has a large list of options that can be specified. Here are a few suggestions for the SAS/GRAPH beginning programmer.

CTEXT = *color* This option specifies the color of the text in the graphs using SAS Color Names. All text in the graphics image will be in this color. That includes titles, footnotes, default header, axis labels, tick mark values, and graph legend. Not covered in this book are the options that may override this choice for TITLEs or FOOTNOTEs.

FTEXT = *font* This option specifies the text font used in the graphs. The selected font will be used in any text in the graphics image. A font that matches a document or presentation can be made. There are some fonts specific to SAS/GRAPH. Here are a few:

ZAPF looks a lot like the Times New Roman font.

ZAPFB is boldface. ZAPFI is italics. ZAPFBI is bold italics.

SWISS is similar to Arial font.

SWISSB is boldface. SWISSI is in italics. SWISSBI is bold italics.

SIMPLEX is the default sans serif SAS/GRAPH font.

SIMPLEXB, SIMPLEXI, SIMPLEXBI are modifications.

GREEK is a sans serif font and, of course, is the Greek alphabet; CGREEK is the Greek option which has serifs.

SAS also has some True Type fonts available, and the availability of these fonts is based on choices made during installation of the SAS software. "Times New Roman" and "Arial" are two such (Microsoft®) fonts. Quotation marks must be used when specifying True Type fonts, such as: FTEXT="Arial".

HTEXT = # There are different units of measurement for height. A unitless measure is the default, and its default value is 1. HTEXT=1.5 will increase the font size by 50%. Enlarging the font may be recommended to maintain legibility if the graphics image will be reduced in size in its final presentation.

A GOPTIONS statement can be written before each SAS/GRAPH procedure if attributes in the graphics images need to be changed. Three important GOPTIONS statements to learn are:

```
GOPTIONS RESET=ALL;
```

This resets all of the graphics options (AXIS (Section 17.3.3), LEGEND (Section 17.3.5), PATTERN (Section 17.2.2), and SYMBOL (Section 17.3.2)) to SAS/GRAPH defaults and resets global statements (TITLE (Section 2.4) and FOOTNOTE (Section 2.4)).

```
GOPTIONS RESET=GOPTIONS;
```

This resets all graphics options but does not reset global statements.

```
GOPTIONS RESET=GLOBAL;
```

This statement resets all global statements but does not reset graphics options.

By now, many SAS programmers probably start their programs with a set of statements, such as,

```
DM 'LOG; CLEAR; ODSRESULTS; CLEAR; ';
TITLE;
FOOTNOTE;
```

After learning a bit about SAS/GRAPH one might consider adding:

```
GOPTIONS RESET=ALL ;
```

or one of the other RESET options when writing programs including SAS/GRAPH statements.

17.2 The GCHART Procedure

17.2.1 Bar Charts

In Chapter 5 the GCHART procedure was introduced, and with some very basic code, bar charts and histograms were produced. These simple tasks demonstrated that SAS/GRAPH has a default set of colors, symbols, and patterns defined. For the simple charting tasks examined at that time, those default settings worked well. In this chapter modifications to these default settings will be demonstrated.

In the GCHART procedure overview in Section 5.1, information about the GROUP= , TYPE= , SUMVAR= , MIDPOINTS= , DISCRETE, LEVELS= , and other options were introduced for the chart statements HBAR and VBAR. In the following objectives, color and pattern control will be incorporated into the bar charts.

HBAR3D and VBAR3D are additional statements in the GCHART procedure that produce three-dimensional bars in the bar chart or histogram. (Just because there is a 3D choice, does not mean 3D is always a good idea!)

Most of the objectives in this chapter will use the *beef3* SAS data set created in Chapter 14. For these objectives *beef3* has been saved as a permanent SAS data set in the *demo* library; thus, the two-level SAS data set name *demo.beef3* is used in the objectives.

OBJECTIVE 17.1: For each producer in *demo.beef3*, construct a vertical frequency bar chart for the category variable Sex where all of the bars appear on one set of axes. That is, there is only one image produced.

```
PROC GCHART DATA=demo.beef3;
VBAR sex / GROUP=producer;
TITLE 'Objective 17.1';
RUN;
```

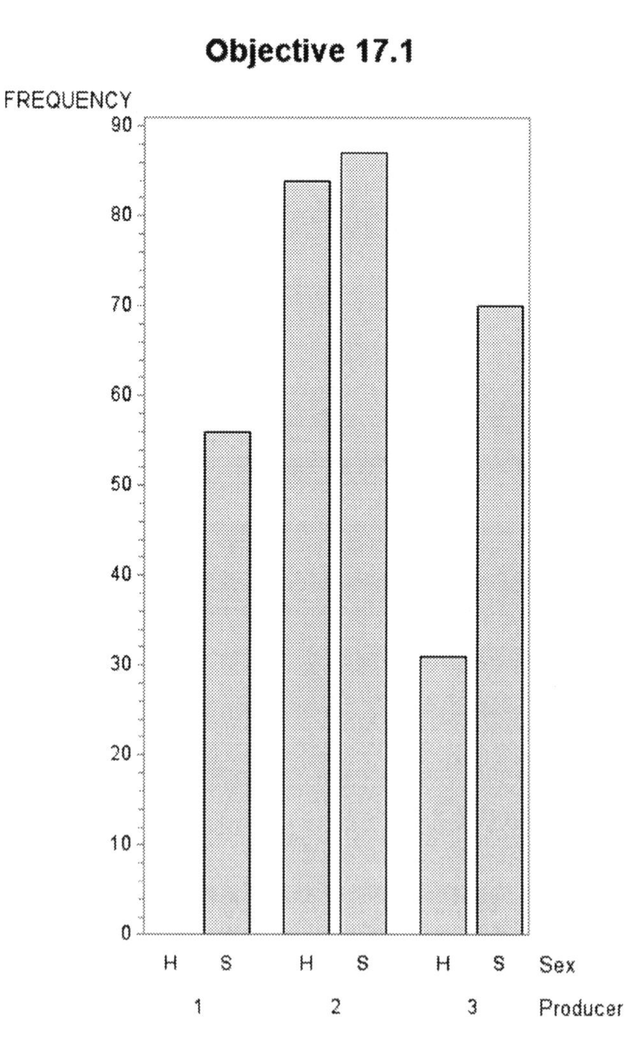

The default bar length is the frequency of the Producer and Sex combinations observed in the *demo.beef3* SAS data set. The default vertical bar chart image has a default color selected for the bars. This book is limited to grayscale images, so the color choice is not evident. In this release of SAS, the bars appear in a medium blue color. The default font used in this grouped bar chart is the default SIMPLEX font. Changing these defaults are topics of this chapter. In Objective 18.4, further enhancing this chart will be done by customizing the 1, 2, 3 values of Producer and the "H" and "S" value for Sex using the FORMAT Procedure and FORMAT statement.

17.2.2 PATTERN Statements

SAS has a default color set for the bars in a bar chart. The color and pattern used to fill the bars is easily controlled in a PATTERN statement. PATTERN statements are numbered, and they are also global statements. That is, there can be multiple PATTERN statements (up to 255), and the choices made in PATTERN statements are active until they are reset or overwritten. The statements that reset the PATTERN statements are:

```
PATTERN; RUN;
```

or

```
GOPTIONS RESET=PATTERN;
```

Either of these statements clears all PATTERN statements and resets colors and patterns to SAS defaults.

```
PATTERN1 COLOR=RED VALUE=SOLID;
```

This statement assigns the solid color red to the bars in a bar chart.

Bars can be a solid color (VALUE=SOLID), or they can simply be outlined in the chosen color by selecting VALUE=EMPTY. For three-dimensional bars (HBAR3D, VBAR3D), VALUE=EMPTY is not recommended by the author. Only two-dimensional bars can be filled with an X pattern (VALUE=X1), or have left sloping lines (VALUE=L1) or right sloping lines (VALUES=R1). Optional line styles for VALUE are X1, …, X5, L1, …, L5, and R1, …, R5.

The following examples illustrate how the image in Objective 17.1 changes with the inclusion of a PATTERN statement. In Figure 17.1A, the bars in the chart can be filled with the X1 pattern instead of a solid color. Figure 17.1B demonstrates the three-dimensional bar chart where the bars are a solid color. Any of the colors from SAS Color Names can be selected; however, since the content of this book appears in grayscale, the examples will select colors such as, black, white, gray, silver, steel, etc.

More than one PATTERN statement can be chosen for a chart or graph, and these are PATTERN1, PATTERN2, etc. . The PATTERNID option in the chart statement (VBAR or HBAR) identifies the chart characteristic that determines when the pattern changes.

```
PATTERNID = <option>
```

PATTERNID=MIDPOINT changes the color and pattern from one bar to the next whether each bar is centered at a default midpoint or a selected midpoint using the DISCRETE, MIDPOINTS= or LEVELS= option.

PATTERNID=GROUP changes the color and pattern from one level of the variable specified in the GROUP option to the next.

PATTERNID=SUBGROUP changes the color and pattern from one level of the variable specified in the SUBGROUP option. The SUBGROUP = *<variable>* on the bar chart statements selects another variable to represent on the bar chart by "stacking" the values of this variable within the bars of the chart variable.

The next objectives will illustrate the MIDPOINT, GROUP, and SUBGROUP options for PATTERNID as well as some of the other options for the HBAR and VBAR statements given in Chapter 5.

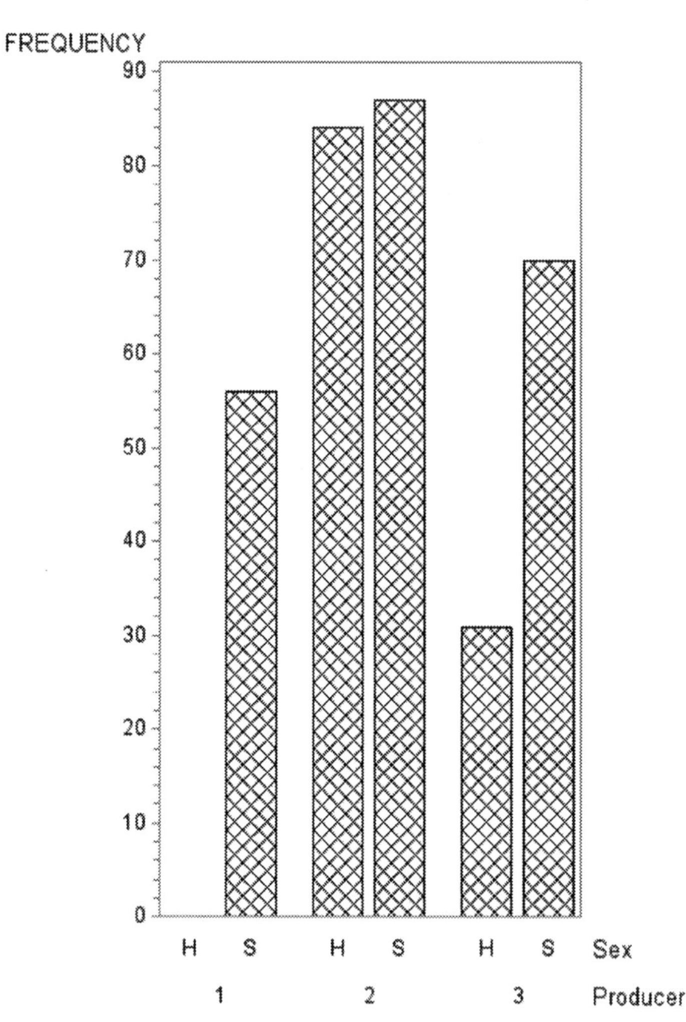

FIGURE 17.1A
PATTERN statement option VALUE=X1.

OBJECTIVE 17.2: Use the *demo.beef3* data for the following bar graphs. Construct all three graphs in a single GCHART procedure using Times New Roman font.

1. Construct a vertical bar chart where the bar height is the mean REA for each Producer. Use a different pattern for each Producer.
2. Modify the first graph so that the bars are grouped by Sex. Again, use a different pattern for each Producer.
3. Modify the second graph so that the bar height is the mean REA for each Sex and the bars are grouped by Producer. Each Producer will have a different pattern.

```
TITLE;
PATTERN;
GOPTIONS FTEXT="Times New Roman";
```

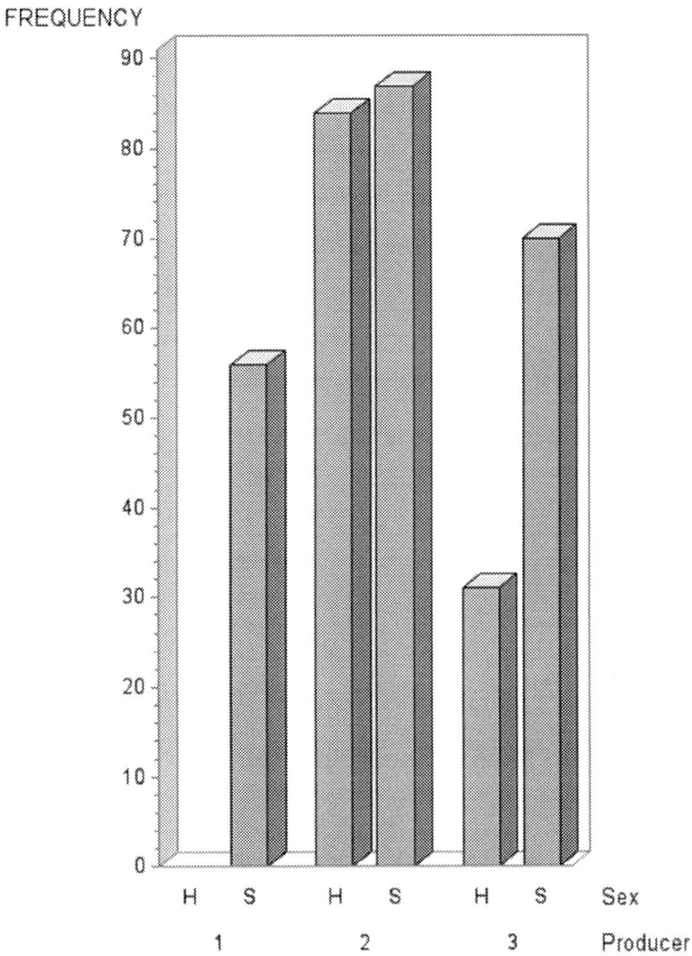

FIGURE 17.1B
Three-dimensional bar chart with COLOR=SILVER VALUE=SOLID.

```
PROC GCHART DATA=demo.beef3;
VBAR producer / DISCRETE TYPE=MEAN SUMVAR=rea PATTERNID=MIDPOINT;
TITLE 'Objective 17.2 - Chart 1';
PATTERN1 COLOR=BLACK VALUE=SOLID;
PATTERN2 COLOR=BLACK VALUE=EMPTY;
PATTERN3 COLOR=BLACK VALUE=X3;
RUN;
VBAR producer / DISCRETE TYPE=MEAN SUMVAR=rea PATTERNID=MIDPOINT
GROUP=sex;
TITLE 'Objective 17.2 - Chart 2';
RUN;
VBAR sex / TYPE=MEAN SUMVAR=rea PATTERNID=GROUP GROUP=producer;
TITLE 'Objective 17.2 - Chart 3';
RUN;
```

A few programming notes:

When creating several titles, footnotes, or patterns in a single SAS session, resetting those global statements, as done in the first two lines of the SAS code for Objective 17.2, is a good practice. In the code for this objective, each of the three levels of Producer are to be represented with different patterns in each of the three charts. It is not necessary to repeat the PATTERN statements unless different selections for COLOR and VALUE are needed in the second and/or third chart. The use of the RUN statement after each chart statement and the corresponding TITLE is necessary to get all of these charts produced in a single GCHART procedure. If only the third RUN statement were included, then only the third chart would be produced.

In Charts 1 and 2, Producer is the chart variable. Since Producer is a discrete numeric variable, it was necessary to include the DISCRETE option. Since Sex is a discrete character variable, the DISCRETE option was not necessary in Chart 3. In each of the charts, the patterns changed for each level of Producer. As the role of the variable Producer changed in Chart 3, the VBAR statement and PATTERNID had to change with it.

Alternatively, these three graphs could be produced using three different blocks of PROC GCHART code. Each PROC GCCHART block would have a single VBAR statement and would contain its own TITLE statement(s).

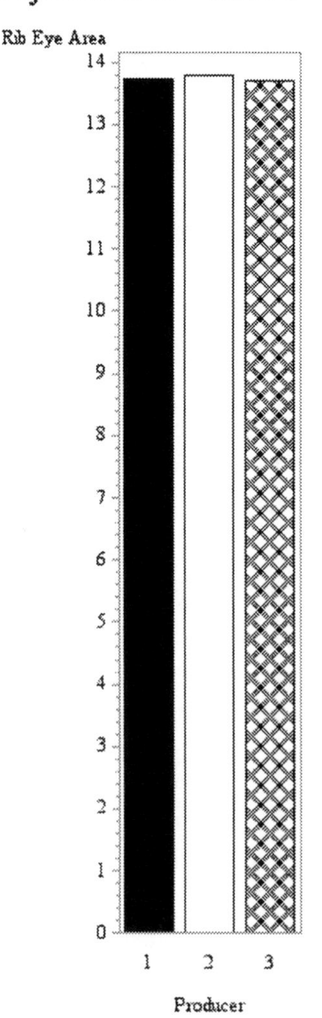

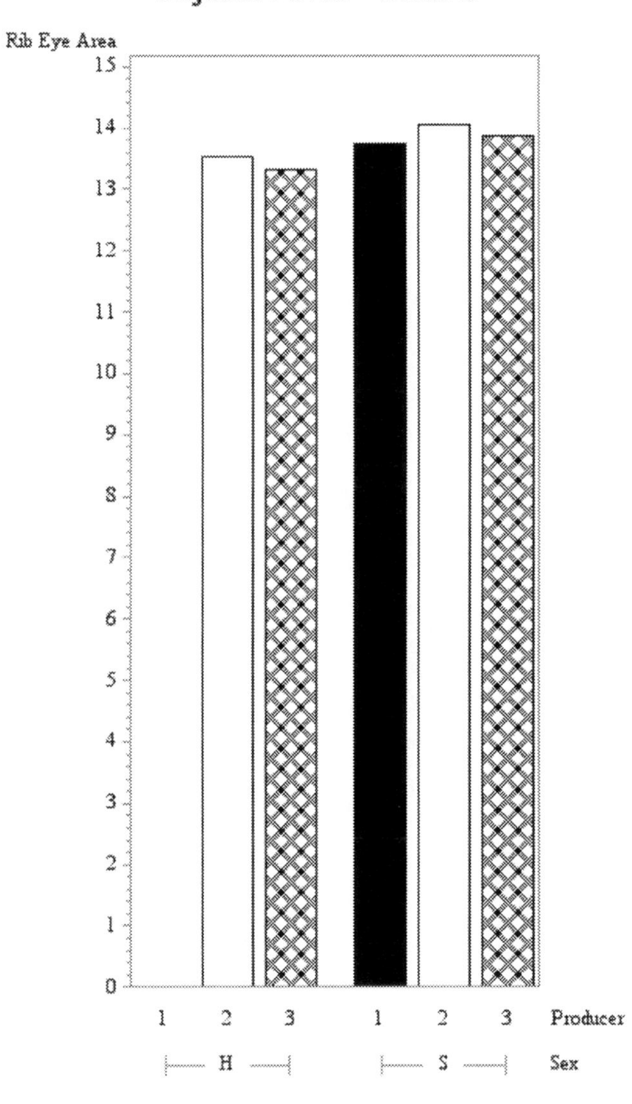
Objective 17.2 - Chart 2
Rib Eye Area
15
14
13
12
11
10
9
8
7
6
5
4
3
2
1
0
1 2 3 1 2 3 Producer
H S Sex

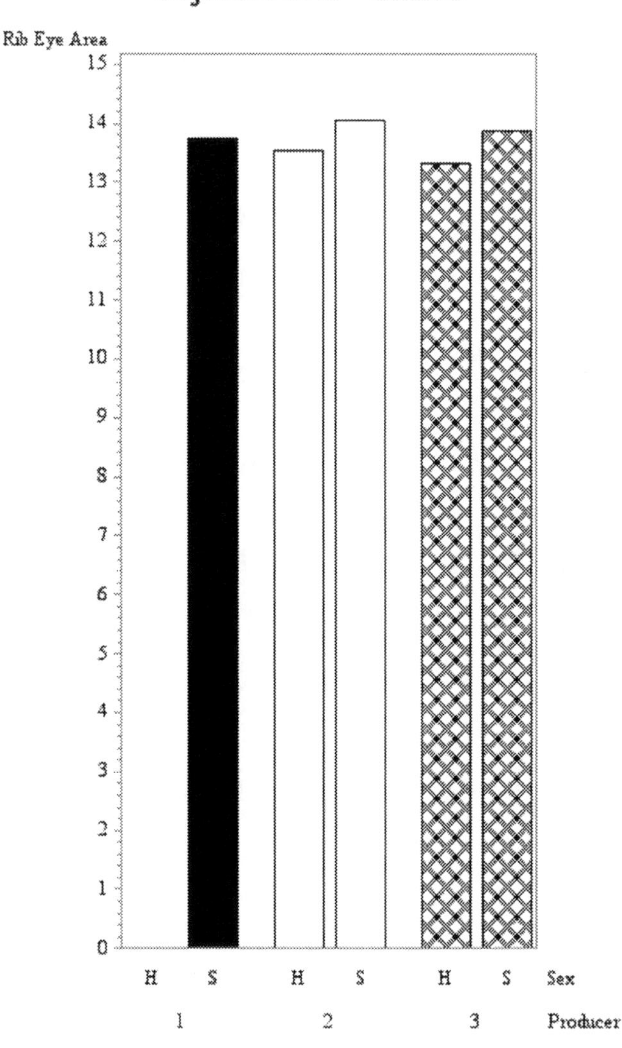

OBJECTIVE 17.3: Using *demo.beef3* chart the frequency distribution of the producers using either a vertical bar chart or a horizontal bar chart. Separate each of the bars into pieces that represent the number of heifers and steers. Change the colors and patterns for the subgroups. Keep a solid bar for the heifers and a bar with an outline only for the steers. Keep the Times New Roman font but increase the size of the font on the chart.

```
TITLE;
PATTERN;
GOPTIONS FTEXT="Times New Roman" HTEXT=1.5;
PROC GCHART DATA=demo.beef3 ;
VBAR producer / DISCRETE SUBGROUP=sex PATTERNID=SUBGROUP;
TITLE 'Objective 17.3';
PATTERN1 COLOR=GRAY VALUE=SOLID ; *heifers;
PATTERN2 COLOR=GRAY VALUE=EMPTY ; *steers;
RUN;
```

As seen in previous charts in this chapter, H is ordered before S alphabetically. Thus, PATTERN1 will apply to the first level (H) of the SUBGROUP variable, then PATTERN2 will apply to the second level (S). In general, the sorted order of the SUBGROUP determines which PATTERNn statement aligns with which level of SUBGROUP. The single-line comments after the PATTERN statements are added for clarity.

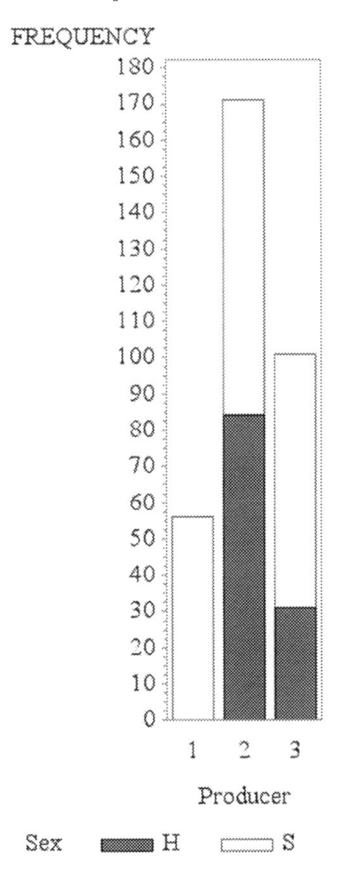

The increase in the font size in this objective versus the font size in Objective 17.2 is due to the HTEXT=1.5 option in the GOPTIONS statement. Here colors could have been selected. When producing this chart in black and white or grayscale, subgroups in solid color bars may be difficult to distinguish. Thus, the choices of SOLID and EMPTY in this objective were made since the variable Sex has two levels.

If the roles of the chart variable and SUBGROUP were switched in Objective 17.3, the vertical bar chart would have two bars, one for heifers and one for steers, and each bar would be divided in up to three pieces since the subgroup variable Producer has three levels. Thus, the SUBGROUP variable and PATTERNID = SUBGROUP would indicate that three PATTERN statements are needed as given here:

```
PROC GCHART DATA=demo.beef3;
VBAR sex / SUBGROUP=producer PATTERNID=SUBGROUP;
PATTERN1 COLOR=GRAY VALUE=SOLID ; *Producer=1;
PATTERN2 COLOR=GRAY VALUE=EMPTY ; *Producer=2;
PATTERN3 COLOR=GRAY VALUE=X3 ;    *Producer=3;
RUN;
```

In Section 16.1.1, the SGPLOT procedure can also produce these "stacked" bars when the HBAR or VBAR statements are used. However, it is the GROUP option on those statements that does this. There is no SUBGROUP option in the SGPLOT procedure.

17.2.3 Pie Charts

In Chapter 5, pie charts were listed as one of the types of charts available in the GCHART procedure. Using the options for the PIE statement in Chapter 5 simple pie charts can be constructed. The use of PATTERN statements presented in this chapter is recommended when producing pie charts as the default colors may not work well in print and/or presentation. For pie charts, only VALUE=SOLID and VALUE=EMPTY can be selected in the PATTERN statements. Among the options on the PIE statement labeling options can also be selected. That is, one can choose whether or not to print the slice values and where the labels are to be located. PIE statement options were listed in Table 5.1. They are explained further in Table 17.1.

These are just a few of the options for pie charts. More advanced programming options allow one to select the order in which the slices appear, the angle of the first slice, color of the text labels, suppressed slices, and exploding (or emphasized) slices. There are also 3D options for pie charts.

OBJECTIVE 17.4: For the SAS data set *demo.beef3*, construct two pie charts for the Producer variable. Use a larger Swiss bold font in these charts.

1. Compute the percent of the data from each Producer. Place the percentage inside the slice and the value of the chart variable outside of the slice. Choose grayscale colors.
2. Compute the frequency of observations from each Producer. Place the frequency inside the slice.

TABLE 17.1

PIE Statement Options

Options	Explanation
DISCRETE	Use if the chart variable is a discrete numeric random variable. MIDPOINTS = *list* or LEVELS = *value* are other options for controlling the number of slices as in bar charts. See Objectives 5.4 and 5.6.
FREQ = *value*	This is necessary if another variable contains a count associated with each level of the chart variable.
NOHEADER	GCHART places a heading such as, "Percent of . . . " under the TITLE lines and above the pie chart. This option disables that header.
SLICE = ARROW | INSIDE | NONE | OUTSIDE	Pick one. This selection positions the slice or midpoint name. OUTSIDE is the default.
TYPE = FREQ | PCT	Pick one. FREQ is the default if no type is specified. TYPE=MEAN with a SUMVAR can be specified, but a pie chart generally is not recommended for representing a mean.
VALUE = ARROW | INSIDE | NONE | OUTSIDE	Pick one. This selection positions the slice value. NONE suppresses it. OUTSIDE is the default.

```
GOPTIONS FTEXT=SWISSB HTEXT=1.5;
PROC GCHART DATA=demo.beef3;
PIE producer / DISCRETE TYPE=PERCENT VALUE=INSIDE SLICE=OUTSIDE;
TITLE 'Objective 17.4 - Pie Chart - Percent';
PATTERN1 COLOR=GRAY VALUE=EMPTY ;
PATTERN2 COLOR=GRAY VALUE=SOLID ;
PATTERN3 COLOR=SILVER VALUE=SOLID;
RUN;
PIE producer / DISCRETE TYPE=FREQ VALUE=INSIDE;
TITLE 'Objective 17.4 - Pie Chart - FREQ';
RUN;
```

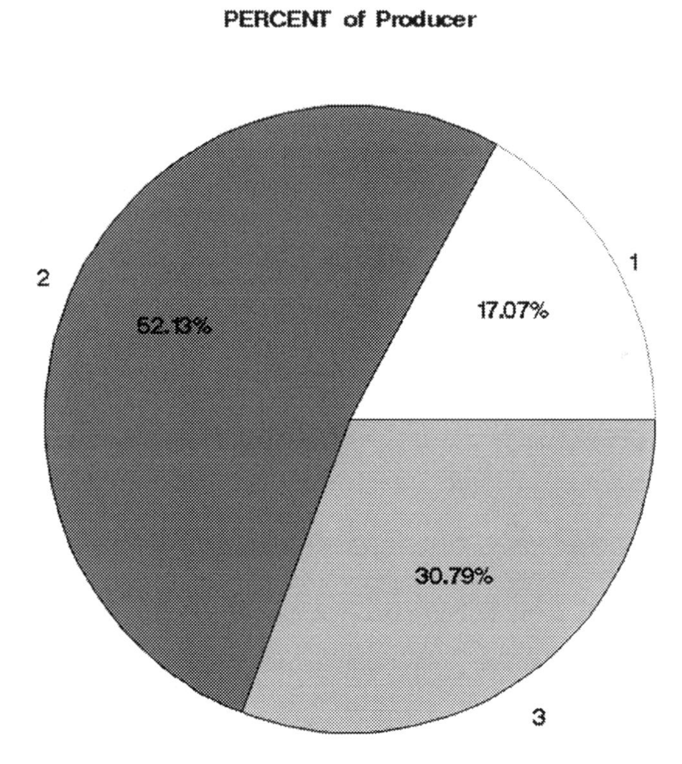

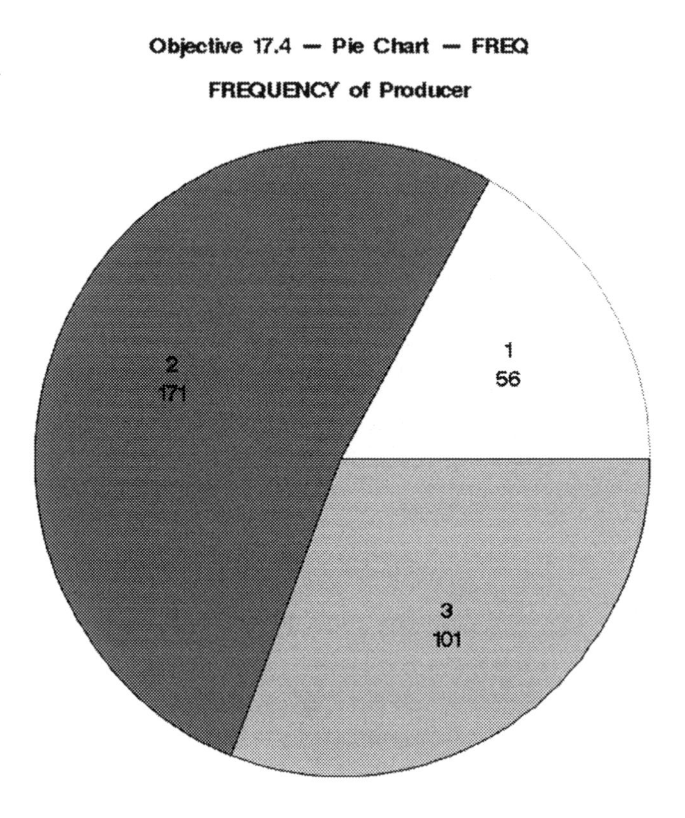

17.3 The GPLOT Procedure

The GPLOT procedure plots numeric variables on an (X, Y) coordinate system. With appropriate options pairs of variables can be plotted separately or overlaid on the same set of axes. Colors, lines, and symbols used in the graphs can be selected in this procedure.

Using the GPLOT procedure three different categories of plots can be produced:

1. Simple (X, Y) plot
2. (X, Y) plot by category
3. Multiple plots per axes

The syntax of the GPLOT procedure is as follows:

PROC GPLOT DATA=*SAS-data-set*;
PLOT (*vertical axis variablelist*) * (*horizontal axis variablelist*) < = *variable*> </*options*>;
SYMBOL1 <*options*>;
⋮
SYMBOLn <*options*>;

In a PLOT statement, the requested list of plots can take various forms. Here are some examples.

PLOT y*x;	generates a plot in which Y is plotted on the vertical axis and X is plotted on the horizontal axis; that is, plot "Y by X".
PLOT a*b a*c; or PLOT a*(b c);	generates two plots in which A is plotted on the vertical axes of both plots. Plot A by B and A by C.
PLOT (a b)*(c d);	generates four plots. A by C, A by D, B by C, B by D.
PLOT a*b=*variable*; or PLOT a*b = *n*;	generates a plot of A by B where the plotting symbol is associated with a third variable or where *n* = the number of the SYMBOLn statement. The SYMBOL statements are overviewed in Section 17.2.2.

Options for the PLOT statement include:

HAXIS=*axisn* specifies the user-defined axis (See Section 17.3.3 AXIS Statements).

VAXIS=*axisn* specifies the user-defined axis (See Section 17.3.3 AXIS Statements).

HREF=*values* draws lines on the plot perpendicular to the horizontal axis at the values specified, for example, HREF = 2 4 6 8. See also AUTOHREF.

VREF=*values* draws lines on the plot perpendicular to the vertical axis at the values specified. See also AUTOVREF.

GRID draws grid lines perpendicular to both axes at each tick mark.

AUTOHREF draws grid lines in the background perpendicular to the horizontal axis only at each tick mark.

AUTOVREF draws grid lines in the background perpendicular to the vertical axis only at each tick mark.

OVERLAY overlays all plots specified in the PLOT statement on one set of axes.

NOLEGEND Some plots will produce a legend by default. NOLEGEND will suppress the legend.

LEGEND=*legendn* specifies the user-defined legend in a LEGENDn statement. The information regarding the LEGENDn statement is in Section 17.3.5.

Many more options for the PLOT statement of the GPLOT procedure can be found in SAS Help and Documentation or SAS Institute, 2016c.

17.3.1 SYMBOL Statements

Line colors, line types, plotting symbols, and other plot enhancing items can be selected using SYMBOL statements. Like the PATTERN statements, SYMBOL statements are numbered (1-255) and are global statements.

Shapes

In a SYMBOL statement, the plotting symbol is selected using the VALUE= option. There are many choices available, but the basics are: NONE, SQUARE, CIRCLE, TRIANGLE, DOT (filled circle), DIAMOND, STAR (looks like an asterisk), and PLUS. PLUS is the default plotting symbol if there are no active SYMBOL statements in the SAS session. A single letter or numeric value can be specified as the plotting symbol if it is enclosed in quotation marks, such as, VALUE = "M" or VALUE="7". The symbols @, #, $ should not be used as these are code for other special plotting symbols.

H = # sets the height of the plotting symbols. The default is H = 1. For larger plotting symbols, numbers larger than one are needed, such as: H = 1.5.

The term "marker" is used for the SGPLOT procedure in Section 16.1.4 for a plotting symbol. In SAS/GRAPH one can examine "marker fonts" or "graphics fonts" in SAS Help and Documentation when a greater selection of plotting symbols is needed.

Colors

The color of the plotting values is selected using the CV=*color* option. Colors are specified using the color-naming scheme referred to as "SAS Color Names" just as described in Section 17.1. COLOR=*color* or C=*color* selects the color of the lines used to connect plotting symbols. If only the CV=*color* option is specified, the plotting value and the selected line style will be the same color.

Line Styles

I = NONE leaves points unconnected as in a scatter plot.

I = JOIN option where "I" stands for "interpolate" connects the plotted values with a line. When "connecting the dots" it may be necessary to SORT by the X (horizontal axis) variable first. There are many more options that can be chosen for I= in the SYMBOL statement, but those will not be presented here.

L = # where # is a number between 1 and 46, chooses a line style to use in the graph. L=1 selects a solid line and is the default line type. L=3 is line of short dashes. All other line styles are dashed or dotted or some combination of both.

W = # sets the width of the line. The default is W = 1. For bolder lines try W = 2.

In all cases, if not enough SYMBOLn statements are specified for the current plot, SAS will use a default SYMBOL statement or an active SYMBOL statement previously defined in another block of statements.

17.3.2 Simple (X, Y) Plot

Only one pair of variables is represented on a two-dimensional set of axes in a simple (X, Y) plot. These plots can be scatterplots or line plots. This is controlled by the choice of options in the SYMBOL statement. This is different from the SGPLOT procedure in Chapter 16 where there are SCATTER and SERIES statements.

OBJECTIVE 17.5: Using the GPLOT procedure and the SAS data set *demo.beef3* produce a scatter plot of CWT versus DMI for Producer 3 only. Use black-filled circles as the plotting symbols. Reset all graphics options first.

```
GOPTIONS RESET=ALL;
PROC GPLOT DATA=demo.beef3;
WHERE producer = 3;
PLOT cwt * dmi ;
SYMBOL1 VALUE=DOT CV=BLACK I=NONE;
TITLE 'Objective 17.5';
RUN;
```

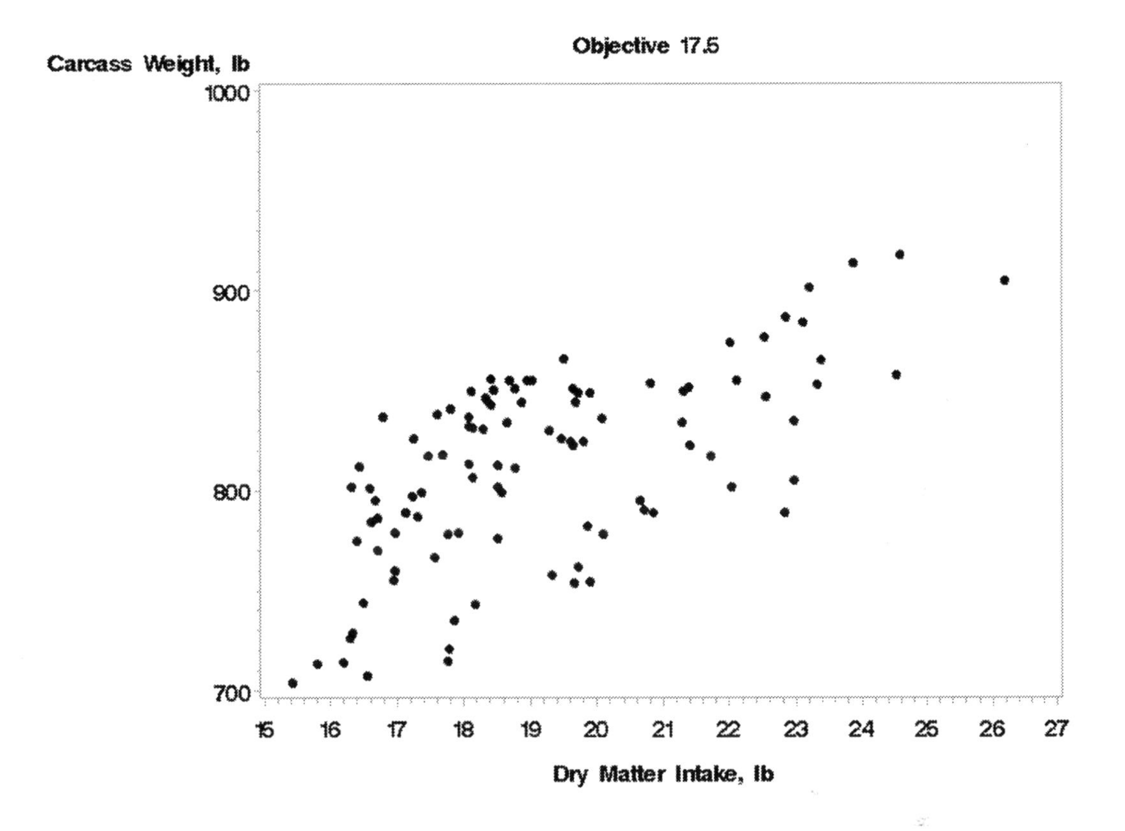

This objective is the same as Objective 16.8 Option 1 using the SGPLOT procedure. For Options 2 and 3 of Objective 16.8 one should see Section 17.3.4 X, Y Plot by Category.

17.3.3 AXIS Statements

AXIS statements are numbered statements that control the appearance of axes in the GPLOT and GCONTOUR (Section 17.5) procedures. AXIS statements are global statements like the TITLE, PATTERN, and SYMBOL statements. An AXIS statement can specify major and minor tick marks, the position of the axis label, and even axis color. There are many options for the AXIS statement. A few specific options are given in Table 17.2.

TABLE 17.2

AXIS Statement Options

Options	Explanation
LABEL = ("*text for label*", A= *angle*) ;	Text for a variable can be included in the AXIS statement if the variable has not been labelled in a DATA step. The label on an axis can be rotated counter-clockwise by the specified *angle* (in degrees).
ORDER = (*values*) or ORDER = (*low to high* BY *increment*)	Values are listed in ascending order with a space delimiter or in descending order with a comma delimiter. Increment is assumed to be 1 unless specified.
COLOR = *color*	Use SAS color names, Section 17.1.

Advanced AXIS options allow one to set logarithmic scales, set the origin, change fonts, change the size of the font, and many more enhancements.

The syntax for the AXIS statement is:

```
AXIS<1 … 99> list options;
```

There can be multiple AXIS statements which is why they are indexed, 1 – 99. On the PLOT statement of the GPLOT procedure, the HAXIS and VAXIS options select the appropriate AXISn definition, such as:

```
PLOT plot-requests/VAXIS = axisn HAXIS = axism;
```

Since AXIS statements are global, they are held in SAS memory until they are reset or rewritten.

17.3.4 (X, Y) Plot by Category

An (X, Y) plot by category is a two-dimensional line plot or scatter plot where information about a third category variable is identified by different plotting symbols (VALUE =) and/or line styles (L = #) using one SYMBOLn statement for each value of the third variable.

OBJECTIVE 17.6: Using the GPLOT procedure and the SAS data set *demo.beef3* produce a scatter plot of CWT versus DMI for Producer 3 only. Rotate the label on the vertical axis 90 degrees, and change the tick marks on the vertical axis to 700, 750, 800, 850, 900, 950, 1000.

Option 1: Use two different symbols to denote heifers and steers.

Option 2: Use only H or S as the plotting symbol to denote heifers and steers. Include both horizontal and vertical grid lines in the background of the plot.

```
AXIS1 ORDER = (700 TO 1000 BY 50) LABEL = (A=90);
PROC GPLOT DATA=demo.beef3;
PLOT cwt * dmi = sex / VAXIS = AXIS1 ;
SYMBOL1 VALUE=CIRCLE   CV=BLACK I=NONE; *heifers;
SYMBOL2 VALUE=TRIANGLE CV=BLACK I=NONE; *steers;
TITLE 'Objective 17.6 - Option 1';
RUN;
PLOT cwt * dmi = sex / VAXIS = AXIS1 GRID;
SYMBOL1 VALUE="H" CV=BLACK I=NONE;
SYMBOL2 VALUE="S" CV=BLACK I=NONE;
TITLE 'Objective 17.6 - Option 2';
RUN;
```

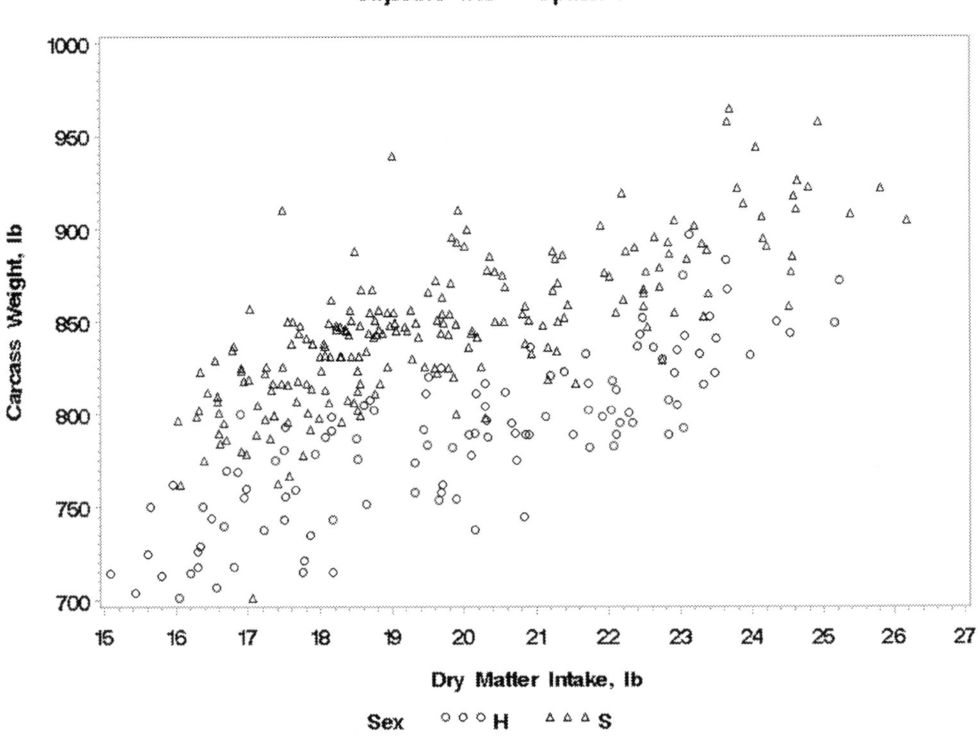
Objective 17.6 — Option 1
Carcass Weight, lb
1000
950
900
850
800
750
700
15 16 17 18 19 20 21 22 23 24 25 26 27
Dry Matter Intake, lb
Sex ○○○ H △△△ S

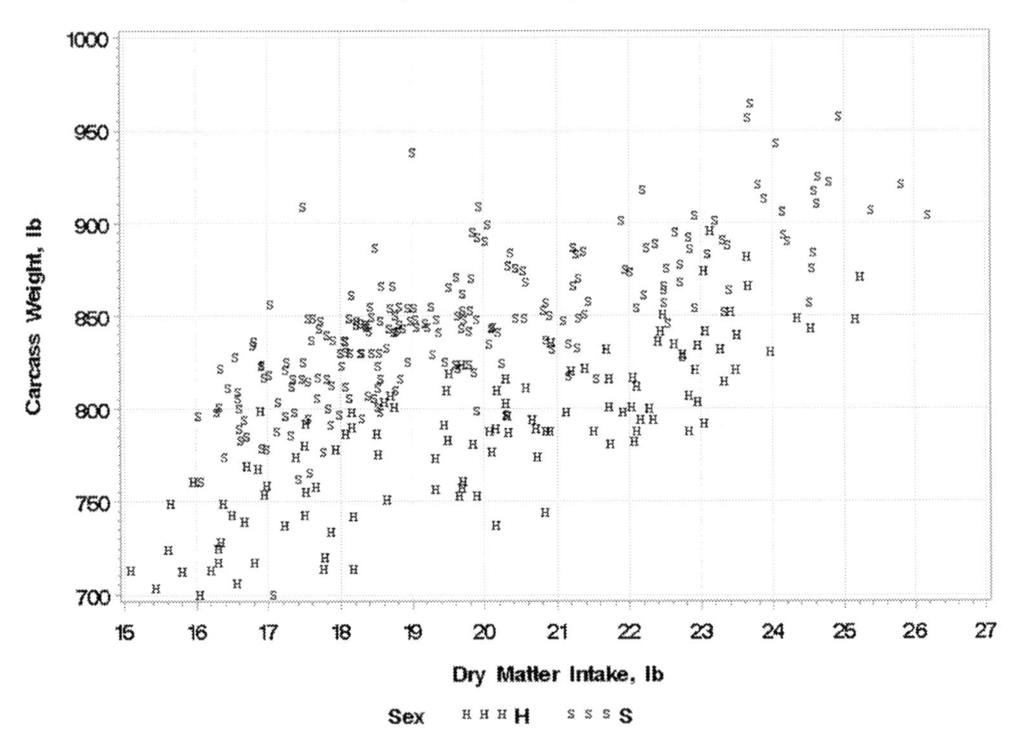
Objective 17.6 — Option 2
Carcass Weight, lb
1000
950
900
850
800
750
700
15 16 17 18 19 20 21 22 23 24 25 26 27
Dry Matter Intake, lb
Sex H H H H S S S S

In both of the options, rotating the vertical axis label 90 degrees (counter clockwise) improves the graph. If the vertical axis variable has a really long label, it can compress the requested plot if the label is not rotated. Additionally, the GRID option produced the background reference lines. Specifying both AUTOHREF and AUTOVREF would achieve this same effect.

OBJECTIVE 17.7: Redo Objective 16.9 using the GPLOT procedure. That is, for the SAS data set *beef3means* from Section 14.3 plot the mean CWT versus Producer identifying the variable Sex as the group variable. Use a line or series plot. This is referred to as an interaction plot for the response variable CWT and the two classification variables Producer and Sex.

```
PROC GPLOT DATA=demo.beef3means;
WHERE _STAT_ = "MEAN";
PLOT cwt * producer = sex / VAXIS = AXIS1 ;
SYMBOL1 VALUE = DOT CV = BLACK I=JOIN L=1 ;
SYMBOL2 VALUE = NONE C = BLACK I=JOIN L=3 ;
TITLE 'Objective 17.7';
RUN;
```

To produce a line plot, plotting symbols did not have to be chosen (VALUE=NONE), but the distinction between the groups may be clearer with the addition of a plotting symbol for at least one of the categories.

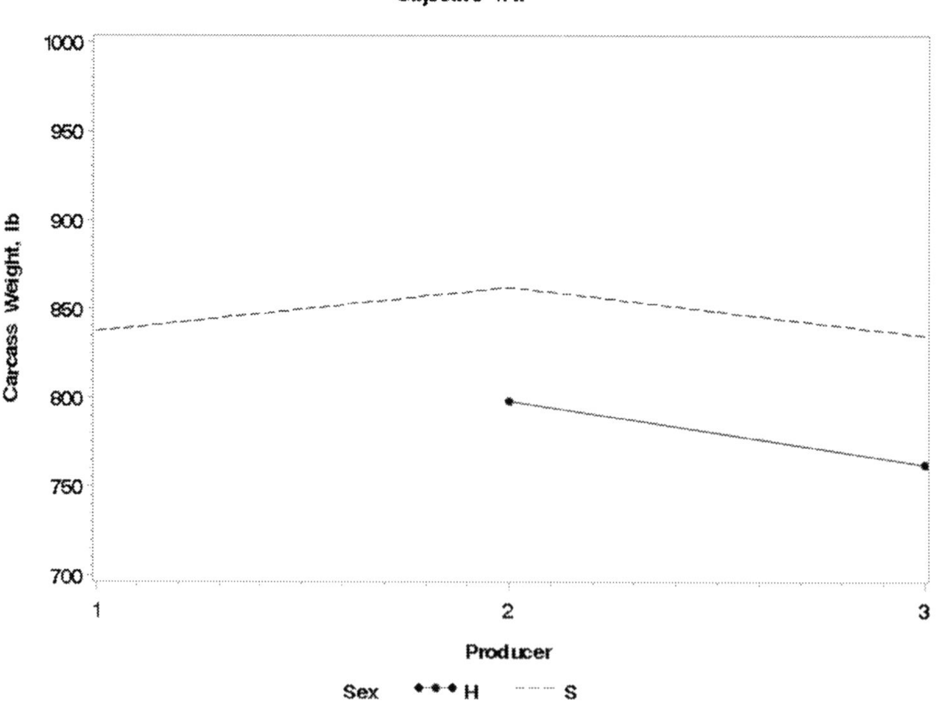

17.3.5 LEGEND Statements

In the plots of Objectives 17.6 and 17.7, a legend was produced in the area outside of the graph and centered below the horizontal axis. In the options for the PLOT statement,

TABLE 17.3

LEGEND Statement Options

<table>
<tr><th>Options</th><th>Explanation</th></tr>
<tr><td>LABEL = (“text for label”) | NONE</td><td>Text for a variable can be included in the LEGEND statement if the variable has not been labelled in a DATA step. More than one word can be used as a LABEL. LABEL=NONE suppresses all labeling of the axis.</td></tr>
<tr><td>POSITION =
(<BOTTOM | MIDDLE | TOP>
< LEFT | CENTER | RIGHT>
<INSIDE | OUTSIDE>)</td><td>For each < A | B | C > option select at most one item from those listed. The default position for the legend is
POSITION=(BOTTOM CENTER OUTSIDE)</td></tr>
<tr><td>ACROSS = number of columns</td><td>If the legend needs to be customized to fit in available space or would appear better with a different column assignment across the graph, then the number of columns needed can be specified. ACROSS can be used alone or with the DOWN option.</td></tr>
<tr><td>DOWN = number of rows</td><td>Specify the number of rows in the legend that would enhance the produced image. DOWN can be used alone or with the ACROSS option.</td></tr>
</table>

NOLEGEND is an option though there are few cases where that would be recommended. Like the AXIS statements, multiple forms for the legend can also be defined. These defined legends are called into the plot in the PLOT statement using the LEGEND = option. The basic syntax for the LEGEND statement is:

```
LEGENDn POSITION = ( specify ) <options> ;
```

The LEGENDn statement(s) where n = 1, …, 99 are global statements in SAS. Legend statement options are found in Table 17.3.

Here are a few examples of LEGEND statements.

```
LEGEND1          POSITION = (TOP INSIDE) LABEL=NONE;
LEGEND2          POSITION = (TOP RIGHT OUTSIDE);
LEGEND3          POSITION = (TOP RIGHT OUTSIDE) ACROSS=1;
LEGEND4          POSITION = (LEFT);
LEGEND5          POSITION = (LEFT) DOWN=4 LABEL=("Gender");
```

LEGEND1 puts the legend at the top of the plot inside the frame of the plot. If LEFT, CENTER, or RIGHT is not specified, the default setting, CENTER, will be used. Similarly, LEGEND4 and LEGEND5 position the legend on the left side of the plot. The default settings of BOTTOM OUTSIDE are assumed since all three position settings were not used in the LEGENDn statements.

OBJECTIVE 17.8: Modify Objective 17.7 by defining the legends *before* the GPLOT procedure. Examine the effects of one or more of the legends. Change to a larger Arial font on the graphs. Since the means for CWT are much smaller than 1000 in Objectives 17.6 and 17.7, create a new AXIS statement where the maximum is 875 pounds.

```
LEGEND1          POSITION = (TOP INSIDE) LABEL=NONE;
LEGEND2          POSITION = ( TOP RIGHT OUTSIDE ) ;
LEGEND3          POSITION = ( TOP RIGHT INSIDE ) ACROSS=1;
LEGEND4          POSITION = (LEFT) ;
LEGEND5          POSITION = (LEFT) DOWN=4 LABEL=("Gender");
```

```
GOPTIONS FTEXT="Arial" HTEXT=1.5 CTEXT=BLACK;
AXIS2 ORDER = (700 TO 875 BY 25);
PROC GPLOT DATA=demo.beef3means;
WHERE _STAT_ = "MEAN";
PLOT cwt * producer = sex / VAXIS = AXIS2 LEGEND=LEGEND3;
SYMBOL1 VALUE = DOT CV = BLACK I=JOIN L=1 ;
SYMBOL2 VALUE = NONE C = BLACK I=JOIN L=3 ;
TITLE 'Objective 17.8 - Legend 3';
RUN;
PLOT cwt * producer = sex / VAXIS = AXIS2 LEGEND=LEGEND5;
TITLE 'Objective 17.8 - Legend 5';
RUN;
```

Programming notes: Only two of the legends are illustrated in this objective, though any or all of them could have been. In LEGEND5 the DOWN=4 option specifies more than the two lines needed for the legend. SAS budgets space for four lines, but only two lines are used. The legend appears "two lines higher" than if the DOWN=2 option was used. When the legend appears outside of the plot and to either the right or the left of the plot, the plot does become compressed as it makes space for the legend. A lengthy label for the variable would compress the plot even more.

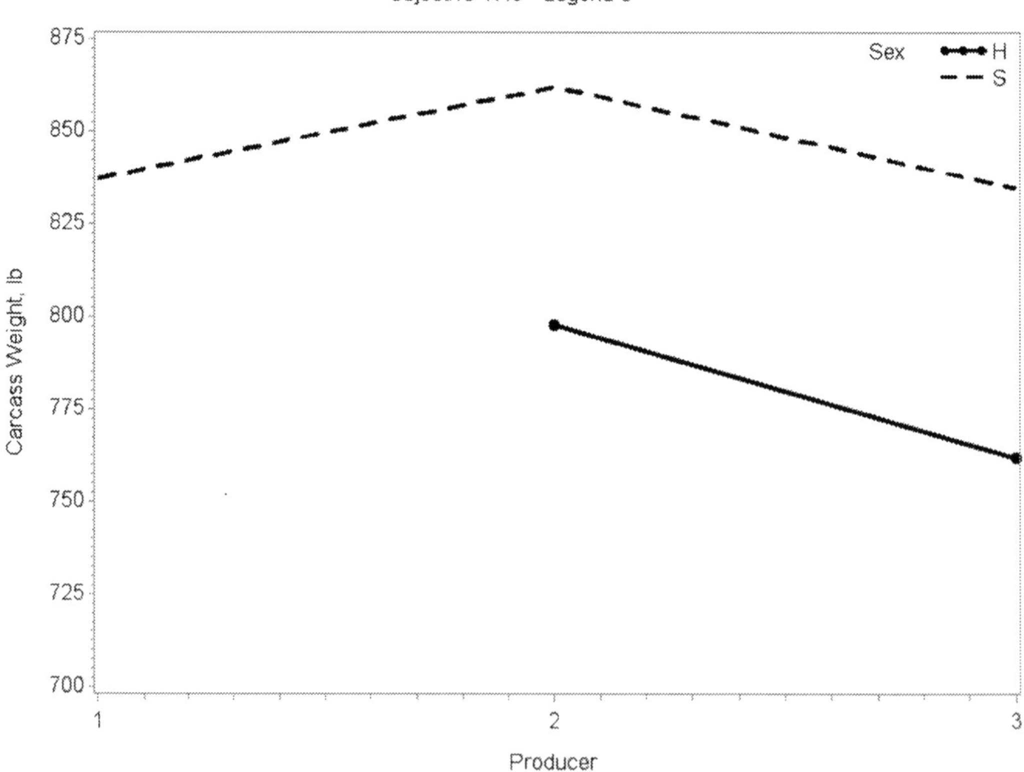

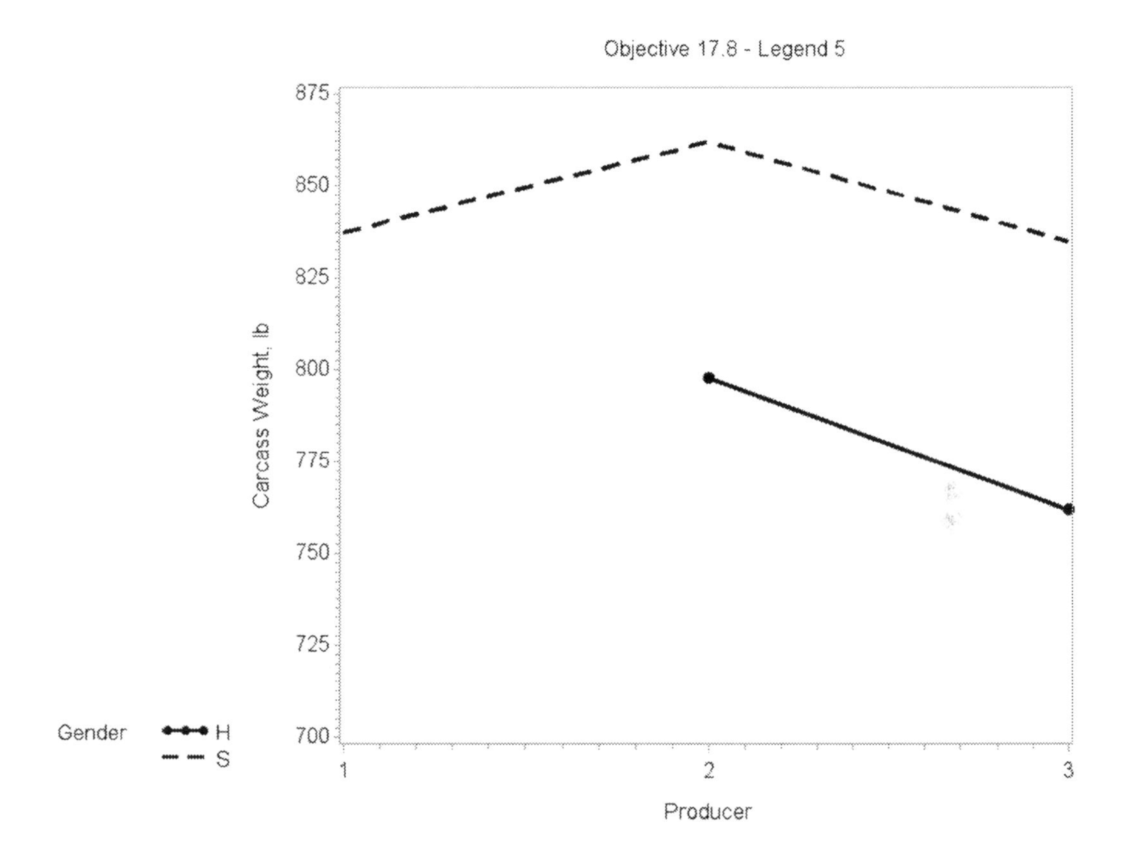

17.3.6 Multiple Plots per Axes

Multiple plots per axes refers to multiple variables represented by the same axes; vertical, horizontal, or both. Consider only one variable on the horizontal axis first. This can be done in two different ways.

First, if the scale or range of variables on this vertical axis is somewhat the same, this can be done in a single PLOT statement. The multiple axis variables are specified in parenthesis and the OVERLAY option is also needed, such as

```
PLOT (variable1 variable2 … variablen) * x/OVERLAY LEGEND=legendn;
SYMBOL1 <options>;
⋮
SYMBOLn <options>;
```

A SYMBOLn statement for each vertical axis variable is needed using this syntax. SYMBOL1 corresponds to variable1, SYMBOL2 corresponds to variable2, and so on. The options on the SYMBOLn statements do not have to be the same. For example, one variable can be plotted as a line graph, and another can be a scatter plot. The OVERLAY option on the PLOT statement is required, or each vertical axis variable will appear on a separate set of axes. A legend is recommended for this plot type.

If a second vertical axis is to appear on the right side of the plot, there is a PLOT2 statement with the same syntax and options as the PLOT statement, and PLOT2 will overlay the images and create a second vertical axis. PLOT2 follows PLOT in the same GPLOT procedure. A legend for either the PLOT or PLOT2 statements or both may be used to distinguish plotting symbols.

OBJECTIVE 17.9: Using the SAS data set *demo.beef3* overlay the scatterplots of ADG by REA and DMI by REA for Producer 3. Do this in two ways. Reset all graphics options. Select ZAPF text.

1. Use a single PLOT statement with the OVERLAY option.
2. Use the PLOT and PLOT2 statements. Rotate the vertical axis labels on both axes.

```
GOPTIONS RESET=ALL FTEXT=ZAPF;
LEGEND1 POSITION = (TOP INSIDE) LABEL=NONE;
AXIS3 LABEL=NONE;
AXIS4 LABEL=(A=90) ;
AXIS5 LABEL=(A=270);
PROC GPLOT DATA=demo.beef3;
WHERE producer = 3 ;
PLOT (adg dmi) * rea / OVERLAY LEGEND=LEGEND1 VAXIS=AXIS3;
SYMBOL1 VALUE=CIRCLE CV=BLACK I=NONE;
SYMBOL2 VALUE=DOT    CV=BLACK I=NONE;
TITLE 'Objective 17.9 - OVERLAY Option';
RUN;
PLOT adg * rea / VAXIS=AXIS4 LEGEND=LEGEND6;
PLOT2 dmi * rea / LEGEND=LEGEND1 VAXIS=AXIS5;
TITLE 'Objective 17.9 - PLOT2 Statement';
RUN;
```

In the first PLOT statement, AXIS3 suppresses the label on the vertical axis otherwise the first vertical axis variable listed (ADG in this instance) will be used to label the vertical axis. In the second plot, both PLOT and PLOT2 have different legends. One can experiment with the placement of these for the best effect.

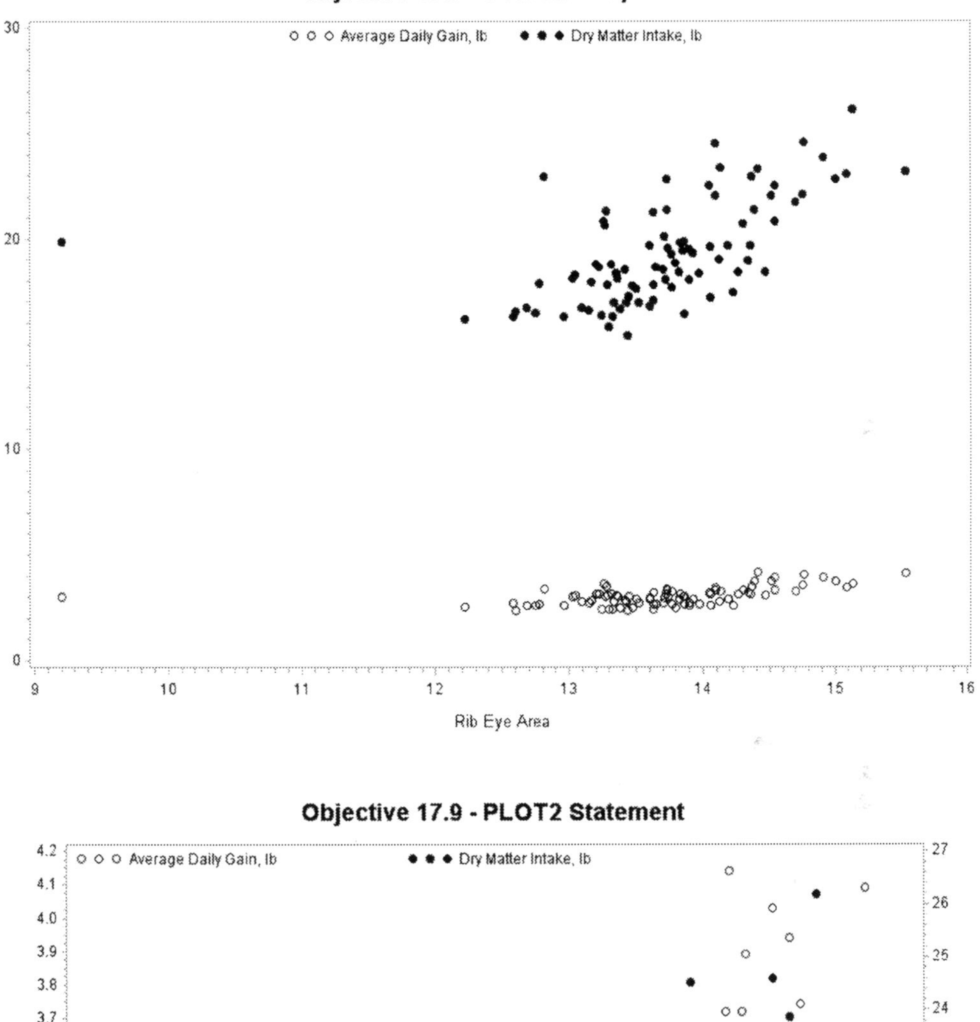
Objective 17.9 - OVERLAY Option
Average Daily Gain, lb
Dry Matter Intake, lb
Rib Eye Area
0
10
20
30
9
10
11
12
13
14
15
16

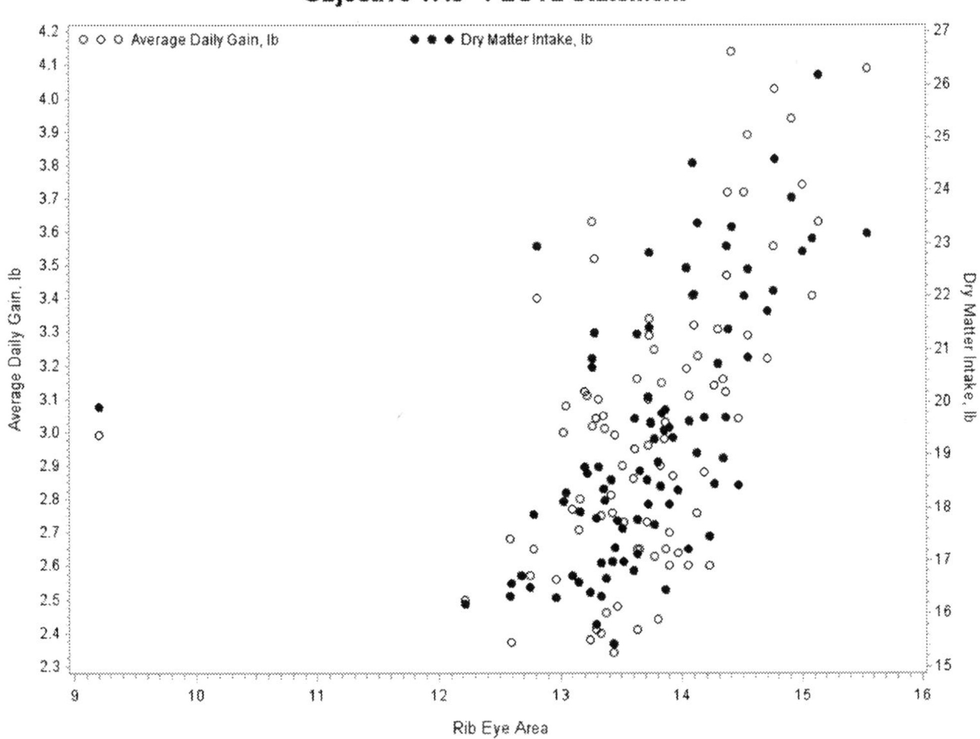
Objective 17.9 - PLOT2 Statement
Average Daily Gain, lb
Dry Matter Intake, lb
Rib Eye Area

In both types of plots, the choices for the legends and vertical axes greatly enhance the plots. Plotting data is one way of checking the data for possible errors also. The REA value close to 9 seems like an outlier. In Section 20.3, other methods of identifying outliers and/or possible data errors will be discussed.

17.4 The G3D Procedure

SAS/GRAPH also has a three-dimensional plotting capability in an (X, Y, Z) coordinate system. Either a scatter plot of the data or a plot of a three-dimensional surface can be requested.

The SAS procedure for this is the G3D procedure.

The syntax of the G3D procedure for a scatter plot is:

PROC G3D DATA=*SAS-data-set;*
SCATTER y * x = z </options>;

The SCATTER statement allows one to produce a three-dimension scatter plot of data. The "z" variable is plotted on the vertical axis, and the (X, Y) plane is the base of the image. Options on the SCATTER statement allow one to select a plotting shape, plotting color, and more.

One or more options on the SCATTER statement that control the plot appearance are:

COLOR="*color*" Stick to the color basics used so far: RED, BLUE, ORANGE, BLACK, GREEN, GRAY, PURPLE, PINK, YELLOW and BROWN. All points plotted will be the color specified. Quotation marks are required, such as COLOR="BLUE".

COLOR=*color-variable* allows for a character variable in the SAS data set to identify multiple colors in a single three-dimensional scatterplot.

NONEEDLE By default G3D draws a needle from the plotting symbol "down" to the x*y plane. NONEEDLE suppresses this.

SHAPE="*shape*" identifies the plotting shape used in the scatterplot. Shapes include: BALLOON, CLUB, CROSS, CUBE, CYLINDER, DIAMOND, FLAG, HEART, PILLAR, POINT, PRISM, PYRAMID, SPADE, SQUARE, and STAR. The NONEEDLE option is available for all symbols except PILLAR and PRISM. All points plotted will be the shape specified. Quotation marks are required, such as SHAPE="CUBE". These SHAPES are different from the values used in the SYMBOL statements of the GPLOT and GCONTOUR procedures.

SHAPE=*symbol-variable* allows for a character variable in the SAS data set to identify multiple symbols on a single three-dimensional scatterplot.

SIZE=*symbol-size* identifies a numeric value which determines the size of all plotting symbols on the graph.

SIZE=*size-variable* allows for a character variable in the SAS data set to identify multiple sizes for plotting symbols on a single scatterplot.

ANNOTATE, ROTATE, TILT are other appearance options that can be specified.

Options that modify one or more of the axes are:

CAXIS=*color* specifies the color of axis lines, tick marks, and grid lines.

CTEXT=*color* specifies the color for text on the axes.

GRID draws reference lines from each tick mark on each axis.

XTICKNUM=*number* specifies the number of major tick marks that are located on a plot's X axis. The default value is 4.

YTICKNUM=*number* specifies the number of major tick marks that are located on a plot's Y axis. The default value is 4.

ZTICKNUM=*number* specifies the number of major tick marks that are located on a plot's Z axis. The default value is 4.

ZMAX=*value* specifies the maximum data value that is displayed on a plot's Z axis. The ZMAX= option can be used to extend the Z axis beyond the observed range. The value that is specified by the ZMAX= option must be greater than that specified by the ZMIN= option. If the ZMAX= option specified is within the range of the z-variable values, the plot's data values are truncated at the ZMAX level specified. This results in a flattening at the largest values of a surface. The default is the observed maximum value of the z-variable. There is no similar maximum option for either the X or Y axes.

ZMIN=*value* specifies the minimum value that is displayed on a plot's Z axis. Defining the ZMIN= value less than the minimum value in the input data set extends the plot's Z axis. Defining the ZMIN= value greater than the minimum value in the input data set displays all Z values in the range of ZMIN-to-ZMAX, and might cause data clipping or flattening of the surface at the ZMIN value. The default is the observed minimum value of the z-variable.

Unfortunately PROC G3D does not support the global AXIS statements in the current version of SAS.

OBJECTIVE 17.10: Use the *demo.beef3* SAS data set for the heifers from Producer 3. Create a three-dimensional scatter plot of CWT, ADG, and DMI where CWT is the z-variable. Use the default plotting symbol and reset all graphing options.

```
GOPTIONS RESET=ALL;
PROC G3D DATA=demo.beef3 ;
WHERE producer = 3 and sex="H";
SCATTER adg * dmi = cwt;
TITLE 'Objective 17.10 - Default G3D Scatter Plot';
RUN;
```

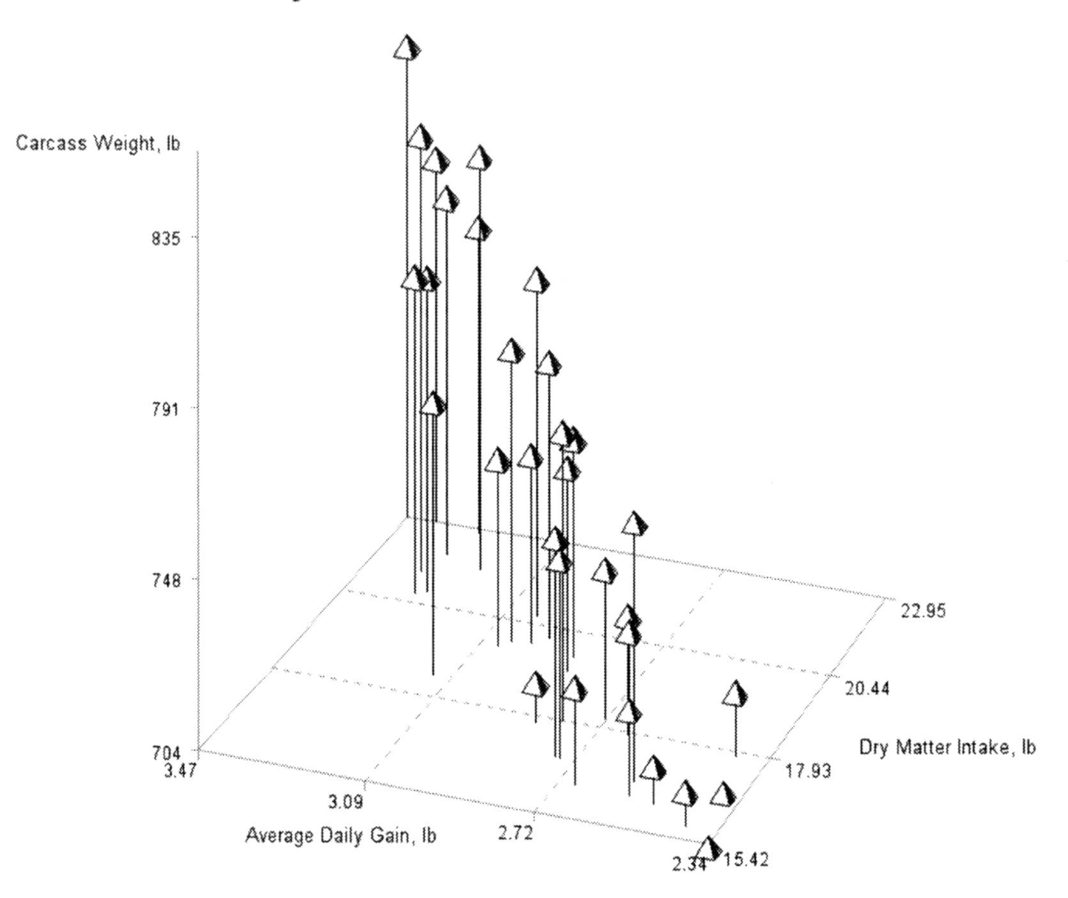

When using the G3D procedure the default symbol is the pyramid and the "needles" are enabled. The use of the "needles" assists with the visualization of the three-dimensional image linking the z-variable to points in the (X, Y) plane. One should also identify the ADG y-axis and the DMI x-axis. In the lower right-hand corner are the minimums for the x- and y-variables, and the minimum value for the z-axis is determined by the smallest observed value of z, 704. In this plot only the heifers were used since the number of observations was smaller. Three-dimensional plots can get very congested very quickly.

OBJECTIVE 17.11: Modify Objective 17.10 by setting the tick marks on the z-axis to be consistent with earlier objectives. That is, set the tick marks at 700 to 1000 in 50-unit increments. Include grid lines on all of the axes. Change the plotting symbol to gray pillars. Produce two plots, one for heifers and one for steers from Producer 3.

```
PROC SORT DATA=demo.beef3;
BY sex;
PROC G3D DATA=demo.beef3 ;
WHERE producer = 3;
BY sex;
SCATTER adg * dmi = cwt / ZMIN=700 ZMAX=1000 ZTICKNUM=7 GRID
                                    SHAPE="PILLAR"   COLOR="GRAY";
TITLE 'Objective 17.11';
RUN;
```

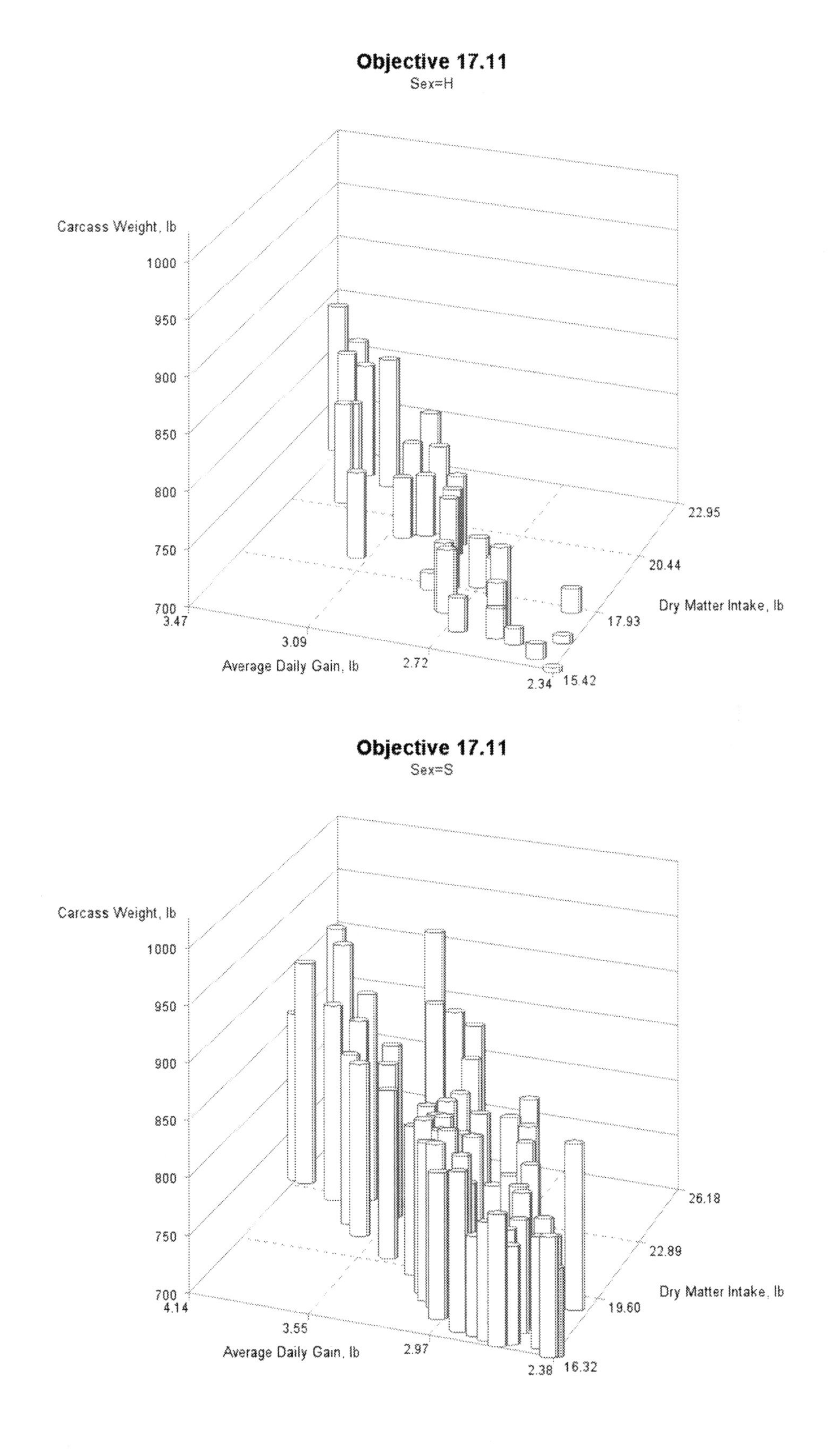
Objective 17.11
Sex=H
Carcass Weight, lb
1000
950
900
850
800
750
700
3.47
3.09
2.72
2.34
Average Daily Gain, lb
15.42
17.93
20.44
22.95
Dry Matter Intake, lb
Objective 17.11
Sex=S
Carcass Weight, lb
1000
950
900
850
800
750
700
4.14
3.55
2.97
2.38
Average Daily Gain, lb
16.32
19.60
22.89
26.18
Dry Matter Intake, lb

Both the PILLAR and PRISM shapes produce the column-like appearance in the graphs. The GRID option (in the author's opinion) aids in the perception of a three-dimensional plot. By keeping the z-axis the same for both heifers and steers, the two groups can be compared more easily.

Partially due to the choice of the PILLAR plotting shape in addition to the larger number of observations of steers to be plotted, the second plot is very congested. So, for very large numbers of observations, the G3D procedure can be used, but one may wish to find reasonable subgroups to plot for more legible plots. In addition to the relationship each variable has with CWT, one can see in both plots the positive association between ADG and DMI in the (X, Y) plane.

The G3D procedure can also be used to plot a three-dimensional surface. A PLOT statement is used rather than a SCATTER statement. In order to graph a surface, one needs many (X, Y) coordinates at which z values are determined. Smooth surfaces are seldom possible for observational data. Surface plots are typically generated for a function in x and y.

Though not always required, one should run a G3GRID procedure prior to generating a surface plot in G3D. The G3GRID procedure places the data in an order that the G3D procedure can more efficiently use to produce a surface. G3GRID does not generate any printed output but simply works on the SAS data set that is to be plotted. The GRID statement in G3GRID should match the PLOT statement of G3D. The simple syntax of these procedures is:

```
PROC G3GRID DATA=SAS-data-set;
GRID y * x = z;

PROC G3D DATA=SAS-data-set;
PLOT y * x = z </options>;
RUN;
```

One or more options on the PLOT statement that control the plot appearance are:

CBOTTOM=color SAS has a default color for the bottom of the surface, so this is optional. Use the SAS color names specified in Section 17.1.

CTOP=color This option selects a color for the top of the surface. See CBOTTOM description.

SIDE requests a side wall of the surface.

XYTYPE=0 | 1 | 2 | 3 Pick one. The plotted surface is represented by lines drawn parallel to one or both the x-axis and y-axis. One can specify the direction of these lines. Both X and Y are displayed by default. The valid values for the XYTYPE= option are as follows:

1. XYTYPE=0 (applicable to Java and ActiveX file types only) No lines are displayed. The plot is displayed as a solid surface.
2. XYTYPE=1 draws lines that are parallel to the x-axis. (1 implies parallel to X, the first letter of "XYTYPE".)
3. XYTYPE=2 draws lines that are parallel to the y-axis. (2 implies parallel to Y, the second letter of "XYTYPE".)
4. XYTYPE=3 draws lines that are parallel to both the x- and y-axes. This is the default setting.

The axis options specified for the SCATTER statement are applicable to the PLOT statement also. (CAXIS=, CTEXT=, GRID, XTICKNUM, YTICKNUM, ZTICKNUM, ZMIN, ZMAX).

OBJECTIVE 17.12: Produce a surface plot for the equation: $z = x^3 + 3xy^2 + 3y^2 - 15x$. To do this, a SAS data set containing many (X, Y, Z) points will first need to be generated. This is done using nested DO loops (Section 15.1). The x and y roles may need to be switched in the G3D procedure to produce the preferred image.

```
DATA twelve;
x = -2.5;
y = -1;
  DO x = -2.5 TO 2.5 BY 0.1;
      DO y = -1 TO 1 BY 0.1;
            z = x*x*x + 3*x*y*y + 3*y*y - 15*x;
            OUTPUT;
      END;
  END;
RUN;
PROC G3GRID DATA=twelve;
GRID y * x = z;

PROC G3D DATA=twelve;
PLOT y * x = z / GRID;
TITLE 'Objective 17.12 - View 1';
RUN;

PROC G3GRID DATA=twelve;
GRID x * y = z;

PROC G3D DATA=twelve;
PLOT x * y = z / GRID XYTYPE=2
                 XTICKNUM=6
                 ZMIN=-25 ZMAX=25 ZTICKNUM=11;
TITLE 'Objective 17.12 - View 2';
RUN;
```

Caution: If the increment (0.1) is too small for the selected starting and stopping values, the DATA step will take a very long time to run. When running nested DO loops for two variables, the number of (X, Y) pairs produced is the product of the number of x values times the number of y values in each of the DO statements. In SAS data set *twelve* above, there are $51 \times 21 = 1071$ points on the (X, Y) plane for which a z value is determined. After running the DATA step generating the values to be plotted, the SAS log will indicate how many observations or lines are in the DATA set, as shown here.

```
NOTE: The data set WORK.TWELVE has 1071 observations and 3 variables.
```

Objective 17.12 - View 1

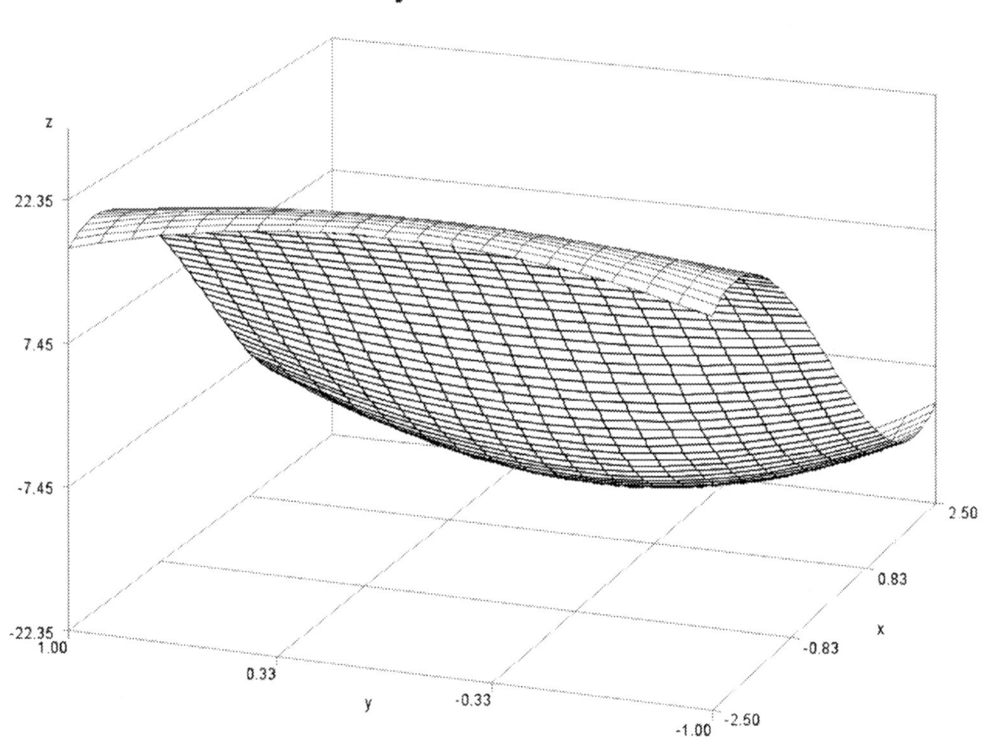

Objective 17.12 - View 2

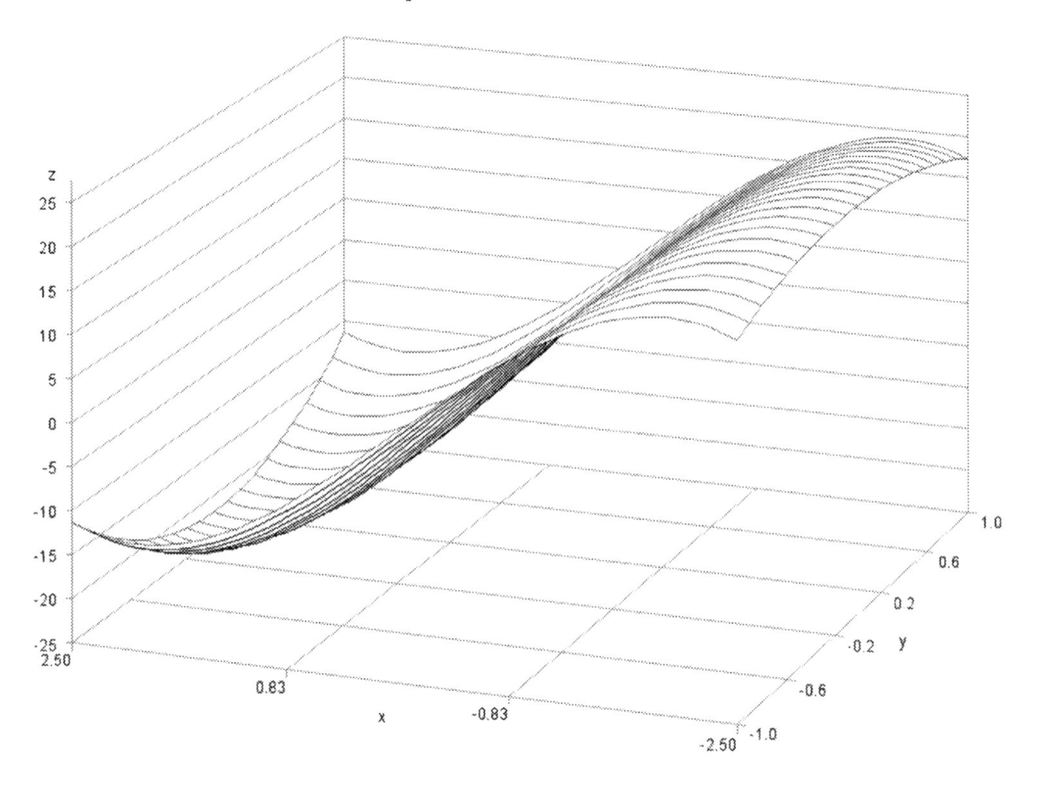

If there are too few data points generated, the image will not look smooth or not look complete. One will need to examine the function to be graphed to determine logical starting values, stopping values, and the increment for both X and Y. These values do not have to be symmetric about 0 nor does the increment need to be the same for both variables. That was just simply the case for the equation in this objective.

In View 1 the default surface pattern is XYTYPE=3. The spacing of the "mesh" on the surface corresponds with the increment size chosen in the DO loops. The GRID option is the only option selected in this view. The orientation of the graph is determined by the x and y roles in the syntax. The default setting is for four tick marks on each axis.

In View 2 the roles of x and y have been switched, thereby changing the orientation of the surface plot. This view also illustrates the modification of two of the axes as well as the XYTYPE option.

17.5 The GCONTOUR Procedure

Frequently, contour plots are presented with or instead of surface plots. These are generated by the GCONTOUR procedure. The simple syntax of the GCONTOUR procedure is:

```
PROC GCONTOUR DATA=SAS-data-set;
PLOT y * x = z </options>;
RUN;
```

The PLOT statement options for the GCONTOUR procedure are in Table 17.4.

There are other advanced settings that will produce solid color areas for the contours, but those methods are not overviewed here.

OBJECTIVE 17.13: Produce a contour plot for the three-dimensional surface in View 2 of Objective 17.12. Place the numeric label on the contour and suppress the legend. Use a solid

TABLE 17.4

PLOT Statement Options for the GCONTOUR Procedure

Options	Explanation
CLEVELS=color(s)	Specify the colors of the contour lines using SAS color names.
LEGEND=LEGEND<1...99> | NOLEGEND	Pick one. NOLEGEND suppresses the legend. See Section 17.3.5 for LEGEND syntax.
LEVELS=value-list	Specify z values for the contour lines, such as: LEVELS = 10 20 30 40 50. If none are specified, SAS will choose the levels.
LLEVELS=line-type-list	Specify the line styles (1 – 46) for each contour. See also Section 17.2.1.
NLEVELS=number-of-levels	If no list of values is specified for contours, NLEVELS can be set for a specific number of contours. SAS will pick the values.
AUTOLABEL	Displays the z value on the contours in the plot. This is a good option when NOLEGEND is requested.
XAXIS = *AXISn* and YAXIS = *AXISn*	Global AXIS statements (Section 17.3.3) are supported by the GCONTOUR procedure.

black line for each contour selected. To do this use a SYMBOL statement with a REPEAT = n option (not covered in Section 17.3.1), where n is the number of contours chosen.

```
PROC GCONTOUR DATA=eleven;
PLOT x * y = z / AUTOLABEL NOLEGEND
                 LEVELS = -20 -15 -10 -5 0 5 10 15 20;
SYMBOL1 I=JOIN C=BLACK L=1 REPEAT= 9;
TITLE 'Objective 17.13 - Contour Plot for View 2';
RUN;
```

REPEAT = 9 was chosen since there were nine values in the LEVELS = option. NLEVELS = 9 could have been chosen, but SAS would pick the levels. NLEVELS = is a reasonable option if one is unsure of the range of z values in the data.

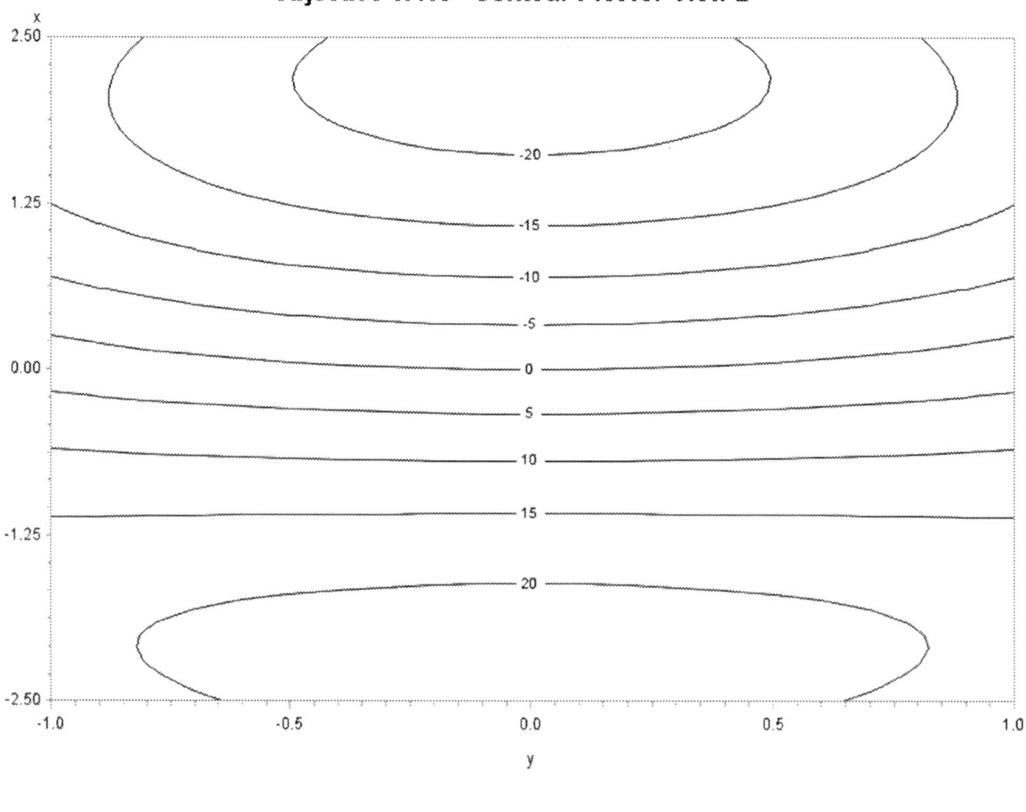

17.6 Saving Graph Images

SAS/GRAPH images appear in the Results Viewer when HTML output is enabled. These images can be saved by right clicking on the image and selecting **Save picture as ...** . The options are to save the image as a .PNG or .BMP image. This information also appeared in Section 5.2.

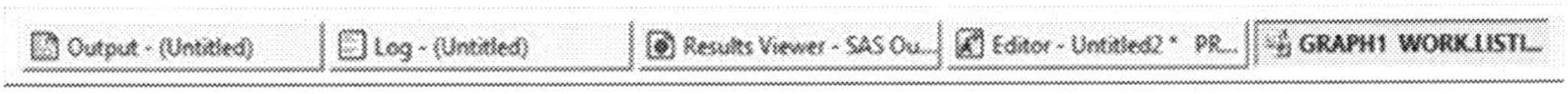

FIGURE 17.2
The GRAPH window that displays SAS/GRAPH when LISTING is enabled.

Additionally, SAS/GRAPH images can also be produced in a Graph window in SAS. To do this, the LISTING Output must be enabled and is done so by inserting

```
ODS LISTING;
```

in a statement before the SAS/GRAPH procedure. ODS LISTING; is a global command (Section 19.1). Once enabled, both the LISTING Output and Graph windows will be enabled until the statement

```
ODS LISTING CLOSE;
```

is submitted, or the SAS session closes.

Once the program is submitted while the Listing Output is enabled, there is an addition of the GRAPH window bar at the bottom of the SAS screen (Figure 17.2).

If the HTML output is still enabled, the image will still appear in the Results Viewer. Selecting the GRAPH window bar at the bottom of the SAS screen will display the image in this new window. When the Listing Output is enabled, the SAS/GRAPH images appear sequentially in the GRAPH window. One can page up and down within the GRAPH window to view created images. Any single image can be saved as an external file.

When a graph is produced in the GRAPH window, it can be edited or saved as a graphics file. Selecting **File – Open Graph** from the pull-down menu or the (right click) pop-up menu will initiate the dialog for editing a graph. Using the SAS Graphics Editor is beyond the scope of this book, but one can *see SAS/GRAPH® 9.4: Reference, Fifth Edition* (SAS Institute, 2016c) for more information. Saving the image for future use is the more likely task needed. To save the image active in the GRAPH window, select **File – Export as Image** from either the pull-down menu or the (right click) pop-up menu. The active image in the GRAPH window can be saved by a user-chosen file name and file type specified. Files types include: bitmap (.bmp), GIF, JPEG, TIF, PS, and PNG and more. There is also a **File – Save as Image** selection available, but this selection requires knowledge of SAS graphics catalogs which are not covered in this book. Thus, the **File – Export as Image** approach is recommended.

Using either of these approaches, these images can then be imported into other word processing and presentation software.

17.7 Chapter Summary

One of the initial motivations for this chapter and the previous chapter was the comment of a researcher who said, "I'll make my graphs in *<other software mentioned>* since SAS doesn't make graphs." The challenge then was to present to the reader SAS/GRAPH procedures in a clear and concise way so that not only is that person convinced that SAS *does* make graphs, but the graphs can be produced with relative ease. Selecting types of graphs, colors, fonts, symbols, and more can be quickly accomplished. The research data is in a SAS data set for the statistical analysis. Information from the analysis can be saved using the OUTPUT statement for analysis procedures or by using additional skills presented in Chapter 19 for creating SAS data sets. So, it makes sense to produce graphs supporting the research using SAS if the data and the analysis are in SAS.

18

Formatting Responses

In SAS there are formats and informats. Formats change how data values are printed in the analysis results or in graphs. Informats convert how the data are stored. In this book, only formats will be introduced. There are SAS formats, which are already defined, and user-defined formats. What does it mean to format responses? As an example, a variable in a SAS data set may have values 1 and 2. These values may be coded values for the responses of "Yes" and "No" to a survey question. With a user-defined format these 1 and 2 values can be printed as "Yes" and "No" in the results or graphs. Additionally, character string responses can be printed with more detailed character strings, such as the value "F" in the data may be printed as "Female". Specifying a given number of decimal places for a numeric variable is controlled by a format. Numeric values can also have templates applied, such as a 10-digit number can have a template for printing a phone number, such as 000-000-0000 or (000) 000-0000.

18.1 The FORMAT Procedure

The FORMAT procedure is used to create user-defined formats. These formats can be applied to the values of selected variables in many SAS procedures. This is done so the appearance of the printed results or SAS/GRAPH applications are improved. A format is different from the LABEL statement used in a DATA step (Section 8.2). LABEL allows the programmer to more fully define the variable name not the actual values the variable can take. The user-defined formats are active during the entire SAS session. When an analysis or graph procedure is used and a format is to be applied to the output, a FORMAT statement in the procedure is required.

The FORMAT procedure produces no printed output and multiple formats can be defined in a single FORMAT procedure. There are more advanced methods of creating formats from SAS data sets, but simple user-defined format creation is presented here.

The simple syntax of the FORMAT procedure is:

```
PROC FORMAT;
PICTURE          name range-1 = 'picture-1'  (<picture1 options>)
                      ⋮
                      range-n = 'picture-n'  (<picturen options>);
VALUE            name range-1 = 'formatted-value-1'
                      ⋮
                      range-n = 'formatted-value-n';
VALUE           $name 'range-1' = 'formatted-value-1'
                      ⋮
                      'range-n' = 'formatted-value-n';
```

There can be multiple PICTURE and/or VALUE statements in a single FORMAT procedure. To apply any format, either user-define or a SAS format, a FORMAT statement is needed. The FORMAT statement is covered in the next section.

PICTURE statement

This format is identified by the value of the required *name* in the PICTURE statement. The format *name* cannot begin or end with numbers. Thus, neither *2check* nor *check12* are valid format names. The PICTURE statement defines a template for printing numbers. *Range-1* through *range-n* specify the numeric values in the data for which the corresponding format will apply. Ranges can be written as: low-high, low – 100, 200 – high, 1-10, and so on. To encompass any and all numeric values of a variable, set *range-1* to OTHER or LOW-HIGH and only a single *range* is needed in the PICTURE statement. The "formatted-value" expressions define the template desired for the numeric responses.

Options for the PICTURE statement do not require a forward slash (/) separating the options from the rest of the statement, but the options are enclosed in parenthesis. A few PICTURE statement options for the beginning programmer are:

PREFIX="*character(s)*"

The PREFIX option is used to specify one or more characters to place in front of the formatted values of a variable. This is useful for placing text, such as, "A", "Chapter", or "Item # " to the left of the formatted values. One could also use PREFIX="$" for monetary values; however, there are existing SAS formats for monetary values that may be recommended over the usage of this prefix.

FILL="*character*"

The FILL option specifies a single value to fill the empty spaces in the formatted values. This is useful for maintaining leading zeros in the responses. FILL cannot be used in conjunction with PREFIX.

MULTIPLIER=*n*

The MULTIPLER option will multiply the numeric value of the variable by n before formatting.

ROUND

The ROUND option will round the number to the nearest integer before formatting.

A 10-digit phone number in the SAS data set (no spaces, hyphens, or other symbols) can be formatted as 000-000-0000 or "(000) 000-0000" using a PICTURE statement:

```
PICTURE phone      other = "000-000-0000";
```

or

```
PICTURE parphone      other = "(000) 000-0000";
```

Another example is U.S. ZIP + 4 postal codes which can be formatted using the statement:

```
PICTURE zip      other="00000+0000";
```

Zeros are typically used in defining these formats, but any digit 0 through 9 could be used. That is, `PICTURE zip other="12345+6789";` will also successfully define the *zip* format. One may need the FILL option for some U.S. ZIP codes that begin with zero:

```
PICTURE zip       other = "00000+0000" FILL = "0";
```

VALUE statement

This statement defines a format for numeric or character values. When the values of the range are numeric, the name of the format is given by *name*. The *name* of the format must begin with a letter and cannot end with a number. Ending the format name with a number is used to indicate the formatted width of a multiple digit number. When the values of the variable are character values, the format is given by *$name*, and the dollar sign "$" must be used in the FORMAT statement that calls this format into use. Then, more specific text or character expressions are set as the definition or "formatted value" in the VALUE statement. The next section gives examples of both types of VALUE statements.

18.2 The FORMAT Statement

The FORMAT procedure creates user-defined formats, and these are applied using a FORMAT statement in the syntax for a procedure. Most of the SAS procedures presented in this book can accommodate a FORMAT statement. Format names are typically followed by a period when they are called or applied in a FORMAT statement. The period syntax is not used when the format name is created in a PICTURE or VALUE statement of the FORMAT procedure.

The syntax of the FORMAT statement is:

FORMAT variable1 format1. - - variablen formatn.;

Not all variables in a procedure need to have a format applied to them. One format can also apply to more than one variable. If a format applies to several variables, each variable must be listed in the FORMAT statement followed by the format which will apply, such as:

```
FORMAT variable1 variable2 variable3 A.;
```

which applies the "A" format to all three variables.

Additionally, there are defined SAS formats that can be applied in a FORMAT statement. SAS formats do not have to be initialized in a FORMAT procedure. One of the simplest SAS formats, controls the number of decimal places. That is, one can specify the number of places a numeric value fills. This is referred to as the width of the number. The number of decimal places, d, can also be controlled. This is done by placing *w.d* after the variable(s) in the FORMAT statement where *w* identifies the width of the numeric value, and *d* identifies the number of decimal places. For example, `FORMAT response 5.2;`

will allow five spaces or a width of five for the *response* variable. Two of those spaces will be to the right of the decimal, one space is for the decimal, and the remaining two spaces are to the left of the decimal.

For U.S. monetary amounts, the SAS format DOLLARw.d prints the numeric value with a dollar sign prefix, d places to the right of the decimal (typically, d=2), and w is the total width of the field. The DOLLARw.d format also places a comma every three digits to the left of the decimal. DOLLARXw.d reverses the roles of the comma and decimal in monetary expressions, such as used in Europe. Similarly, there are EUROw.d and EUROXw.d formats for the monetary euro. The COMMAw.d format is the same as the DOLLARw.d format except there is no dollar sign prefix. There is also a SAS format for U.S. Social Security Numbers. It is SSN.

Both user-defined and SAS formats are illustrated in the following program. TITLE statements (Section 2.4), LABEL statement (Section 8.2), and line comments (Section 20.1) are used to annotate parts of the program and output. SAS data set *one* contains the following variables: identification (ID), experience (exp), salary in thousands of dollars, region of the United States, Gender, Yes/No responses for BS, MS, PhD degrees, and a Group classification of Engineering, Business, and Science.

```
DATA one;
INPUT ID exp Salary Region Gender BS MS PhD Group $;
salary2 = salary*1000; *convert salary in $thousands to dollars;
LABEL exp="Work Experience" salary="Salary ($thousands)"
                  region="Region of US";
DATALINES;
000171831 8.5   41.5  1  1 1 1 2 E
077889999 10    53.4  3  2 1 1 1 E
111223333 13    65.0  4  . 1 1 2 B
222334444 20    75.0  2  3 2 1 1 S
;
RUN;

PROC FORMAT;
   *** templates for the ID and Salary variables ***:
PICTURE a OTHER = '000-00-0000' (FILL="0");
PICTURE b OTHER = '000.00';
PICTURE c OTHER = '000000' (PREFIX='$');

   *** create salary intervals for original salary variable ***;
VALUE salfmt low-49.9 = 'Below 50,000'
             50.0-59.9 = '50,000 - 60,000'
             60-high   = '60,000 and over';

   *** Identify numerical regions to parts of the US ***;
VALUE regfmt 1 = 'Northwest'
             2 = 'Central'
             3-4= 'Southern';

...*** Note that a format can assist with missing or miscoded values***;
VALUE gen    1 = 'Male'
             2 = 'Female'
             . = 'Missing value'
             OTHER = 'Miscoded';
```

```
*** formats can translate numeric values into another language if needed ***;
VALUE degree 1 = 'Yes' 2 = 'No';
VALUE degrus 1 = 'Da'  2 = 'Nyet';   *Yes/No Russian;
VALUE ynspan 1 = 'Si'  2 = "No";   *Yes/No Spanish;

* $ is needed to create a format when the responses are character strings *;
VALUE $grp                  "E" = "Engineering"
                            "B" = "Business"
                            "S" = "Science";
RUN;
```

Programming Notes: Once the previous syntax is submitted, all of the picture, value, and character value formats are defined for the active SAS session. The "fmt" part of the salfmt and regfmt is a naming convention the author uses to assist in identifying user-defined formats in program syntax. Each PICTURE and VALUE statement is a single statement and is punctuated with a semicolon only after all of the definitions are created. Arranging the definitions or levels of the format in multiple lines as done with *salfmt, regfmt,* and *gen* formats is another of the author's preferences selected for ease of reading. These PICTURE and VALUE statements can continue on the same line as done for the formatting of the Yes and No responses with the *degree, degrus,* and *ynspan* formats. In addition to defining the formats, the syntax serves an annotation role for the programmer.

OBJECTIVE 18.1: Print the SAS data set *one* formatting the ID's with leading zeros and keeping two decimal places for the salary variable.

```
PROC PRINT DATA=one;
FORMAT id a. salary b.;
TITLE 'Objective 18.1';
RUN;
```

Objective 18.1

Obs	ID	exp	Salary	Region	Gender	BS	MS	PhD	Group	salary 2
1	000017-1831	8.5	41.50	1	1	1	1	2	E	41500
2	077-88-9999	10.0	53.40	3	2	1	1	1	E	53400
3	111-22-3333	13.0	65.00	4	.	1	1	2	B	65000
4	222-33-4444	20.0	75.00	2	3	2	1	1	S	75000

The *a* format for the ID requests that leading zeros be included. When there were three leading zeros (one adjacent to the – delimiter in the format) the delimiter was filled with a zero also, as seen in Observation 1.

OBJECTIVE 18.2: Modify the PRINT procedure in Objective 18.1. Use the SAS SSN format for the ID, *salfmt* for the salaries, and format all of the remaining variables selecting from the user-defined formats. Include the LABEL option on the PROC PRINT. Position *salary* and *salary2* so that they are in adjacent columns.

```
PROC PRINT DATA=one LABEL;
VAR id exp salary salary2 region gender bs ms phd group;
FORMAT id SSN. salary salfmt. salary2 c. region regfmt. gender gen.
       bs ms phd degree. group $grp.;
TITLE 'Objective 18.2';
RUN;
```

Objective 18.2

Obs	ID	Work Experience	Salary ($thousands)	salary 2	Region of US	Gender	BS	MS	PhD	Group
1	000-17-1831	8.5	Below 50,000	$41500	Northwest	Male	Yes	Yes	No	Engineering
2	077-88-9999	10.0	50,000 - 60,000	$53400	Southern	Female	Yes	Yes	Yes	Engineering
3	111-22-3333	13.0	60,000 and over	$65000	Southern	Missing value	Yes	Yes	No	Business
4	222-33-4444	20.0	60,000 and over	$75000	Central	Miscoded	No	Yes	Yes	Science

The SAS format *SSN* manages the leading zeros in the first observation better that the user-defined *a* format in Objective 18.1. A format can create intervals from the data such as done for the salary variable using the *salfmt* format. Usage of the text "low" and "high" in the VALUE statement for *salfmt* is convenient in that the programmer does not have to know the extreme values for the variable to create these intervals. Only the user-defined *gen* format included the missing and miscoded levels, but this could be done for any of the variables. These two options may assist the programmer in locating possible typographical errors in the data.

Additionally, in this objective all three variables bs, ms, and phd are consecutively listed with a single user-defined format *degree* entered in the FORMAT statement after these variables. This concept is extremely useful when many variables have the same set of responses such as the "yes/no" questions in this example. Another example is a data set containing responses to several 5-point Likert scale survey questions where the data values are 1, 2, 3, 4, 5 representing levels of agreement with the question posed. The numeric values are certainly much more easily entered in the data set than the longer character strings. Using the FORMAT procedure, one can create the format for these responses. That is,

```
PROC FORMAT;
VALUE likert    1 = "Strongly Disagree"
                2 = "Disagree"
                3 = "Neutral"
                4 = "Agree"
                5 = "Strongly Agree";
```

And then in a procedure, such as FREQ or REPORT or GCHART, the *likert* format can be used for all survey questions of this type by using the FORMAT statement in the procedure. As an example:

```
FORMAT q6 q7 q8 q9 q10 q20 q21 q22 q23 likert.;
```

All of the "q-variables" listed prior to the *likert* format are responses to survey questions that have these five possible answers. The format will be used in the output results or graph rather than the numeric values 1, …, 5

OBJECTIVE 18.3: Use the SAS formats SSN, DOLLARw.d, and numeric format w.d for the id, salary2, and salary variables, respectively. Print only these variables with their labels.

```
PROC PRINT DATA=one LABEL NOOBS;
VAR id salary2 salary;
FORMAT id SSN. Salary2 DOLLAR9.2 salary 7.3 ;
TITLE 'Objective 18.3';
RUN;
```

Objective 18.3

ID	salary2	Salary ($thousands)
000-17-1831	$41500.00	41.500
077-88-9999	$53400.00	53.400
111-22-3333	$65000.00	65.000
222-33-4444	$75000.00	75.000

In the DOLLARw.d format, the value of w must take into account all of the digits, the decimal point, and the dollar sign. If one had specified DOLLAR8.2, the dollar sign is not printed. In general, if the value of w specified in the SAS format is too small, SAS will make some truncation or not apply the format if w is severely incorrect. There will be a notice in the log indicating this misspecification of the value of w:

```
NOTE: At least one W.D format was too small for the number to be printed.
The decimal may be shifted by the "BEST" format.
```

More than just the PRINT procedure, the benefits of using formats can be seen in graphics and other tables of results. The next two objectives demonstrate these benefits.

OBJECTIVE 18.4: Redo the vertical frequency bar chart in Objective 17.1 of the *demo.beef3* using formats to enhance the graph. Specifically, format the variable Sex for heifers and steers, and the information that Producer 1 is the Rocking K Ranch, Producer 2 is Superior Beef Co., and Producer 3 is Royal Beef Producers. Produce the graph using grayscale color choices. The library containing the file *beef3.sas7bdat* may need to be redefined for this SAS session.

```
LIBNAME demo 'G:\CLGoad\';

PROC FORMAT;
VALUE $sexfmt  "H"="Heifers"  "S" = "Steers";
VALUE pfmt 1="Rocking K Ranch" 2="Superior Beef Co." 3="Royal Beef
Producers";

GOPTION RESET=ALL; *restore the default graph settings;
PROC GCHART DATA=demo.beef3;
VBAR sex / GROUP=producer;
TITLE 'Objective 18.4';
PATTERN1 VALUE=SOLID COLOR=GRAY;
FORMAT sex $sexfmt. producer pfmt.;
RUN;
```

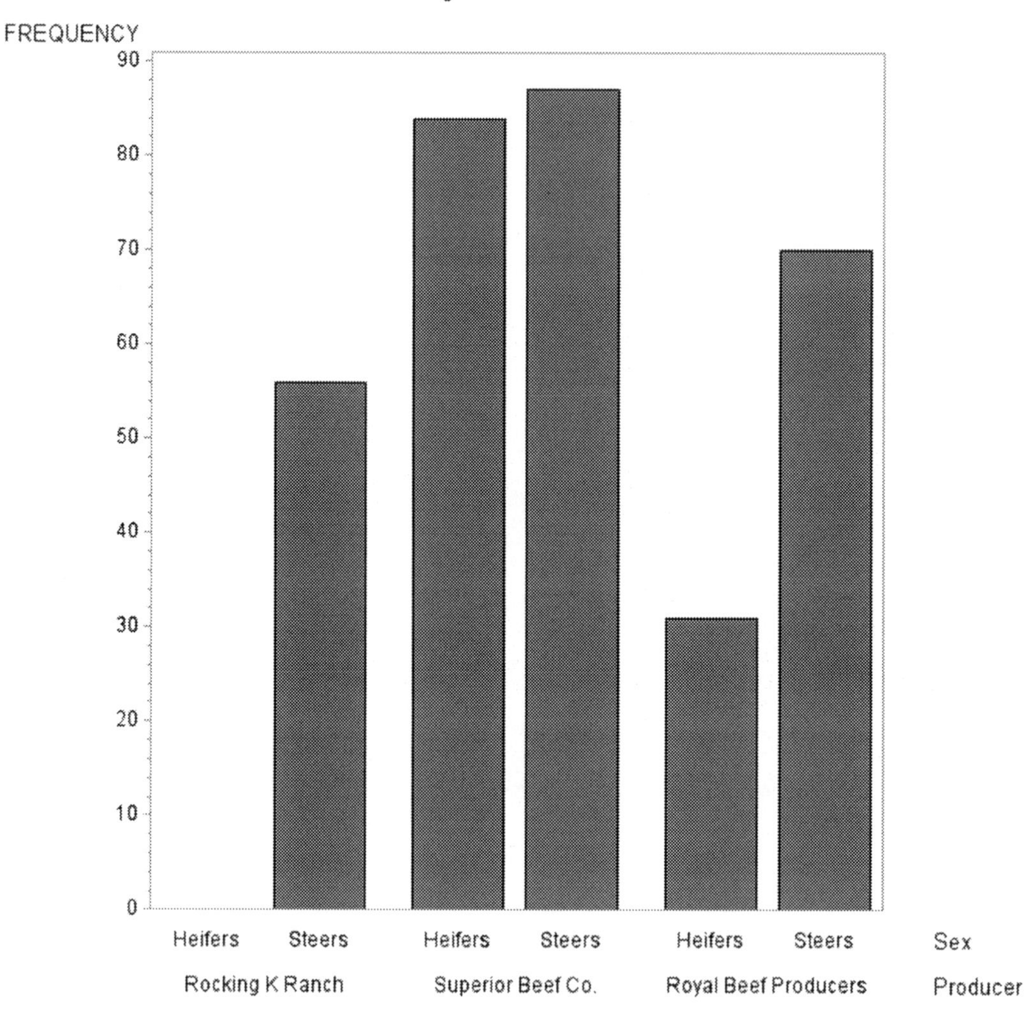

OBJECTIVE 18.5: Compute the mean and range for each of the sexes and producers in the SAS data set *demo.beef3* using the MEANS Procedure with a CLASS statement (Section 3.5) for the response variables DMI and REA. Apply the user-defined formats from the previous objective to the variables Sex and Producer.

```
PROC MEANS DATA=demo.beef3 MEAN RANGE;
CLASS sex producer;
VAR dmi rea;
TITLE 'Objective 18.5';
FORMAT sex $sexfmt. producer pfmt.;
RUN;
```

Objective 18.5

The MEANS Procedure

Sex	Producer	N Obs	Variable	Label	Mean	Range
Heifers	Superior Beef Co.	84	DMI	Dry Matter Intake, lb	20.3860179	10.1442000
			REA	Rib Eye Area	13.5437389	2.9444000
	Royal Beef Producers	31	DMI	Dry Matter Intake, lb	18.9123806	7.5336000
			REA	Rib Eye Area	13.3055000	5.5000000
Steers	Rocking K Ranch	56	DMI	Dry Matter Intake, lb	18.8512929	8.2838000
			REA	Rib Eye Area	13.7459778	3.6294000
	Superior Beef Co.	87	DMI	Dry Matter Intake, lb	20.5432908	9.7858000
			REA	Rib Eye Area	14.0627761	3.4620000
	Royal Beef Producers	70	DMI	Dry Matter Intake, lb	19.3271414	9.8579000
			REA	Rib Eye Area	13.8805367	2.5142000

In the table of results for this objective, the user-defined formats for Sex and Producer are used in the first two columns. Without the FORMAT statement, the original response values for these variables would be used in these columns. The variables DMI and REA were labeled in Objective 14.1, and those labels appear in this table of results also. A slightly different table results when the CLASS statement is changed to:

```
CLASS producer sex;
```

18.3 Chapter Summary

Long or repetitive text values do not have to be entered in a SAS data set. The FORMAT procedure can be used to manage those values. SAS formats and user-defined formats can enhance the output produced by graphics procedures and analysis procedures. There are many, many SAS formats available, and perusing *SAS Help and Documentation* is recommended to pick up a few more tips. There are a number of SAS formats just for time and date. In this chapter the addition of formats was limited to SAS procedures. There are applications where formats can also be applied in a DATA step, but this is NOT recommended to the beginning SAS programmer as it affects data management skills. For now, a few simple guidelines to create and apply formats in SAS can result in better quality graphics and tables of results.

19

Output Delivery System (ODS)

In addition to the suite of graphs seen in many of the procedures overviewed in this book, the Output Delivery System (ODS) can suppress or include the printing of selected tables in the Results Viewer or Output Listing. ODS has the capability to recover any of the tables of the results in the output and make them into SAS data sets. Whereas the results generated appear in the Results Viewer are in HTML format, all or part of the output can be directed to a PDF, RTF, or HTML file, if desired. These are just a few of the third-party file types that are overviewed in this chapter.

19.1 Enabling Graphics, HTML, and Listing Output

When a SAS session is initiated, there are one or two default destinations for results to appear. In Figure 1.2, there is a dialog window shown when SAS first opens. In this dialog, a programmer can select Output Changes and select whether HTML format or a simple text format is desired default destination for the results. Typically, the SAS programmer closes this dialog window and gets to work. However, this window has an option to set up preferences for the SAS programmer. If this dialog window has been suppressed, one can also set the default setting for the output by selecting **Tools – Options – Preferences** from the pull-down menus. Then one selects the **Results** tab in the Preferences dialog window in Figure 19.1. **Create HTML** and **Use ODS Graphics** are the typical default selections.

In this **Results** dialog, one can set whether the HTML output (the Results Viewer used throughout this book) is the default setting for the program output or the Listing Output (located using the **Output** window bar at the bottom of the SAS screen, Figure 1.3) or both by checking the boxes to the left of the options. Because ODS Graphs do not appear in the Listing Output, it is recommended to leave the default setting as HTML. One can temporarily enable the Listing Output and disable the HTML Output with two simple SAS statements.

In this dialog (Figure 19.1), one also notes that **Use ODS Graphics** is typically enabled as a default setting. As early as Chapter 3, ODS Graphics were an essential part of a statistical analysis, and these graphics have appeared in several chapters throughout the book. This is because **Use ODS Graphics** is enabled by default. (In Chapter 16 *Statistical Graphics Procedures*, all of the output produced are ODS Graphics!) It is not recommended to "turn off" these graphics for the default settings. When needed in special cases, one can easily enable/disable ODS Graphics with a simple SAS statement.

If the default settings for the reader are different than those indicated here, it is recommended that HTML Output and ODS Graphics are enabled. Selecting **OK** stores the desired default settings for the results and graphs in **Preferences**.

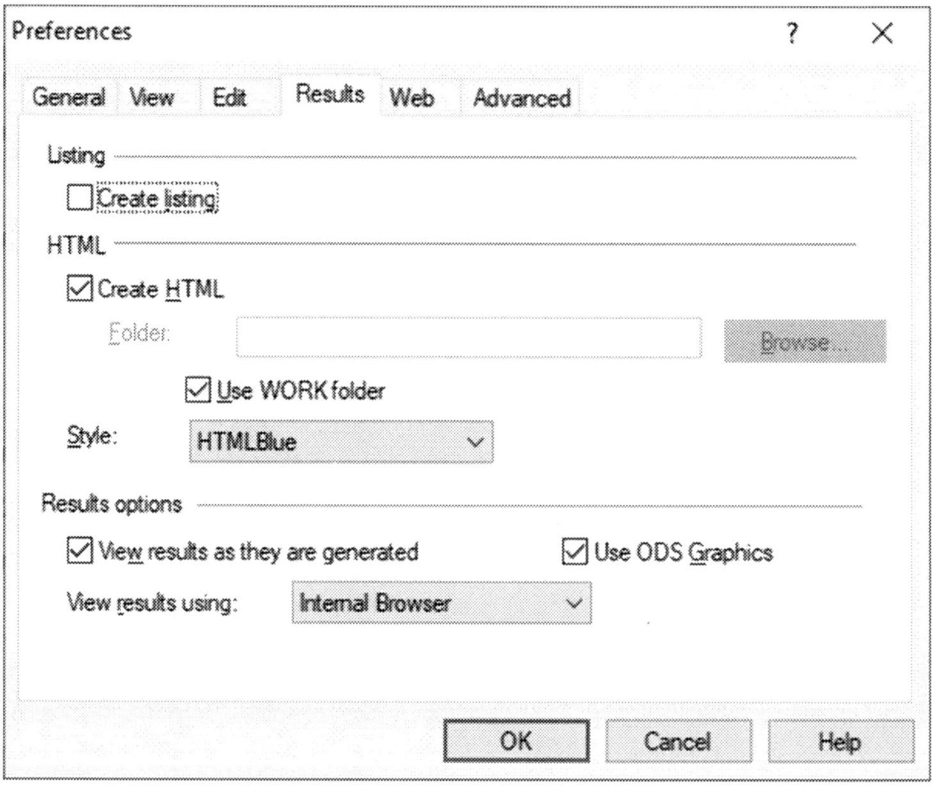

FIGURE 19.1
The options for Results Preferences (**Tools – Options – Preferences**).

Should ODS Graphics need to be disabled or "turned off" in a SAS session the SAS statement is:

```
ODS GRAPHICS OFF;
```

This statement disables ODS Graphics for the remainder of the SAS session or until ODS Graphics are enabled using the SAS statement:

```
ODS GRAPHICS ON;
```

ODS Graphics requires that the HTML Output (the Results Viewer) be enabled. If HTML Output is not enabled at the SAS session start-up, it is enabled by the SAS statement:

```
ODS HTML;
```

and is disabled by the SAS statement:

```
ODS HMTL CLOSE;
```

Similarly, if the Listing Output is needed, the SAS statement:

```
ODS LISTING;
```

enables the Listing Output (appearing in the Output window) for the SAS session or until the Listing Output is disabled in the current SAS session. To disable the Listing Output, the SAS statement is:

```
ODS LISTING CLOSE;
```

Each of these ODS statements for Graphics, HTML, and LISTING are global commands. For example, once the Listing Output is enabled, all subsequent SAS programs and procedures submitted during that SAS session will have Output Listings produced until the statement

```
ODS LISTING CLOSE;
```

is submitted, or SAS is closed.

19.2 ODS Table and Graph Names

Whenever a procedure that produces HTML or Listing Output is submitted, the results are a series of tables of information. Each of these tables can be printed, suppressed, output to a SAS data set, or written to an external file in HTML or another format. SAS refers to each of these tables by a SAS-assigned name. There is a different set of ODS Table Names for every procedure producing SAS output. Similarly, there is an ODS Graphic name for each graphics image produced. It is not feasible to memorize or know them all. Two ways to determine the names of these tables and graphics are by:

1. Consulting SAS Help and Documentation for the procedure and examining the lists of ODS Table Names or ODS Graphics.
2. "Tracing" the output of a submitted SAS program to obtain the ODS Table Names and ODS Graphics.

19.2.1 Using SAS Help and Documentation

A list of ODS Table Names can be found in SAS Help and Documentation for most procedures. After accessing SAS Help and Documentation, select the **Index** tab and enter the procedure for which the ODS names are needed in the search blank, such as PROC UNIVARIATE. Selecting the **Details** tab for the procedure displays the available topics, and one would look for **ODS Table Names** and/or **ODS Graphics** and select the item. The screen capture marking these selections that lead to the ODS items for the UNIVARIATE procedure is shown in Figure 19.2.

If "ODS Table Names" is selected from the "Details" of the UNIVARIATE procedure, then a list of ODS Table Names and the associated syntax options to produce those tables are listed. This list is shown in Figure 19.3. For example, the ODS Table Name "BasicMeasures" is produced by default in the output of the UNIVARIATE procedure, while using the CIBASIC option on the PROC UNIVARIATE statement will produce the output table with the ODS Table Name, "BasicIntervals". The output from the CIBASIC option was overviewed in Objective 3.2.

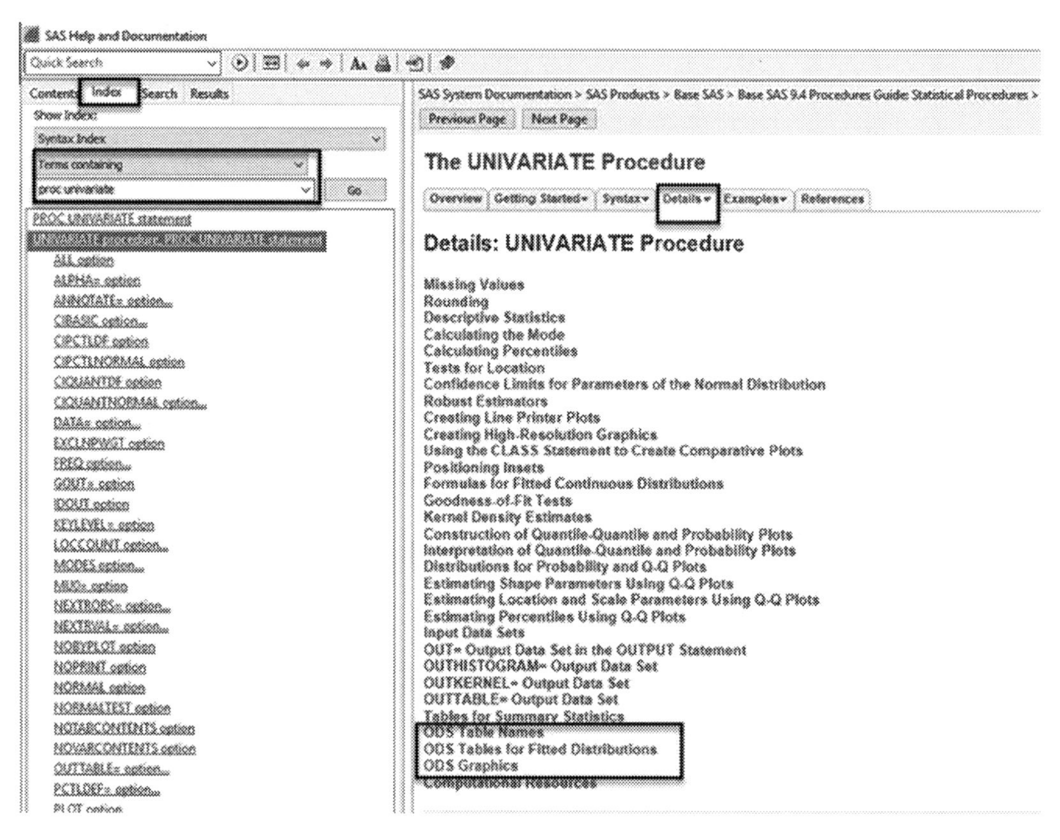

FIGURE 19.2
Finding the ODS Table Names and ODS Graphics for the UNIVARIATE procedure.

Additionally, ODS Graphics are similarly listed in SAS Help and Documentation. That content is not shown here. Once these ODS references are obtained, the remaining sections of this chapter demonstrate some of the many ways these are used.

19.2.2 Tracing SAS Output

When a SAS program is submitted, the output can be traced for ODS Table Names and ODS Graphics by including the following SAS statement:

```
ODS TRACE ON;
```

This SAS statement can be inserted into a SAS program or submitted as a single-line SAS program from the Editor. When a program is submitted, in addition to the usual feedback in the SAS Log there is a list of ODS Table Names and ODS Graphics **available for that current submission from the Editor**. The trace can appear in the Output Listing rather than the SAS Log if the SAS statement is:

```
ODS TRACE ON/LISTING;
```

ODS TRACE ON; is a global SAS statement. That is, once the trace is enabled, all subsequent SAS programs and procedures submitted during that SAS session will be traced until the

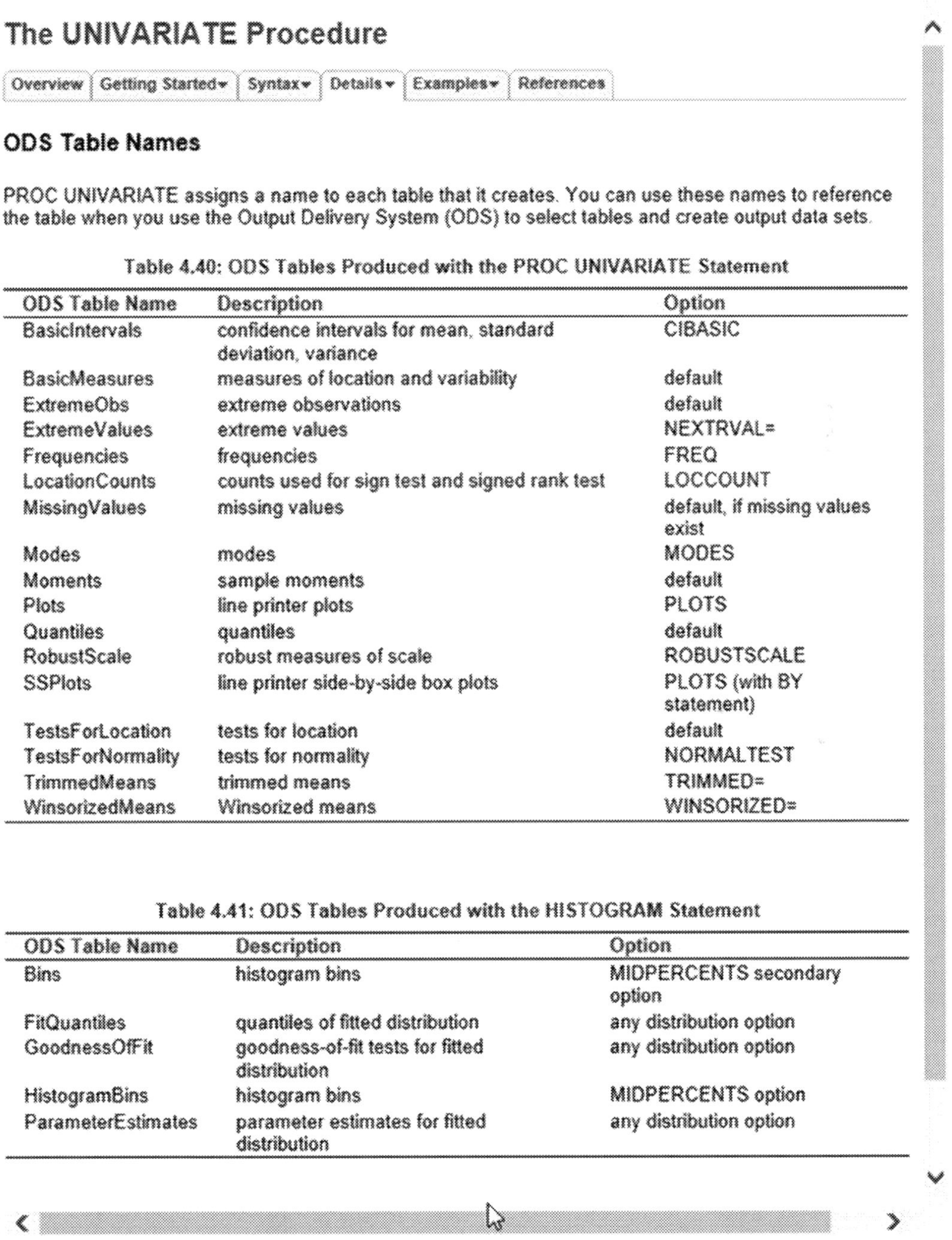

The UNIVARIATE Procedure

Overview | Getting Started | Syntax | Details | Examples | References

ODS Table Names

PROC UNIVARIATE assigns a name to each table that it creates. You can use these names to reference the table when you use the Output Delivery System (ODS) to select tables and create output data sets.

Table 4.40: ODS Tables Produced with the PROC UNIVARIATE Statement

ODS Table Name	Description	Option
BasicIntervals	confidence intervals for mean, standard deviation, variance	CIBASIC
BasicMeasures	measures of location and variability	default
ExtremeObs	extreme observations	default
ExtremeValues	extreme values	NEXTRVAL=
Frequencies	frequencies	FREQ
LocationCounts	counts used for sign test and signed rank test	LOCCOUNT
MissingValues	missing values	default, if missing values exist
Modes	modes	MODES
Moments	sample moments	default
Plots	line printer plots	PLOTS
Quantiles	quantiles	default
RobustScale	robust measures of scale	ROBUSTSCALE
SSPlots	line printer side-by-side box plots	PLOTS (with BY statement)
TestsForLocation	tests for location	default
TestsForNormality	tests for normality	NORMALTEST
TrimmedMeans	trimmed means	TRIMMED=
WinsorizedMeans	Winsorized means	WINSORIZED=

Table 4.41: ODS Tables Produced with the HISTOGRAM Statement

ODS Table Name	Description	Option
Bins	histogram bins	MIDPERCENTS secondary option
FitQuantiles	quantiles of fitted distribution	any distribution option
GoodnessOfFit	goodness-of-fit tests for fitted distribution	any distribution option
HistogramBins	histogram bins	MIDPERCENTS option
ParameterEstimates	parameter estimates for fitted distribution	any distribution option

FIGURE 19.3
ODS Table Names for the UNIVARIATE procedure and the programming options that produce the tables.

statement `ODS TRACE OFF;` is submitted, or SAS is closed. ODS TRACE OFF; can be submitted as part of a SAS program or submitted as a single-line SAS program from the Editor.

One possible recommendation is to place ODS TRACE ON; at the top of a program to be traced and place ODS TRACE OFF; after the final RUN statement and before the QUIT statement at the end of the SAS program. In this way, the trace is not left enabled thereby creating lengthier SAS Log or Output Listings for subsequent program submissions. No RUN statement is needed after ODS TRACE OFF; but no error will result if an additional RUN statement is placed before the QUIT statement.

OBJECTIVE 19.1: Find the ODS Table Names and ODS Graphics for the UNIVARIATE procedure used in the analysis in Objective 3.2. Trace the procedure in the Log window. (The SAS data set *instruction* shown was originally given in Objective 3.1.)

```
DATA instruction;
INPUT program $ score @@;
DATALINES;
A 71 A 82 A 88 A 64 A 59 A 78 A 72
A 81 A 83 A 66 A 83 A 91 A 79 A 70
B 65 B 88 B 92 B 76 B 87 B 89 B 85
B 90 B 81 B 91 B 78 B 81 B 86 B 82
B 73 B 79
;
ODS TRACE ON;
PROC UNIVARIATE DATA=instruction CIBASIC ALPHA=0.01 NORMAL;
CLASS program;
VAR score;
HISTOGRAM score / NORMAL;
TITLE 'Objective 19.1 - Tracing Objective 3.2';
RUN;
ODS TRACE OFF;
QUIT;
```

Here the UNIVARIATE procedure is submitted, and each of the two levels of the class variable *program* are summarized. The ODS TRACE statement identifies the available ODS Table Names in the output. When submitted, this SAS Log identifies the ODS Table Names in the "Name" line: *Moments, BasicMeasures, BasicIntervals, TestsForLocation, TestsForNormality, Quantiles, ExtremeObs, Histogram, ParameterEstimates, GoodnessOfFit,* and *FitQuantiles* for each level of the class variable. The heading for the first table in the HTML Output for this UNIVARIATE procedure is Moments. Typically, the ODS Table Name will very closely match the table heading in the output. When a table has a multi-word heading, the ODS Table Name will run the words together or may truncate some of the words as shown here.

For this objective, only the first three tables of results in the Results Viewer are shown. The results of the trace of "Program = A" in the SAS Log are aligned with the corresponding tables in the Results Viewer. Since a CLASS statement was used in the UNIVARIATE procedure the results of the trace for Program B are repeated in the SAS Log but not shown here. ODS Graphics can be requested in the UNIVARIATE procedure, and the HISTOGRAM statement is one example of this. The trace also reveals the ODS Graphic *Histogram* for this image.

SAS Log

```
Output Added:-------------
Name:       Moments
Label:      Moments
Template:   base.univariate.Moments
Path:       Univariate.score.A.Moments
-------------
```

```
Output Added:
-------------
Name:       BasicMeasures
Label:      Basic Measures of Location and
Variability
Template:   base.univariate.Measures
Path:    Univariate.score.A.BasicMeasures
-------------
```

```
Output Added:
-------------
Name:       BasicIntervals
Label:      Basic Confidence Limits
Template:   base.univariate.ConfLimits
Path:   Univariate.score.A.BasicIntervals
-------------
```

☐

```
Output Added:
-------------
Name:       Histogram
Label:      Panel 1
Template: base.univariate.Graphics.
CompHistogram
Path:   Univariate.score.Histogram.
Histogram
-------------
```

☐

SAS HMTL Output

Objective 19.1 - Tracing Objective 3.2

The UNIVARIATE Procedure
Variable: score
program = A

Moments

N	14	**Sum Weights**	14
Mean	76.2142857	**Sum Observations**	1067
Std Deviation	9.40685978	**Variance**	88.489011
Skewness	−0.2825394	**Kurtosis**	−0.7691737
Uncorrected SS	82471	**Corrected SS**	1150.35714
Coeff Variation	12.3426464	**Std Error Mean**	2.51408903

Basic Statistical Measures

Location		Variability	
Mean	76.21429	**Std Deviation**	9.40686
Median	78.50000	**Variance**	88.48901
Mode	83.00000	**Range**	32.00000
		Interquartile Range	13.00000

Basic Confidence Limits Assuming Normality

Parameter	Estimate	99% Confidence Limits	
Mean	76.21429	68.64116	83.78742
Std Deviation	9.40686	6.21107	17.96323
Variance	88.48901	38.57738	322.67770

☐

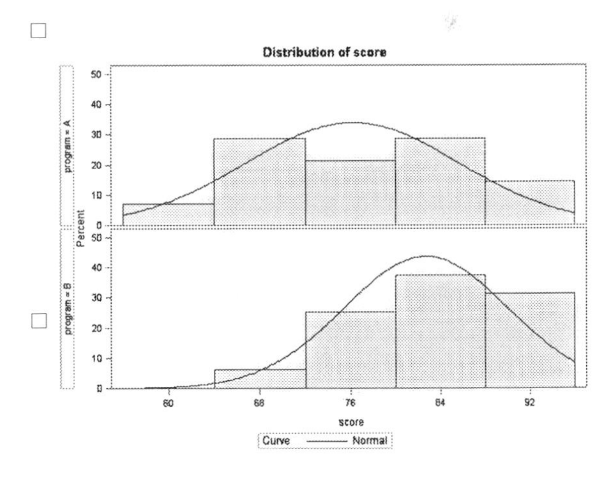

☐

Additionally, the results of the trace can be directed to the Output Listing **only if** the Output Listing has been enabled, such as:

```
ODS LISTING;
ODS TRACE ON / LISTING;
PROC UNIVARIATE DATA=instruction CIBASIC ALPHA=0.01 NORMAL;
CLASS program;
VAR score;
HISTOGRAM score / NORMAL;
TITLE 'Objective 19.1 - Tracing Objective 3.2';
RUN;
ODS TRACE OFF;
```

The resulting Listing Output will place the trace result above the generated output as shown here for the first table of results, *Moments*:

```
                    Objective 19.1 - Tracing Objective 3.2
                          The UNIVARIATE Procedure
                              Variable: score
                                program = A
Output Added:
-------------
Name:       Moments
Label:      Moments
Template:   base.univariate.Moments
Path:       Univariate.score.A.Moments
-------------
                                  Moments
            N                          14    Sum Weights                 14
            Mean               76.2142857    Sum Observations          1067
            Std Deviation      9.40685978    Variance             88.489011
            Skewness           -0.2825394    Kurtosis             -0.7691737
            Uncorrected SS          82471    Corrected SS         1150.35714
            Coeff Variation    12.3426464    Std Error Mean       2.51408903
```

This, of course, would continue for all tables of results in the Listing Output.

The advantage of using ODS TRACE statements in the SAS Editor is that without consulting SAS Help and Documentation, one can obtain the ODS Table Names and ODS Graphics easily while programming. The disadvantage of using ODS TRACE statements is that the trace only identifies the ODS Table Names and ODS Graphics resulting from the submitted program. To determine what additional syntax will generate selected ODS Table Names or ODS one must search SAS Help and Documentation as shown in Figures 19.2 and 19.3.

Three important things to remember when using ODS TRACE statements are:

- If the Listing Output is not enabled, the trace will default to the SAS Log.
- If there are ODS Graphics produced by a procedure, they will not appear in the Listing Output but the trace information will be given for those ODS Graphics in the Output Listing. In Objective 19.1 then, the *Histogram* trace information appears in the Listing Output though the histogram itself does not.
- If a BY statement is used for any procedure, the trace will repeat for each level of the BY variable just like it did for the CLASS statement of the UNIVARIATE procedure in Objective 19.1. The author's recommendation is to "turn the trace off" once the desired ODS Table Names or ODS Graphics names are obtained. When a procedure has a BY variable with several values or levels, the trace can become quite

lengthy. If say, the BY variable has 100 values or levels, the ODS TRACE would yield 100 "traces" of the submitted procedure. A lengthy trace can slow down the performance time of the submitted SAS program.

19.3 Controlling the Content of the Output

In Section 19.1, the SAS statements for disabling all HTML or Listing Output were given. These statements can be modified to select or omit portions of the output. When submitting a SAS procedure, some or all of the tables in the output may not be needed. Perhaps a procedure is run so that an output SAS data set is created, and neither the HTML Output nor Listing Output are needed. Not all procedures have a NOPRINT option such as PROC MEANS and PROC UNIVARIATE. Submitted programs or procedures would produce less unnecessary output if the output were suppressed wholly or in part. Using the ODS HTML statement, syntax options for excluding or selecting ODS Tables and ODS Graphics are given.

```
ODS HTML        EXCLUDE|SELECT        ALL| NONE | ODSlist;
```

`EXCLUDE|SELECT`
Only one of these actions can appear in this statement.

`ALL| NONE | ODSlist`
And only one of these actions can appear in this statement. For the *ODSlist* one can choose multiple items from ODS Table Names or ODS Graphics.

EXCLUDE ALL suppresses all output for the procedure. This is equivalent to SELECT NONE. This syntax is useful if the output SAS data sets are needed but the procedure's output in the Results Viewer is not.

EXCLUDE NONE enables all the output, and SELECT ALL is the equivalent syntax. This is also used to restore the complete output when all or a part of it was suppressed in a prior procedure.

Likewise, ODS LISTING has the options EXCLUDE | SELECT and ALL| NONE | *ODSlist*.

Some procedures generate multiple tables in the output. It may not be desirable or necessary to print all of the possible output tables for a procedure. One can select one or more tables or graphics for inclusion in the output or exclude one or more tables or graphics from the output.

The general syntax for the selecting tables to be included in the HTML Output is:

```
ODS HTML SELECT ODSlist;
```

The general syntax for excluding tables from the HTML Output is:

```
ODS HTML EXCLUDE ODSlist;
```

In the *ODSlist*, one or more ODS Table Names can appear and need only be separated by a space (no commas). Either SELECT a few tables or EXCLUDE a few tables, whichever is a more efficient method. The chosen ODS statement would occur within the block of statements associated with a SAS procedure. If the ODS Table Names are not known, the program would need to be first traced or the SAS Help and Documentation would have to be consulted to obtain that information.

OBJECTIVE 19.2: Modify the program in Objective 19.1. Produce only the Tests for Normality and the stacked histograms in the HTML Output. Use the trace information from Objective 19.1 to identify the ODS names needed.

```
PROC UNIVARIATE DATA=instruction CIBASIC ALPHA=0.01 NORMAL;
CLASS program;
VAR score;
HISTOGRAM score / NORMAL;
TITLE 'Objective 19.2 - SELECT Output';
ODS HTML SELECT TESTSFORNORMALITY HISTOGRAM;
RUN;
```

A "trace" of the program in Objective 19.1 identified the ODS names *TestsForNormality* and *Histogram* as the items to be selected for the HTML Output. Once the selection of the ODS information is made in the program, it is recommended that the trace be "turned off".

Objective 19.2 - SELECT Output

The UNIVARIATE Procedure
Variable: score
program = A

Tests for Normality

Test	Statistic		p Value	
Shapiro-Wilk	W	0.965895	Pr < W	0.8175
Kolmogorov-Smirnov	D	0.146708	Pr > D	>0.1500
Cramer-von Mises	W-Sq	0.04613	Pr > W-Sq	>0.2500
Anderson-Darling	A-Sq	0.251411	Pr > A-Sq	>0.2500

The UNIVARIATE Procedure
Variable: score
program = B

Tests for Normality

Test	Statistic		p Value	
Shapiro-Wilk	W	0.938759	Pr < W	0.3341
Kolmogorov-Smirnov	D	0.12384	Pr > D	>0.1500
Cramer-von Mises	W-Sq	0.041509	Pr > W-Sq	>0.2500
Anderson-Darling	A-Sq	0.312564	Pr > A-Sq	>0.2500

The UNIVARIATE Procedure

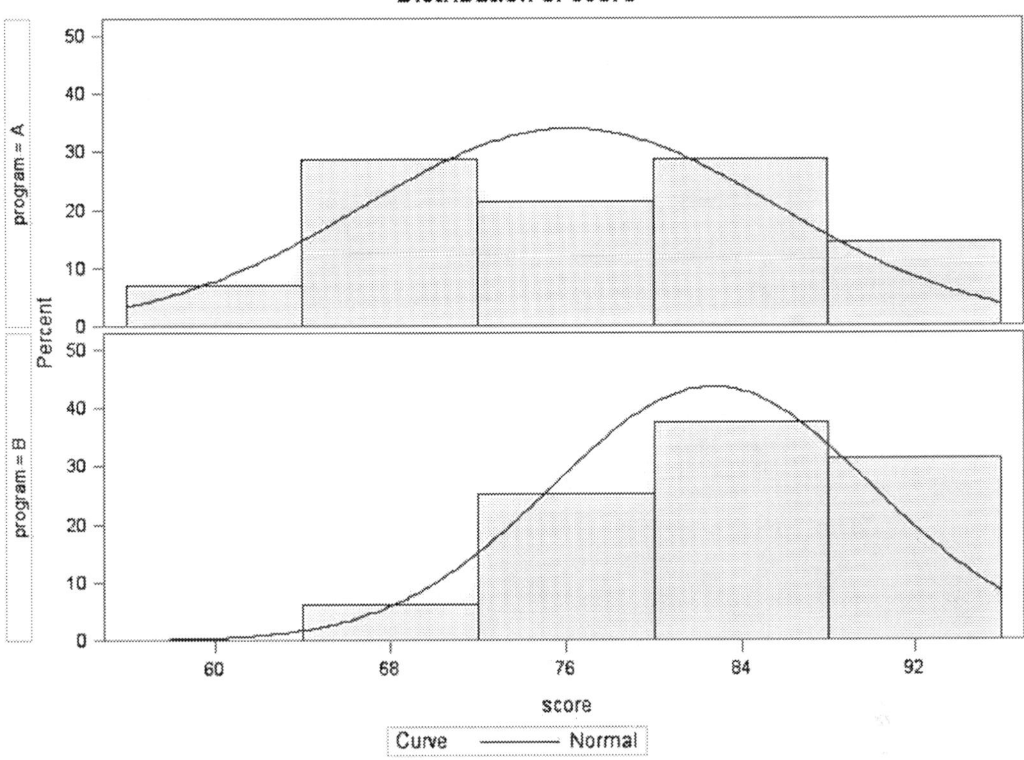

Using ODS to select only the items necessary for the analysis, the HTML Output for the UNIVARIATE procedure is much shorter than usual, containing only those items requested.

OBJECTIVE 19.3: Modify the program in Objective 19.2 so that the table of Extreme Observations is the single item excluded from the HTML Output.

```
PROC UNIVARIATE DATA=instruction CIBASIC ALPHA=0.01 NORMAL;
CLASS program;
VAR score;
HISTOGRAM score / NORMAL;
TITLE 'Objective 19.3 - EXCLUDE Output';
ODS HTML EXCLUDE EXTREMEOBS;
RUN;
```

The UNIVARIATE procedure HTML Output for this procedure is not shown here since it appears in Objective 3.2 with the exception of the table of Extreme Observations, of course.

It is also possible to have both the HTML and Listing Output active at the same time and to select different ODS Table Names or ODS Graphics in the same program or procedure. For example, in Objective 19.2 tables of information can be directed to the Listing Output while the Tests for Normality and the Histogram are excluded from the Listing Output and are the only items selected for the HTML Output by including two or three ODS statements. If the Listing Output is not active, one would need to enable it first by:

```
ODS LISTNG;
```

Within the block of code for the UNIVARIATE procedure, two more ODS statements are needed:

```
ODS LISTING EXCLUDE TESTSFORNORMALITY HISTOGRAM;
ODS HTML SELECT TESTSFORNORMALITY HISTOGRAM;
```

If the Listing Output were not enabled before submitting the procedure, the following error message in the SAS Log is obtained.

```
ERROR: The LISTING destination is not active; no select/exclude lists are
available.
```

A similar error message would result if the HTML "destination" was not active.

Since HISTOGRAM is an ODS Graphic and is not produced in the Listing Output anyway, it is not necessary to EXCLUDE it from the Listing Output. In general, those items selected for one destination do not have to be the same as those excluded from the other destination as they were in this illustration.

19.4 Creating SAS Data Sets from the Output

One of the many attractive applications of ODS is the ability to recover any of the tables of output results as either a temporary or permanent SAS data set. Using the methods discussed in Section 19.2 to identify the ODS Table Names produced, the SAS data sets can be created within the block of code for a procedure by including another ODS statement. The general syntax for this is:

ODS OUTPUT *ODStablename=SAS-data-set;*

where *SAS-data-set* can be either a one- or two-level SAS data set name (Chapter 12).

The ODS OUTPUT statement is placed within a procedure in order to recover the procedure's output as a SAS data set. Multiple SAS data sets can be defined in a single ODS OUTPUT statement.

OBJECTIVE 19.4: Using the SAS program code in Objective 19.2, include the statement that would create the SAS data set *work.basicstats* from the Moments information and *work.meantests* from the Tests for Locations. Print the two resulting SAS data sets. Print the SAS-assigned variables for *work.basicstats* and the SAS-assigned variable labels for *work.meantests*. Close the Listing Output first.

```
ODS LISTING CLOSE;
PROC UNIVARIATE DATA=instruction CIBASIC ALPHA=0.01 NORMAL;
CLASS program;
VAR score;
```

```
HISTOGRAM score / NORMAL;
ODS HTML SELECT TESTSFORNORMALITY HISTOGRAM;
ODS OUTPUT MOMENTS=basicstats TESTSFORLOCATION=meantests;
TITLE 'Objective 19.4';
RUN;
PROC PRINT DATA=basicstats;
TITLE2 'The Recovered Moments Table';
PROC PRINT DATA=meantests LABEL;
TITLE2 'The Recovered Tests for Location Table';
RUN;
QUIT;
```

Tests for Normality and Histograms are produced as in the Results Viewer by the UNIVARIATE procedure and ODS HTML SELECT statements, but these are not shown in the HTML output in this objective since that information is presented in Objective 19.2. The Moments and Tests for Location were not printed in the Results Viewer as part of the UNIVARIATE procedure, but that information can still be recovered as a SAS data set. This implies that the full procedure "runs", but only what is selected in the ODS HTML statement appears in the Results Viewer.

Objective 19.4

The Recovered Moments Table

Obs	VarName	program	Label1	cValue1	nValue1	Label2	cValue2	nValue2
1	score	A	N	14	14.000000	Sum Weights	14	14.000000
2	score	A	Mean	76.2142857	76.214286	Sum Observations	1067	1067.000000
3	score	A	Std Deviation	9.40685978	9.406860	Variance	88.489011	88.489011
4	score	A	Skewness	−0.2825394	−0.282539	Kurtosis	−0.7691737	−0.769174
5	score	A	Uncorrected SS	82471	82471	Corrected SS	1150.35714	1150.357143
6	score	A	Coeff Variation	12.3426464	12.342646	Std Error Mean	2.51408903	2.514089
7	score	B	N	16	16.000000	Sum Weights	16	16.000000
8	score	B	Mean	82.6875	82.687500	Sum Observations	1323	1323.000000
9	score	B	Std Deviation	7.32774408	7.327744	Variance	53.6958333	53.695833
10	score	B	Skewness	−0.891007	−0.891007	Kurtosis	0.69363001	0.693630
11	score	B	Uncorrected SS	110201	110201	Corrected SS	805.4375	805.437500
12	score	B	Coeff Variation	8.86197319	8.861973	Std Error Mean	1.83193602	1.831936

Objective 19.4

The Recovered Tests for Location Table

Obs	VarName	program	Test for Location	Label for Test Statistic	Test Statistic	P Value Description	P Value	Mu0 Value
1	score	A	Student's t	t	30.31487	Pr > |t|	<.0001	0
2	score	A	Sign	M	7	Pr >= |M|	0.0001	0
3	score	A	Signed Rank	S	52.5	Pr >= |S|	0.0001	0
4	score	B	Student's t	t	45.13667	Pr > |t|	<.0001	0
5	score	B	Sign	M	8	Pr >= |M|	<.0001	0
6	score	B	Signed Rank	S	68	Pr >= |S|	<.0001	0

OBJECTIVE 19.5: Using ODS statements, suppress all of the output from the UNIVARIATE procedure of Objective 19.4 and recover the Basic 99% Confidence Intervals data set named *work.ci99* and print it. The CIBASIC and ALPHA=0.01 options on the PROC UNIVARIATE statement are needed in order to obtain the requested SAS data set.

```
PROC UNIVARIATE DATA=instruction CIBASIC ALPHA=0.01 NORMAL;
CLASS program;
VAR score;
HISTOGRAM score / NORMAL;
ODS HTML SELECT NONE;   *or ODS HTML EXCLUDE ALL;
ODS OUTPUT BASICINTERVALS=ci99;
TITLE 'Objective 19.5';
RUN;
PROC PRINT DATA=ci99;
TITLE2 '99% CIs for Mean, Std Dev, and Variance';
ODS HTML SELECT ALL;
RUN;
```

In the ODS HTML statement of the UNIVARIATE procedure, NONE of the output is selected for inclusion in the Results Viewer even though tests for normality and a histogram were requested. This could also have been done with the ODS statement:

```
ODS HTML EXCLUDE ALL;
```

SELECT NONE and EXCLUDE ALL are extreme options. The first procedure after a procedure where all of the results are suppressed and for which output is needed in the Results Viewer, an ODS statement is needed to enable the output once again. Thus,

```
ODS HTML SELECT ALL;
```

was included in the PRINT Procedure immediately following the UNIVARIATE procedure.

Objective 19.5

99% CIs for Mean, Std Dev, and Variance

Obs	VarName	program	Parameter	Estimate	LowerCL	UpperCL
1	score	A	Mean	76.21429	68.64116	83.78742
2	score	A	Std Deviation	9.40686	6.21107	17.96323
3	score	A	Variance	88.48901	38.57738	322.67770
4	score	B	Mean	82.68750	77.28931	88.08569
5	score	B	Std Deviation	7.32774	4.95530	13.23103
6	score	B	Variance	53.69583	24.55503	175.06027

One could modify the information in these SAS data sets for many other uses. For example, if the 99% confidence intervals for each of the parameters (mean, standard deviation, and variance) could be recovered in a new SAS data set for each parameter, such as *work.ci99_mean, work.ci99_sd,* and *work.ci99_var.* This can be done in a new DATA step (Sections 4.2, 4.4, and 15.2.), and each of the new SAS data sets must be initialized in the DATA statement (Objective 15.3).

```
DATA ci99_mean ci99_sd ci99_var;
SET ci99;
IF PARAMETER="Mean" THEN OUTPUT ci99_mean;
IF PARAMETER="Std Deviation" THEN OUTPUT ci99_sd;
IF PARAMETER="Variance" THEN OUTPUT ci99_var;
RUN;
```

The SAS Log identifies the three newly created SAS data sets. Two observations result in each of the three SAS data sets since there are only two Programs, A and B, in the *instruction* SAS data set.

```
NOTE: There were 6 observations read from the data set WORK.CI99.
NOTE: The data set WORK.CI99_MEAN has 2 observations and 6 variables.
NOTE: The data set WORK.CI99_SD has 2 observations and 6 variables.
NOTE: The data set WORK.CI99_VAR has 2 observations and 6 variables.
```

19.5 Creating External Files from the Output

In addition to recovering parts of the output in a SAS data set, all or part of the tables of results and graphs can be saved in an external file. External file types shown here are limited to Portable Document Format (PDF), Rich Text Format (RTF), and Hypertext Markup Language (HTML) files. Of course, the ability to copy and paste information from the Editor, Log, Output, and Results Viewer into another software for the purpose of producing documents or presentations has been available for all of the tasks covered thus far in this book. For lengthier output, the results can be "sent" to an external file using ODS Table Names and ODS Graphics selections presented earlier in this chapter.

Creating a PDF destination file of the SAS results will be considered first. A minimum of two ODS statements are needed to do this.

The syntax is:

```
ODS PDF FILE = "file path and filename";
```

to open the communication and

```
ODS PDF CLOSE;
```

to close the communication with the destination file. This closing line must appear **after** the RUN statement for the final procedure to have results sent to the destination file. If it appears before that RUN statement, the communication with the destination file will close prematurely and none of the final procedure's results will be "sent" to the destination file. These are global statements, so omitting the second (closing) statement will leave the communication with the destination file open. Any additional procedures or programs submitted will have results sent to the destination file until the ODS PDF CLOSE statement is encountered.

All of the results from a series of SAS procedures between these two statements will be saved in the PDF destination file. The FILENAME statement (Section 8.1) can also be used in conjunction with the first ODS statement. The following two-line syntax is equivalent to the first ODS PDF statement:

```
FILENAME fileref "file path and filename";
ODS PDF FILE = fileref;
```

In both instances, *file path and filename* identifies the destination or folder for the newly created file, and *filename* should include the ".pdf" extension. Omission of the file extension will cause an error when the one later tries to open this destination file.

The SELECT and EXCLUDE options on the ODS statements shown in Section 19.3 can be utilized here. That is, the statement

```
ODS PDF        SELECT | EXCLUDE        ALL | NONE | ODSlist;
```

can be included "inside" the SAS Procedure(s) to identify those ODS Table Names and ODS Graphics that are to be included in the PDF file created by this submitted program.

For RTF files, the syntax is like that for the PDF files. That is, the syntax is:

```
ODS RTF FILE = "file path and filename";
```

to open the communication (or the two-line syntax with the FILENAME statement) and

```
ODS RTF CLOSE;
```

to close the communication.

The *filename* must have the ".rtf" extension. Results can similarly be selected for or excluded from the RTF destination file by the SAS statement:

```
ODS RTF        SELECT | EXCLUDE        ALL | NONE | ODSlist
```

which can be included "inside" the SAS Procedure(s).

And finally, HTML files can also be created using ODS statements. Using similar syntax to name the destination HTML file:

```
ODS HTML FILE=" file path and filename";
```

to open the communication and

```
ODS HTML CLOSE;
```

to close the communication with the destination file. This also disables any further output from being displayed in the Results Viewer. To enable the HTML Output for subsequent submissions from the Editor, one would need to submit ODS HTML; as a single line or as one of the first lines in the subsequent SAS program.

And, as the previous two file types require an extension, the *filename* must have the ".html" extension. Results can similarly be selected for or excluded from the HTML destination file by the SAS statement:

```
ODS HTML       SELECT | EXCLUDE        ALL | NONE | ODSlist;
```

which can be included "inside" the SAS Procedure(s). The contents of the Results Viewer will match the contents of the HTML file.

HTML 4.0 files are the present form of HTML files SAS creates. There are multiple versions or types of HTML files and other third-party file types supported by SAS with the ODS tool. SAS Help and Documentation will provide the most up-to-date list of file types available.

OBJECTIVE 19.6: Modify the syntax in Objective 19.4. Allow all of the UNIVARIATE procedure's results in the Results Viewer, create permanent SAS data sets *basicstats* and *meantests* in a SAS library *ch19*, and select the Tests for Normality and Histogram for a destination PDF file *Obj19_6.pdf*. Print the permanent SAS data sets in the Results Viewer but do not print them in *Obj19_6.pdf*.

```
LIBNAME ch19 "G:\CLGoad";
ODS PDF FILE="G:\CLGoad\Obj19_6.pdf";
PROC UNIVARIATE DATA=instruction CIBASIC ALPHA=0.01 NORMAL;
CLASS program;
VAR score;
HISTOGRAM score / NORMAL;
ODS HTML SELECT ALL;
ODS PDF SELECT TESTSFORNORMALITY HISTOGRAM;
ODS OUTPUT MOMENTS=ch19.basicstats TESTSFORLOCATION=ch19.meantests;
TITLE 'Objective 19.6';
RUN;
ODS PDF CLOSE;
PROC PRINT DATA=ch19.basicstats;
TITLE2 'The Recovered Moments Table';
PROC PRINT DATA=ch19.meantests LABEL;
TITLE2 'The Recovered Tests for Location Table';
RUN;
QUIT;
```

This objective demonstrates that multiple ODS operations can occur within the same program or procedure. Enabling all of the results to print in the Results Viewer allows the programmer the opportunity to confirm results on screen while selected items are saved to PDF. The *filepath* for the library and the PDF file do not have to be the same. The information in the Results Viewer is repetitive of earlier objectives and is not shown here. Examining the SAS Log however, does confirm the creation of the *ch19* SAS library (line 133 note), the PDF file creation (notes after lines 134 and 144), and the two permanent SAS data sets (line 143 notes).

```
133   LIBNAME ch19 "G:\CLGoad";
NOTE: Libref CH19 was successfully assigned as follows:
      Engine:        V9
      Physical Name: G:\CLGoad
134   ODS PDF FILE="G:\CLGoad\Obj19_6.pdf";
NOTE: Writing ODS PDF output to DISK destination "G:\CLGoad\Obj19_6.pdf",
printer "PDF".
135   PROC UNIVARIATE DATA=instruction CIBASIC ALPHA=0.01 NORMAL;
136   CLASS program;
137   VAR score;
138   HISTOGRAM score / NORMAL;
139   ODS HTML SELECT ALL;
140   ODS PDF SELECT TESTSFORNORMALITY HISTOGRAM;
141   ODS OUTPUT MOMENTS=ch19.basicstats TESTSFORLOCATION=ch19.meantests;
```

```
142  TITLE 'Objective 19.6';
143  RUN;
NOTE: The data set CH19.MEANTESTS has 6 observations and 8 variables.
NOTE: The data set CH19.BASICSTATS has 12 observations and 8 variables.
NOTE: PROCEDURE UNIVARIATE used (Total process time):
      real time           0.90 seconds
      cpu time            0.23 seconds
144  ODS PDF CLOSE;
NOTE: ODS PDF printed 3 pages to G:\CLGoad\Obj19_6.pdf.
145
146  PROC PRINT DATA=ch19.basicstats;
147  TITLE2 'The Recovered Moments Table';
NOTE: There were 12 observations read from the data set CH19.BASICSTATS.
NOTE: PROCEDURE PRINT used (Total process time):
      real time           0.07 seconds
      cpu time            0.00 seconds
148  PROC PRINT DATA=ch19.meantests LABEL;
149  TITLE2 'The Recovered Tests for Location Table';
150  RUN;
NOTE: There were 6 observations read from the data set CH19.MEANTESTS.
NOTE: PROCEDURE PRINT used (Total process time):
      real time           0.05 seconds
      cpu time            0.01 seconds
151
152  QUIT;
```

In the program above, the ODS OUTPUT statement does not influence what tables are included in the HTML output nor in the destination file. ODS OUTPUT only specifies the parts of the output to be saved as SAS data sets.

19.6 Chapter Summary

ODS is certainly one of the "cool tools" in SAS. ODS enhances the programmer's ability to manage output from various procedures. Many of the controls require the programmer to open and close the ODS action (TRACE, HTML, PDF, RTF, LISTING), and other ODS statement options are applicable to the current procedure (SELECT, EXCLUDE, OUTPUT). There are many ODS capabilities not covered here, such as page formatting in a destination file, additional text blocks, or display methods. This chapter introduces a few of the basic ODS capabilities. For more coverage of ODS, the reader is encouraged to investigate *SAS® 9.4 Output Delivery System: Procedures Guide* (SAS Institute, 2016b) or the list of topics in SAS Help and Documentation within the SAS product.

20

Miscellaneous Topics

In this chapter, a few additional SAS tools are presented. These items did not seem to fit in any of the earlier chapters, and each is not big enough to form a stand-alone chapter.

20.1 Usage of a Double-dash to Specify a Range of Variables

The usage of a double-dash "- - " in programming statements can efficiently represent a list of several variables. One needs to learn when or how to do this correctly though. When using the double-dash "- -" in the Editor, it does not matter if there is zero, one, or more spaces between the dashes (or hyphens). A single space is used in this book so that it is evident that exactly two dashes are used. SAS DATA step information from Sections 2.2 and 4.1 is used to illustrate the usage of the double-dash. The first illustration begins with the following DATA step:

```
DATA survey;
INPUT ID q1 q2 q3 q4 q5 q6 q7 q8 q9 q10;
DATALINES;
⋮
```

For the *survey* SAS data set, the researcher creates the numeric variables: survey ID, and one variable for each of the ten questions on the survey where q# denotes the question number. When reading "q1 - - q10" one comprehends "questions 1 through 10"; however, in SAS all variables must first be defined before the double-dash can be used. If the INPUT statement were written: INPUT ID q1 - - q10; SAS could make no sense of this since variables q2, q3, q4, q5, q6, q7, q8, and q9 do not yet exist. That is, SAS is not going to fill in what the reader perceives. Thus, in the *survey* DATA step, all of the variable names must be individually entered. Once the *survey* SAS data set is successfully created, then the double-dash notation can be used in procedures. An example using the MEANS Procedure is:

```
PROC MEANS DATA=survey;
VAR q1 - - q6 q10;
RUN;
```

The VAR statement is equivalent to:

```
VAR q1 q2 q3 q4 q5 q6 q10;
```

The default statistics for the requested variables will be produced. When the number of variables to be included is large, the double-dash is certainly a convenience. The order of the variables (or columns) in the SAS data set is important when this double-dash is used. The SAS data set *murphy* is created here to illustrate variable order.

```
DATA murphy;
INPUT a b c d Group $ pH Rate;
L = log(b);
f = (c + d)/2;
DATALINES;
⋮
```

The order of the variables or columns in the SAS data set *murphy* is:

a	b	c	d	Group	pH	Rate	L	f

The PRINT Procedure (Section 2.3) or ViewTable (Section 8.3) can confirm the order of the columns or variables. Again, using the MEANS Procedure as an example, if one were computing summary statistics for variables a, b, c, d, and f, the syntax is:

```
PROC MEANS DATA=murphy;
VAR a b c d f;
RUN;
```

Additionally,

```
VAR a - - d f;
```

would also accomplish this task. If the following VAR statement been used

```
VAR a - - f;
```

an error would have resulted since Group is a character variable and not numeric, and the MEANS Procedure is attempting to execute for all of the variables in the *murphy* SAS data set since *a* is the first variable created (or initialized), and *f* is the last variable initialized. If Group had also been numeric (no $ on the INPUT statement), then extraneous summary statistics for Group, pH, Rate, and L would be computed.

20.2 Annotating Programs

When programming in SAS, explanatory comments can and should be added to the program. There are many reasons for annotating programs. Two of the primary purposes are: 1. To aid oneself in identifying parts of the program and their purpose and 2. To aid another person to read a program and comments and quickly gain understanding of the program's objectives. There are two ways to write comments: single-line comments and block comments. In the Enhanced Editor, comments appear in a different color than the SAS statements, so it is easier to identify comment text.

20.2.1 Single-Line Comments

A single line in the SAS program can be designated as a comment. If the first character in a SAS statement is an asterisk (*) , SAS interprets this as a comment line. That is, the first character of the SAS statement is an asterisk. This is not necessarily the first character in the line of the program. Each single-line comment ends with a semicolon. Almost any text in can be used in a comment line. Single or double quotes cannot be used in single-line comments.

Example 1:

A Single-Line Comment

```
     * program name: survey.sas *;
DATA one ;
INFILE . . . ;
INPUT;
⋮
```

Several single-line comments can appear consecutively in a program.

Example 2:

Stacked Single-Line Comments

```
* Program name: survey.sas ;
* Project number: A4X7-400 ;
* Project leader: Dr. Y. B. Boring ;
DATA two;
SET a b c ;
⋮
```

Only one asterisk is necessary at the left of the comment. Comments do not have to end with an asterisk and a semicolon, just a semicolon. Multiple asterisks can be used. This is usually done out of the programmer's style preference. Multiple asterisks tend to catch the eye and draw attention to necessary support information. It is not a programming necessity.

Example 3:

Illustrations of Different Styles of (Single Line) Commenting a Program

```
*********************************;
*** Program name: survey.sas  ***;
*** Project number: A4X7-400  ***;
*** Client: Roger Barnes      ***;
*********************************;
    *Analysis of initial health of participants ;
PROC MEANS DATA=client.health;
VAR wt ht hr ;

*PROC FREQ DATA=client.health; *BY gender;
*TABLES heart diabetes -- thyroid;
RUN;
QUIT;
```

The usage of several asterisks is the programmer's style for creating a brief header for the program. The FREQ procedure has each line "commented out". This has the effect of skipping that procedure. Though the BY statement occurs on the same line as the PROC FREQ statement, it must also be "commented out". Programmers may "turn off" parts of their program using comment syntax while constructing and editing other parts of the program.

20.2.2 Block Comments

Multiple lines of text can be used to annotate the program using the symbols /* and */ to identify the block comment. These symbols can be thought of as bookends – they must occur as a pair, and one set of block comment symbols cannot nest within another set. /* "turns the program off" until */ is encountered, and the program is "turned on" again. These symbols can be used to block out a portion of a line or SAS statement, only one line, or several lines of the program. When programs become lengthy and the programmer does not wish for part of the program to execute while working on another section, a block comment can "turn off" a portion of the program until the programmer is ready for that part of the program to run again and removes the block comment symbols.

Example 4:

Illustrations of Block Commenting a Program

```
/*   Program name: survey.sas
Project number: A4X7-400
Client: Roger Barnes       */
    /* Analysis of initial health of participants */
PROC MEANS DATA=client.health;
VAR wt ht hr ;
/*
PROC FREQ DATA=client.health; BY gender;
TABLES heart diabetes -- thyroid;
*/
RUN;
QUIT;
```

This example is the program code of Example 3 except all of the single-line comments have been replaced with equivalent block comments. This example also illustrates the different positions in which the comment symbols can appear. The blank lines some programmers leave between blocks of code can be insertion points for the block comment symbols, or the symbols can occur in the same line as text.

Example 5:

Illustration of Both Styles of Commenting a Program

```
/* *******************************
*** Program name: survey.sas   ***
*** Project number: A4X7-400   ***
*** Client: Roger Barnes       ***
******************************** */
    *Analysis of initial health of participants ;
PROC MEANS DATA=client.health;
VAR wt /* ht hr */ ;
```

```
/*
PROC FREQ DATA=client.health; BY gender;
TABLES heart diabetes -- thyroid ;
*/
RUN;
QUIT;
```

This example uses both single-line and block comment techniques. Additionally, in this example, a block comment within a SAS statement is demonstrated. In the VAR statement of the MEANS Procedure, the block comment allows the programmer to temporarily "turn off" or skip the analysis of two of the variables (ht and hr) while building or debugging a program. The advantage to this type of commenting is that by deleting the commenting symbols the analysis of the original set of variables is restored.

These are the general rules for commenting programs. Each programmer adopts a style of his or her own. Whatever the style, one must be certain that it conforms to the rules outlined here.

20.3 Strategies for Checking for Data Errors Using SAS Procedures

For data sets that are small, it is generally fairly easy to proofread the DATA step of the SAS program or the external data file containing the data. As data sets become larger, finding data errors becomes increasingly more difficult. Using some of the procedures in this book, a few strategies for detecting errors will be shown.

When the *beef* data set was introduced in Table 11.2, it was noted that all data was from a single producer, and all of the animals were steers. The *beef3* SAS data set, introduced in Objective 14.1, combined data from two other producers for both steers and heifers with the *beef* data. Whenever a SAS data set contains class variables, the FREQ Procedure can be used to examine the number of levels of each of the class variables. In this *beef3* data set, the class variables are Producer and Sex. A good place to start is to examine each of the class variables. The permanent SAS data set *demo.beef3* is again used in this chapter. The observed frequencies for the variables Producer and Sex in both one-way and two-way frequency tables are computed.

```
LIBNAME demo 'G:\CLGoad';

PROC FREQ DATA=demo.beef3;
TABLES producer sex producer*sex / NOROW NOCOL NOPERCENT;
RUN;
```

Programming Note: The options on the TABLES statement are not necessary for this task. Their usage was preferred since tables of only the observed frequencies are produced.

The FREQ Procedure

Producer	Frequency	Cumulative Frequency
1	56	56
2	171	227
3	101	328

Sex	Frequency	Cumulative Frequency
H	115	115
S	213	328

Frequency	Table of Producer by Sex			
	Producer	Sex		
		H	S	Total
	1	0	56	56
	2	84	87	171
	3	31	70	101
	Total	115	213	328

Researchers generally should know how much data they have. Therefore, checking the frequencies in the one-way tables is a good idea. In the two-way table, what catches the eye is that there are no heifers from Producer 1. A good follow-up investigation should take place to verify this. Quite possibly, the data for heifers was omitted for some reason. In this case, this producer did not process any heifers at this time. Additionally, one should look at the frequency values of the Producer and Sex variables in the one-way tables. Are these correct?

The following one-way frequency tables illustrate possible typographical errors that may occur in the data.

Producer	Frequency	Cumulative Frequency
1	56	56
2	171	227
3	100	327
6	1	328

Sex	Frequency	Cumulative Frequency
H	113	113
h	2	115
S	157	272
Steer	56	328

In the first table for Producer, Producer 6 is observed with one observation. This is likely a typographical error since there is no Producer 6, perhaps it is the result of striking the 6 next to the 3 on the ten-key number pad. In the second table the "h" values would likely also be typographical errors, and the two indications for steers (S and Steer) are also problematic. Further investigation indicates that Producer 1 initially reported "Steer" rather than "S" as the value for Sex. The frequency of "Steer" is 56, the same as the number of animals from Producer 1 further confirming the programmer's suspicion. The typographical errors of Producer 6 and Sex "h" can be traced back to the data set using a simple PRINT Procedure using a WHERE statement. That is,

```
PROC PRINT DATA=demo.beef3;
WHERE producer=6 or Sex="h";
RUN;
```

Obs	DMI	ADG	REA	CWT	BackFat	Producer	Sex
68	18.1517	3.04	13.9347	791.000	0.50944	2	h
193	16.0322	2.36	.	701.000	.	2	h
266	18.3597	3.05	13.3536	845.000	0.52961	6	S

The observation numbers in *demo.beef3* can assist the programmer in locating the error in the SAS data set. If the SAS data set was created using the IMPORT Procedure, the observation numbers may also assist in finding the line of data in the original Microsoft Excel file so that a correction can be made.

The MEANS or UNIVARIATE Procedures can also be used to identify possible errors. The usage of a CLASS statement, such as,

```
CLASS producer sex;
```

for the *demo.beef3* SAS data set, could also have identified the "h" and "Steer" data errors identified by the FREQ procedure. For the continuous variables, one can examine the extreme observations. Here, the MEANS Procedure investigates the first cattle data set, *beef*, introduced in Chapter 11.

```
PROC MEANS DATA=beef;
VAR dmi adg rea cwt backfat;
RUN;
```

The MEANS Procedure

Variable	N	Mean	Std Dev	Minimum	Maximum
DMI	56	18.8512959	2.1086015	16.2730501	24.5568749
ADG	56	3.0705357	0.3558315	1.5300000	3.8400000
REA	54	13.7459805	0.7017480	12.3436893	15.9731481
CWT	56	837.3976183	39.8765986	701.0000000	965.0000000
BackFat	54	0.5091691	0.0380773	0.3945631	0.5838961

The SAS programmer and/or researcher should examine the ranges of the continuous variables. Do these ranges make sense? For this application, low values of 1.53 for ADG and 701.0 for CWT concern the researcher. ADG values are typically larger than 2, and typically CWT is also larger. Using the PRINT Procedure, one can identify the lines of data for these values. The researcher can further investigate whether or not these are correct.

```
PROC PRINT DATA=beef;
WHERE adg < 2;
RUN;
```

Obs	BackFat	REA	ADG	DMI	CWT
43	0.39456	12.3437	1.53	17.0606	701

In this case, there is only one animal with an ADG smaller than 2, and it is the same animal with CWT of 701. From here the programmer and/or researcher can review the original data record for accuracy. In this case, the observation was correct, and the animal was indeed much smaller than the others in this group from Producer 1.

Examining the minimum and maximum values for the numeric response variables can also identify typographical errors. For example, suppose CWT had a low value of 83.27. This value is much too small. The correct value is 832.7, and a typographical error placing the decimal point in the wrong place can be located and corrected. Minimum and Maximum can also be affected by the incorrect recording of missing values, such as placing a zero when the value is missing. The sample sizes, N, in the MEANS Procedure results should also be reviewed. In the Results of the MEANS procedure for *beef*, unequal sample sizes were evident. There are only 54 observations of REA and BackFat. This could be the result of a character typographical error, or it simply and correctly means that the measurement was not taken.

Graphs can also be used to identify some of these extreme values which may be indicative of an error in the SAS data set. Using either of the SGPLOT or GPLOT Procedures, plotting variables on an X, Y plane may be helpful in identifying possible data errors. For example,

```
PROC SGPLOT DATA=beef;
SCATTER Y=backfat X=cwt;
RUN;
```

or

```
PROC GPLOT DATA=beef;
PLOT backfat*cwt;
SYMBOL1 VALUE=CIRCLE CV=BLACK I=NONE;
RUN;
```

The resulting images look very much like the scatter plots produced in Objectives 11.5 and 11.6 by the CORR Procedure. The suspect data value appears in the lower left corner of the scatter plots in these objectives. Additionally, the suspect data value is visualized in the histograms of the scatter plot matrix in Objective 11.6.

These are just a few ways introductory SAS procedures can be used to investigate a data set for possible errors. A researcher knows how much data they have collected, and thus should be able to confirm the values of sample size from these procedures, and the values of the character variables. Likewise, knowing that the responses for a variable should be between *a* and *b*, can be quickly checked using any of the procedures that produce minimum and maximum values in the results. The procedures shown here assist in identifying errors that are more extreme. Smaller errors, such as entering 12.5 instead of 12.4 for the REA variable, are, of course, very difficult to detect.

20.4 DATA Step Information – Indicator Function

Within the DATA step new variables can be created or existing variables modified using variable assignment statements, as shown in Chapter 4. IF – THEN or IF – THEN – ELSE statements were also introduced to perform these sorts of operations with a DATA step. Here, an indicator function is presented.

In general, an indicator function is a function that takes a value of 1 when a specified condition is met, and a value of 0 otherwise. The indicator function is a SAS condition (Section 4.2) enclosed in parentheses. This is illustrated in the following simple SAS program.

```
DATA one;
INPUT Group $ w x y z;
example1 = (group = "A");
example2 = (INT(w/2) = w/2 );
example3 = (x >= 30);
example4 = ( (group ne "A") OR (y < 100) ) ;
example5 = ( (group ne "A") AND (y < 100) );
example6 = z + 5*(group = "A") + 10*(group = "B");
DATALINES ;
A  17  35  104  79
A  19  32   90  92
B  16  39  101  89
B  21  40   95  85
C  12  29   88  81
C  16  27   84  83
;
PROC PRINT DATA=one NOOBS;
RUN;
QUIT;
```

Group	w	x	y	z	example1	example2	example3	example4	example5	example6
A	17	35	104	79	1	0	1	0	0	84
A	19	32	90	92	1	0	1	1	0	97
B	16	39	101	89	0	1	1	1	0	99
B	21	40	95	85	0	0	1	1	1	95
C	12	29	88	81	0	1	0	1	1	81
C	16	27	84	83	0	1	0	1	1	83

For the new variable *example1*, a value of 1 is assigned to those observations in group A, and 0 is assigned to those outside of group A. To do this using IF – THEN – ELSE statements, one could have written

```
IF group = "A" then example1 = 1; ELSE example1 = 0;
```

Without the ELSE statement, when the group was different from A, the value of *example1* would have been missing.

The variable *example2* identifies with a 1 those values of w that are even numbers. If the greatest integer, INT ( ) function, of the quantity w divided by 2 is the same as w divided by 2, then the value of w must be even. The value of w divided by 2 was not assigned to a new variable in the SAS data set *one*, but it could have been. Like *example1*, the *example2* variable assignment could have been completed using IF – THEN – ELSE statements.

```
IF INT(w/2) = w/2 THEN example2 = 1; ELSE example2 = 0;
```

The variable *example3* is a dichotomous variable that identifies when the value of x is at least 30. Equivalently, IF – THEN – ELSE statements could have been used:

```
IF x >= 30 THEN example3 = 1; ELSE example3 = 0;
```

In the assignment statements creating the variables *example4* and *example5,* it is shown that more than one condition can be specified within an indicator function. The differences in the values of the two variables are due to the differences between the logical OR and the logical AND in the indicator function. OR implies that either or both of the conditions must be true to assign a 1; AND implies that both conditions must be true to assign a 1.

The final variable, *example6,* is the value of the variable z only for group C; for group A *example6* is z + 5; and for group B *example6* is z + 10.

The syntax for the indicator functions for these examples tends to be simpler than the IF – THEN – ELSE syntax.

The following example demonstrates three different approaches for creating categorical data from a continuous response variable.

For the SAS data set *demo.beef3,* three categories of size based on the CWT of the animals can be created using 1. IF – THEN syntax, 2. Indicator functions, and 3. FORMAT Procedure and a FORMAT Statement. The FREQ Procedure is used to compare the outcomes of the three.

```
DATA twenty4;
SET demo.beef3;
**** 1. IF-THEN-ELSE ****;
IF cwt LE 825 THEN size1 = 1;
IF 825 < cwt LE 875 THEN size1=2;
IF 875 < cwt THEN size1=3;

*** 2. Indictor Function ***;
size2 = (cwt LE 825) + 2*(825 < cwt LE 875) + 3*(875 < cwt);

*** 3. FORMAT to create intervals ***;
PROC FORMAT;
*for size1 and size2;
VALUE sfmt                 1 = "Small"
                           2 = "Medium"
                           3 = "Large";
VALUE sizefmt     low - 825 = "Small"
                  825.001 - 875 = "Medium"
                  875.001 - high= "Large";

PROC FREQ DATA=twenty4;
TABLES size1 size2 cwt ;
FORMAT size1 size2 sfmt. Cwt sizefmt.;
RUN;
QUIT;
```

The FREQ Procedure

size1	Frequency	Percent	Cumulative Frequency	Cumulative Percent
Small	157	47.87	157	47.87
Medium	122	37.20	279	85.06
Large	49	14.94	328	100.00

size2	Frequency	Percent	Cumulative Frequency	Cumulative Percent
Small	157	47.87	157	47.87
Medium	122	37.20	279	85.06
Large	49	14.94	328	100.00

Carcass Weight, lb

CWT	Frequency	Percent	Cumulative Frequency	Cumulative Percent
Small	157	47.87	157	47.87
Medium	122	37.20	279	85.06
Large	49	14.94	328	100.00

Clearly each of the three approaches yields the same intervals. The first two approaches require the creation of a new variable, and the third approach is successfully accomplished with a user-defined format.

20.5 Chapter Summary

These topics were not an exact fit to any of the previous chapters, but that does not diminish their importance. Annotating programs is a habit SAS programmers should get into when starting out. These annotations can provide clarity when reviewing a program at a later date or when sharing the program with others on a project team. A few quick processes for checking for data errors were also overviewed. As one's programming skills grow, the tools for detecting data errors will also grow. If the data are in error, then it is not possible to perform a correct analysis. Throughout this book there have been multiple approaches to complete some of the analyses and graphing tasks. The indicator function is another one of these approaches to accomplish the task of creating categorical variable levels, and it is one of those SAS "tools" that is surprisingly useful in DATA step operations.

In general, this book has used notation, terminology, and font selection consistent with SAS Institute publications so that the reader can comfortably read program syntax and support documentation used within those resources. With the instruction and practice of this book's objectives, the ability to read SAS documentation, and the ability to interpret SAS error messages, the reader should now be a more confident SAS Programmer.

References

Agresti, A. (2007), *An Introduction to Categorical Data Analysis*, 2nd edition, New York: John Wiley & Sons.

Agresti, A. (2018), *An Introduction to Categorical Data Analysis*, 3rd edition, New York: John Wiley & Sons.

Chambers, J. M., Cleveland, W. S., Kleiner, B., and Tukey, P. A. (1983), *Graphical Methods for Data Analysis*, Belmont, CA: Wadsworth International Group.

Conover, W. J. (1999), *Practical Nonparametric Statistics*, 3rd edition, New York: John Wiley & Sons.

D'Agostino, R. B. and Stephens, M., eds. (1986), *Goodness-of-Fit Techniques*, New York: Marcel Dekker.

Draper, N. R. and Smith, H. (1998), *Applied Regression Analysis*, 3rd edition, New York: John Wiley & Sons.

Freund, R. J., Wilson, W. J., and Mohr, D. L. (2010), *Statistical Methods*, 3rd edition, New York: Academic Press.

IBM Corp. Released 2017. IBM SPSS Statistics for Windows, Version 25.0. Armonk, NY: IBM Corp.

Kutner, M. H., Nachtsheim, C. J., and Neter, J. (2004), *Applied Linear Regression Models*, 4th edition, New York: McGraw-Hill/Irwin.

Microsoft Corporation. (2016). *Microsoft Access*.

Microsoft Corporation. (2016). *Microsoft Excel*.

Montgomery, D. C., Peck, E. A., and Vining, G. G. (2012), *Introduction to Linear Regression Analysis*, 5th edition, New York: John Wiley & Sons.

Myers, R. H. (2000), *Classical and Modern Regression with Applications*, Duxbury Press.

Ott, R. L. and Longnecker, M. (2016), *An Introduction to Statistical Methods and Data Analysis*, 7th edition, Boston, MA: Cengage Learning.

Peck, R. and Devore, J. (2011), *Statistics: The Exploration & Analysis of Data*, 7th edition, Boston, MA: Cengage Learning.

Royston, J. P. (1992), "Approximating the Shapiro-Wilk W Test for Nonnormality," *Statistics and Computing*, 2, 117–119.

SAS Institute Inc. (2013a), *Base SAS® 9.4 Procedures Guide: Statistical Procedures*, 2nd edition. Cary, NC: SAS Institute Inc.

SAS Institute Inc. (2013b), *SAS® 9.4 Statements: Reference*. Cary, NC: SAS Institute Inc.

SAS Institute Inc. (2016a), *SAS® 9.4 DS2 Language Reference*, 6th edition. Cary, NC: SAS Institute Inc.

SAS Institute Inc. (2016b), *SAS® 9.4 Output Delivery System: Procedures Guide*, 3rd edition. Cary, NC: SAS Institute Inc.

SAS Institute Inc. (2016c), *SAS/GRAPH® 9.4: Reference*, 5th edition. Cary, NC: SAS Institute Inc.

Satterthwaite, F. E. (1946), "An Approximate Distribution of Estimates of Variance Components," *Biometrics Bulletin*, 2, 110–114.

Shapiro, S. S. and Wilk, M. B. (1965), "An Analysis of Variance Test for Normality (Complete Samples)," *Biometrika*, 52, 591–611.

StataCorp. 2019. *Stata Statistical Software*: Release 16. College Station, TX: StataCorp LP.

Index

NOTE: Page numbers in **bold** and *italics* refer to tables and figures, respectively

D

L

M

N

T